VARGIC'S

CURIOUS COSMIC

COMPENDIUM

MICHAEL JOSEPH

UK | USA | Canada | Ireland | Australia
India | New Zealand | South Africa

Michael Joseph is part of the Penguin Random House group of companies
whose addresses can be found at global.penguinrandomhouse.com

Penguin
Random House
UK

First published 2019
001

Printed in China

A CIP catalogue record for this book is available from the British Library

ISBN: 978–0–718–18526–8

www.greenpenguin.co.uk

MIX
Paper from
responsible sources
FSC® C018179

Penguin Random House is committed to a
sustainable future for our business, our readers
and our planet. This book is made from Forest
Stewardship Council® certified paper.

VARGIC'S

CURIOUS COSMIC COMPENDIUM

SPACE, THE UNIVERSE AND
EVERYTHING WITHIN IT

MICHAEL JOSEPH
an imprint of
PENGUIN BOOKS

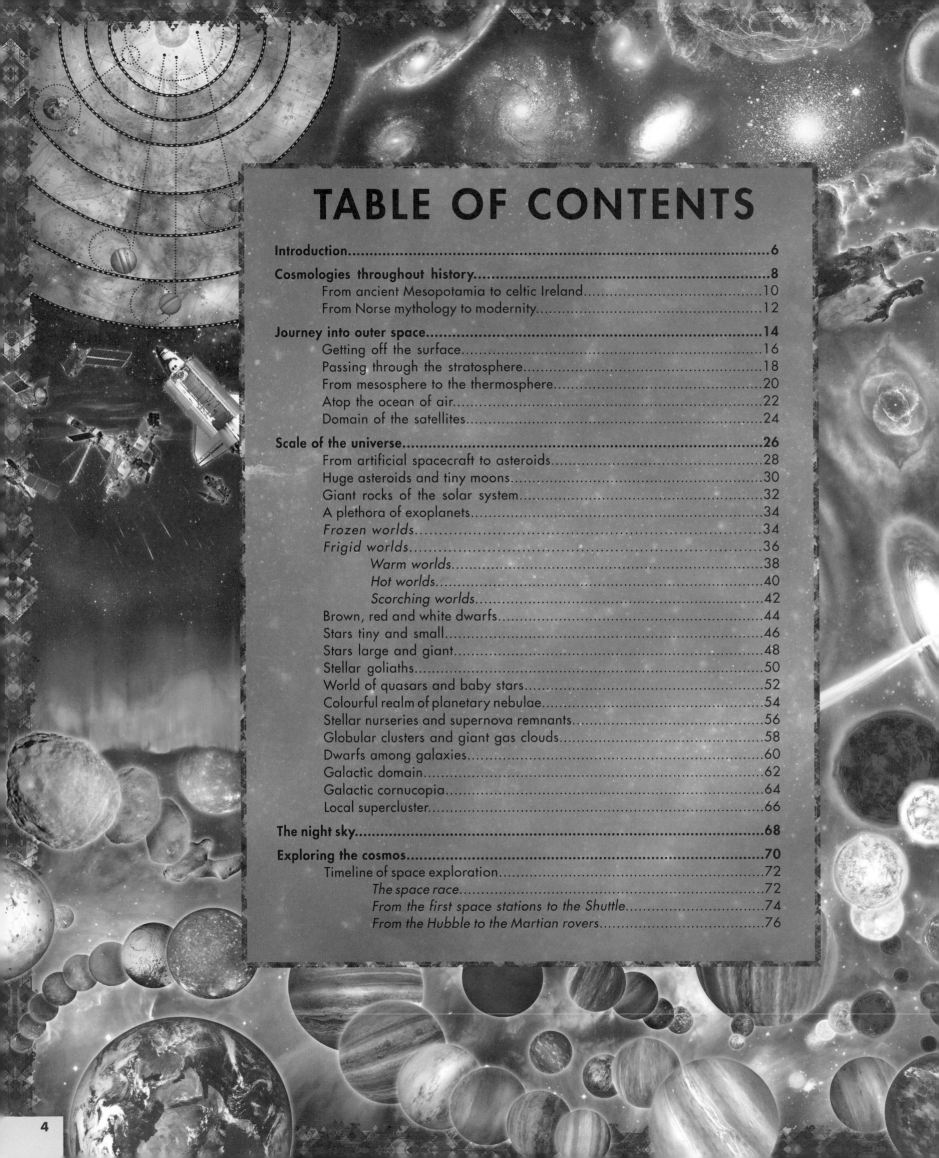

TABLE OF CONTENTS

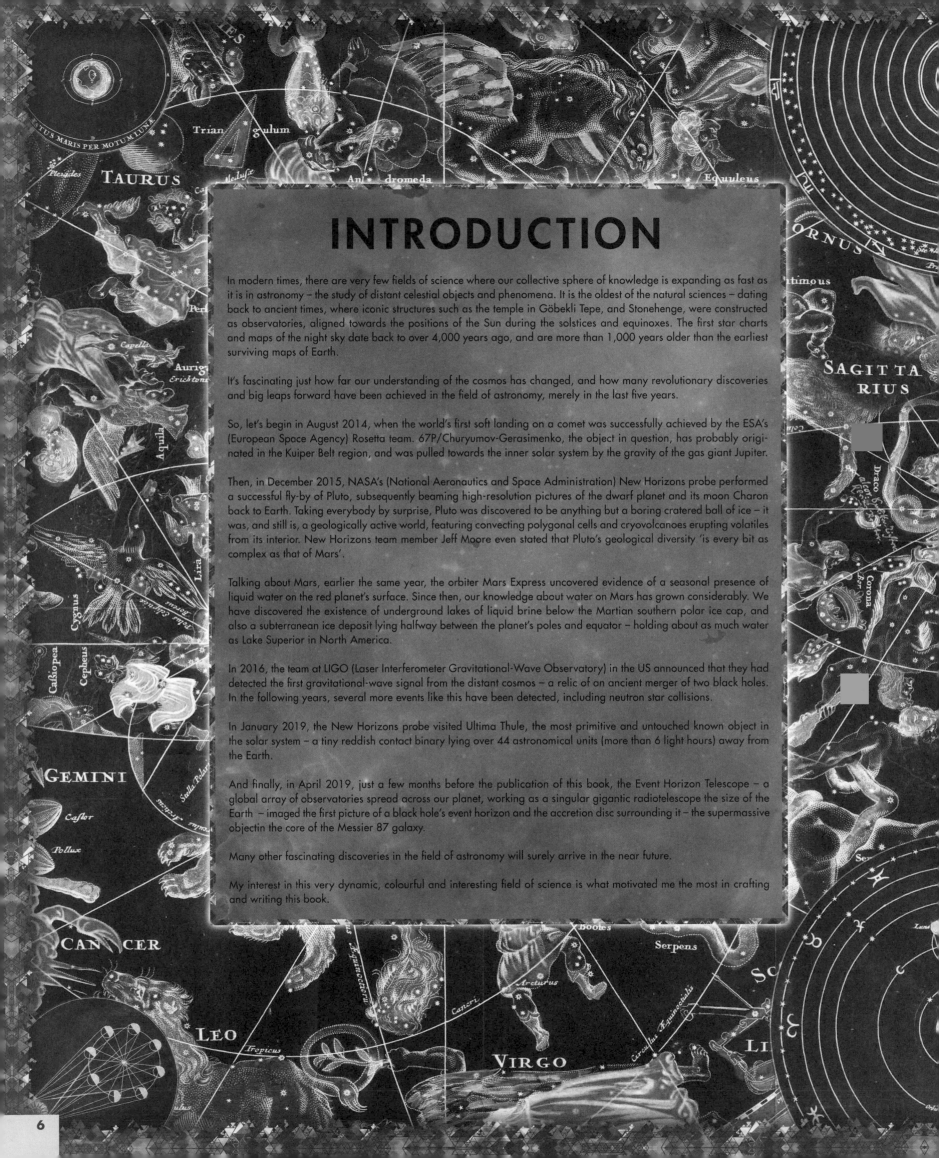

INTRODUCTION

In modern times, there are very few fields of science where our collective sphere of knowledge is expanding as fast as it is in astronomy – the study of distant celestial objects and phenomena. It is the oldest of the natural sciences – dating back to ancient times, where iconic structures such as the temple in Göbekli Tepe, and Stonehenge, were constructed as observatories, aligned towards the positions of the Sun during the solstices and equinoxes. The first star charts and maps of the night sky date back to over 4,000 years ago, and are more than 1,000 years older than the earliest surviving maps of Earth.

It's fascinating just how far our understanding of the cosmos has changed, and how many revolutionary discoveries and big leaps forward have been achieved in the field of astronomy, merely in the last five years.

So, let's begin in August 2014, when the world's first soft landing on a comet was successfully achieved by the ESA's (European Space Agency) Rosetta team. 67P/Churyumov-Gerasimenko, the object in question, has probably originated in the Kuiper Belt region, and was pulled towards the inner solar system by the gravity of the gas giant Jupiter.

Then, in December 2015, NASA's (National Aeronautics and Space Administration) New Horizons probe performed a successful fly-by of Pluto, subsequently beaming high-resolution pictures of the dwarf planet and its moon Charon back to Earth. Taking everybody by surprise, Pluto was discovered to be anything but a boring cratered ball of ice – it was, and still is, a geologically active world, featuring convecting polygonal cells and cryovolcanoes erupting volatiles from its interior. New Horizons team member Jeff Moore even stated that Pluto's geological diversity 'is every bit as complex as that of Mars'.

Talking about Mars, earlier the same year, the orbiter Mars Express uncovered evidence of a seasonal presence of liquid water on the red planet's surface. Since then, our knowledge about water on Mars has grown considerably. We have discovered the existence of underground lakes of liquid brine below the Martian southern polar ice cap, and also a subterranean ice deposit lying halfway between the planet's poles and equator – holding about as much water as Lake Superior in North America.

In 2016, the team at LIGO (Laser Interferometer Gravitational-Wave Observatory) in the US announced that they had detected the first gravitational-wave signal from the distant cosmos – a relic of an ancient merger of two black holes. In the following years, several more events like this have been detected, including neutron star collisions.

In January 2019, the New Horizons probe visited Ultima Thule, the most primitive and untouched known object in the solar system – a tiny reddish contact binary lying over 44 astronomical units (more than 6 light hours) away from the Earth.

And finally, in April 2019, just a few months before the publication of this book, the Event Horizon Telescope – a global array of observatories spread across our planet, working as a singular gigantic radiotelescope the size of the Earth – imaged the first picture of a black hole's event horizon and the accretion disc surrounding it – the supermassive object in the core of the Messier 87 galaxy.

Many other fascinating discoveries in the field of astronomy will surely arrive in the near future.

My interest in this very dynamic, colourful and interesting field of science is what motivated me the most in crafting and writing this book.

6

Deep space, astronomy and rocketry have been some of my greatest fascinations since childhood. From the age of twelve, I have spent many winter evenings looking at the night sky with a telescope, observing double stars, craters on the Moon and the moons of Jupiter, with satellites quickly whizzing by every once in a while. One particularly dark evening in eastern Slovakia, I even managed to spot Neptune near the star Deneb Algedi in the constellation Capricornus. This dim blue dot was only discovered to be the eighth planet of the solar system in the late 19th century.

As with my previous work over the years, including my first book, *Vargic's Miscellany of Curious Maps*, the main focus of this book is the highly visual graphics (although it still contains a lot of interesting info in the text as well). On the pages that follow, you will find thousands of detailed original illustrations, aiming to showcase the many different and colourful facets of the cosmos, and our role in exploring it.

This work is split into several distinct, internally consistent chapters, a number of which are designed to seamlessly stretch throughout dozens of pages, intended to unfold in front of your eyes as continuous scrolls.

In the book, I will take you step-by-step on a lavishly illustrated journey across our universe. Firstly, in Cosmologies Throughout History, I aim to examine and visually depict the variety of diverse and fascinating understandings of the forming of the universe, held by our ancestors from many different cultures. Then we will unravel the Earth's atmosphere with a brief history of how humanity has managed to surpass the countless challenges presented by nature in our journey towards reaching the weightlessness of the Earth's orbit.

The next chapter, the Scale of the Universe, is a journey that will carry you from orbit through the various asteroids and small moons of the solar system to planets and exoplanets and red, brown and white stellar dwarfs, into the domain of the largest stars, accretion discs of quasars and great interstellar nebulae – the cradles of stars. With each page on a far larger scale to the previous one, we will gradually zoom out to one of the largest structures of the cosmos we know about – the local supercluster of galaxies.

The chapter that follows deals with the human exploration of the cosmos. The Timeline of Space Exploration is a chronicle of the most important achievements in both unmanned and manned spaceflight, from the launch of Sputnik 1 to the planned missions into outer space starting in the 2020s and continuing through to the late 2030s. The next section is focused on the evolution of rocket science over the last seventy years. It includes highly detailed illustrations of more than ninety-five orbital and suborbital rockets from fifteen countries, beginning with the famous V2 missile, and ending with some of the most ambitious rocket models that are currently in the development stage, such as NASA's Space Launch System, Jeff Bezos's New Glenn and the United Launch Alliance's Vulcan rocket. A quick addendum then showcases (to scale) the most powerful Earth and space-based observatories – the main instruments we use to learn about the distant universe.

The subsequent chapter is a brief showcase of the landscapes and most iconic geographical features of the five large rocky objects in the inner solar system – Mercury, Venus, Earth, the Moon and Mars – in a series of maps, constructed using the best-quality orbital and radar imagery available.

Finally, the last segment of the book is dedicated to the Timeline of the Universe, with thirty-six pages that will take you on a grand voyage from the first picosecond of the universe through all 13.5 billion years of our universe's existence, to the events that the near and far future have in store for us, the Earth and the cosmos as a whole. It is complemented by dedicated graphics about topics such as elementary particles, the classification of galaxies, the formation and evolution of the solar system, the major orbital cycles our planet Earth goes through during hundreds of thousands of years, and even how our night sky has changed during the geologically very recent past.

All in all, I hope you will enjoy reading this book at least as much as I enjoyed crafting and writing it – and that it will make you better appreciate and understand the complex and fascinating inner workings of the universe that we have found ourselves in.

Ancient Meso-American 'Tree of Life', located in the centre of the world

Depiction of the universe and the world 'tree Yggdrasil, inspired by the Norse sagas

Ancient Indian depiction of the universe as 'Vishvarupa', a cosmic form of the god Vishnu

Reconstruction of Ptolemy's 2nd-century AD map of the world. Contrary to popular belief, Earth was known to be a globe at least since the 5th century BC

COSMOLOGIES THROUGHOUT HISTORY

Since the dawn of humanity, people have always strived to make sense of the universe around them, questioning what the world's origins are, what it is composed of, and asking whether it all has a deeper, underlying purpose. This drive is what eventually led to the emergence of countless different religious, philosophical and, ultimately, scientific schools of thought.

From prehistory, humans certainly took notice of the changing length of day and night throughout the year, and the corresponding changes in seasons, caused by the periodic cycles of the Sun's motion across the night sky. Having little else to do in the evenings, our ancestors would certainly have spent hours staring in amazement at the pristine night sky, trying to find patterns in the stars. They undoubtedly took notice of the non-stationary nature of the Moon and the planets, and understood that the world, which could often seem chaotic and arbitrary, had to have some underlying and unchanging rules and laws that even the distant heavenly bodies were subject to. To understand it more easily, our ancestors started to associate the various aspects of nature and the cosmos with gods and goddesses, often human or animal in form – and as the gods commanded, the heavens obeyed.

Understanding the laws of nature and the heavenly cycles became especially important after the transition from a nomadic hunter-gatherer lifestyle to a more stable agricultural way of life, as a badly timed sowing or harvest could mean starvation and death. Some of the world's oldest surviving structures, such as the Neolithic stone pillars in Göbekli Tepe and earthwork enclosures found all over Europe, were ancient observatories, used to mark the position of the Sun, Moon and the stars for centuries. At the same time, they were also centres of religious worship, as people believed that in order to maintain the natural order, the gods needed to be regularly worshipped and appeased by rituals and sacrifices – both animal and human.

The following chapter showcases and summarizes twenty of the most iconic historical models of the universe from many diverse cultures across the world, showing how they evolved over time – from the highly symbolic cosmologies of the ancient Egyptians, Babylonians and Native Americans, into the far more precise and mathematically complex models of classical antiquit, and, ultimately, into modern times.

All early models of the cosmos have one thing in common – they hold the Earth to be stationary and unmoving, placing it in the centre of the universe, while the heavenly bodies swiftly move around it every day. This is hardly surprising, as the centrifugal forces created by the Earth's rotation around its axis and its orbit around the Sun are far too weak to be perceived by humans, and can only be detected with the help of sensitive instruments.

The credit for the oldest comprehensive cosmological model belongs to the people of ancient Mesopotamia (Sumerians, Akkadians and later Babylonians). Their view of the world had a profound influence on ancient Hebrews as well. The Earth was believed to be a flat disc supported by tremendous pillars, sitting in the middle of a primeval ocean, which surrounded the world on all sides. It was kept from flooding the Earth by a heavenly vault, which also separated the earthly realm from the realms of the gods. Under the surface, there lay an underworld, the resting place for the souls of all mortals.

13th-century illustration showing God in the process of creating the cosmos

Ancient Egyptian goddess Nut forming the vault of the sky, supported by Shu, the god of air

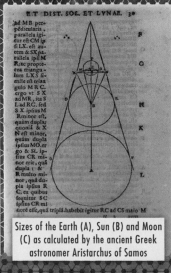
Sizes of the Earth (A), Sun (B) and Moon (C) as calculated by the ancient Greek astronomer Aristarchus of Samos

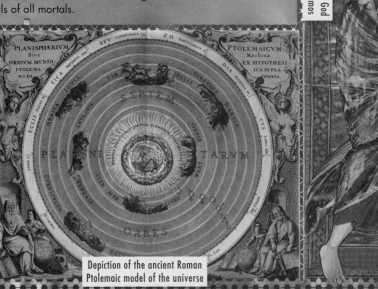
Depiction of the ancient Roman Ptolemaic model of the universe

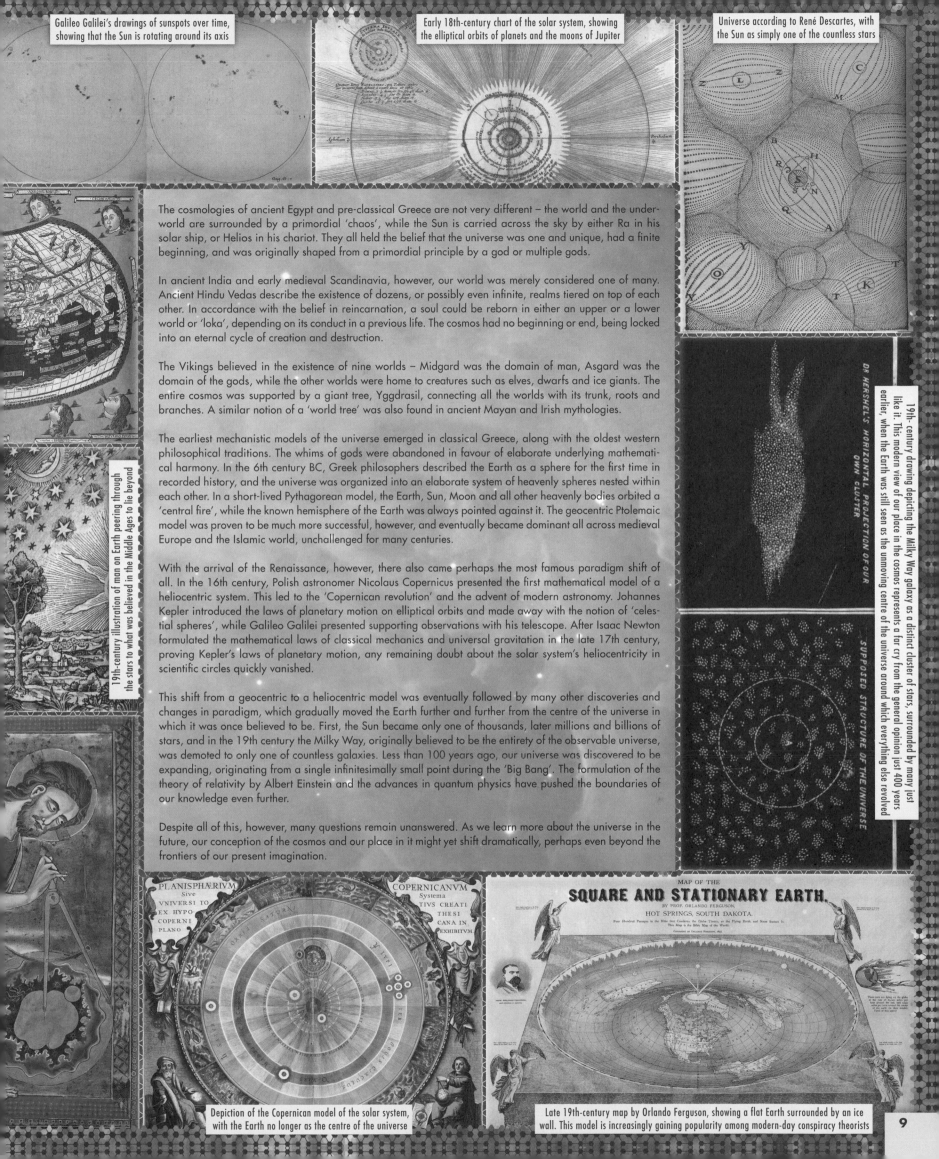

The cosmologies of ancient Egypt and pre-classical Greece are not very different – the world and the under-world are surrounded by a primordial 'chaos', while the Sun is carried across the sky by either Ra in his solar ship, or Helios in his chariot. They all held the belief that the universe was one and unique, had a finite beginning, and was originally shaped from a primordial principle by a god or multiple gods.

In ancient India and early medieval Scandinavia, however, our world was merely considered one of many. Ancient Hindu Vedas describe the existence of dozens, or possibly even infinite, realms tiered on top of each other. In accordance with the belief in reincarnation, a soul could be reborn in either an upper or a lower world or 'loka', depending on its conduct in a previous life. The cosmos had no beginning or end, being locked into an eternal cycle of creation and destruction.

The Vikings believed in the existence of nine worlds – Midgard was the domain of man, Asgard was the domain of the gods, while the other worlds were home to creatures such as elves, dwarfs and ice giants. The entire cosmos was supported by a giant tree, Yggdrasil, connecting all the worlds with its trunk, roots and branches. A similar notion of a 'world tree' was also found in ancient Mayan and Irish mythologies.

The earliest mechanistic models of the universe emerged in classical Greece, along with the oldest western philosophical traditions. The whims of gods were abandoned in favour of elaborate underlying mathematical harmony. In the 6th century BC, Greek philosophers described the Earth as a sphere for the first time in recorded history, and the universe was organized into an elaborate system of heavenly spheres nested within each other. In a short-lived Pythagorean model, the Earth, Sun, Moon and all other heavenly bodies orbited a 'central fire', while the known hemisphere of the Earth was always pointed against it. The geocentric Ptolemaic model was proven to be much more successful, however, and eventually became dominant all across medieval Europe and the Islamic world, unchallenged for many centuries.

With the arrival of the Renaissance, however, there also came perhaps the most famous paradigm shift of all. In the 16th century, Polish astronomer Nicolaus Copernicus presented the first mathematical model of a heliocentric system. This led to the 'Copernican revolution' and the advent of modern astronomy. Johannes Kepler introduced the laws of planetary motion on elliptical orbits and made away with the notion of 'celestial spheres', while Galileo Galilei presented supporting observations with his telescope. After Isaac Newton formulated the mathematical laws of classical mechanics and universal gravitation in the late 17th century, proving Kepler's laws of planetary motion, any remaining doubt about the solar system's heliocentricity in scientific circles quickly vanished.

This shift from a geocentric to a heliocentric model was eventually followed by many other discoveries and changes in paradigm, which gradually moved the Earth further and further from the centre of the universe in which it was once believed to be. First, the Sun became only one of thousands, later millions and billions of stars, and in the 19th century the Milky Way, originally believed to be the entirety of the observable universe, was demoted to only one of countless galaxies. Less than 100 years ago, our universe was discovered to be expanding, originating from a single infinitesimally small point during the 'Big Bang'. The formulation of the theory of relativity by Albert Einstein and the advances in quantum physics have pushed the boundaries of our knowledge even further.

Despite all of this, however, many questions remain unanswered. As we learn more about the universe in the future, our conception of the cosmos and our place in it might yet shift dramatically, perhaps even beyond the frontiers of our present imagination.

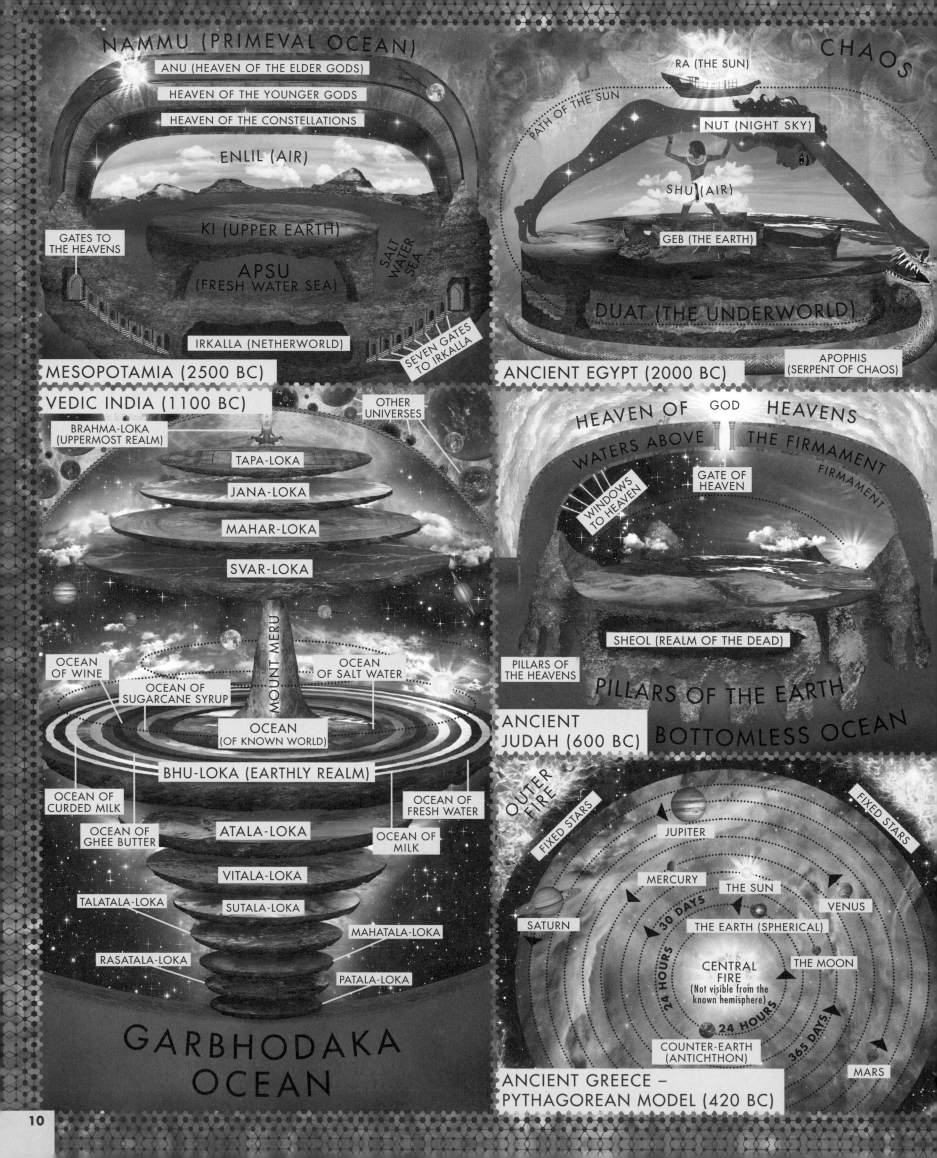

MESOPOTAMIA (2500 BC)

NAMMU (PRIMEVAL OCEAN)

ANU (HEAVEN OF THE ELDER GODS)

HEAVEN OF THE YOUNGER GODS

HEAVEN OF THE CONSTELLATIONS

ENLIL (AIR)

KI (UPPER EARTH)

GATES TO THE HEAVENS

APSU (FRESH WATER SEA)

SALT WATER SEA

IRKALLA (NETHERWORLD)

SEVEN GATES TO IRKALLA

ANCIENT EGYPT (2000 BC)

CHAOS

RA (THE SUN)

PATH OF THE SUN

NUT (NIGHT SKY)

SHU (AIR)

GEB (THE EARTH)

DUAT (THE UNDERWORLD)

APOPHIS (SERPENT OF CHAOS)

VEDIC INDIA (1100 BC)

OTHER UNIVERSES

BRAHMA-LOKA (UPPERMOST REALM)

TAPA-LOKA

JANA-LOKA

MAHAR-LOKA

SVAR-LOKA

MOUNT MERU

OCEAN OF WINE

OCEAN OF SALT WATER

OCEAN OF SUGARCANE SYRUP

OCEAN (OF KNOWN WORLD)

OCEAN OF FRESH WATER

BHU-LOKA (EARTHLY REALM)

OCEAN OF CURDED MILK

OCEAN OF MILK

OCEAN OF GHEE BUTTER

ATALA-LOKA

VITALA-LOKA

TALATALA-LOKA

SUTALA-LOKA

MAHATALA-LOKA

RASATALA-LOKA

PATALA-LOKA

GARBHODAKA OCEAN

ANCIENT JUDAH (600 BC)

HEAVEN OF GOD HEAVENS

WATERS ABOVE THE FIRMAMENT

FIRMAMENT

GATE OF HEAVEN

WINDOWS TO HEAVEN

SHEOL (REALM OF THE DEAD)

PILLARS OF THE HEAVENS

PILLARS OF THE EARTH

BOTTOMLESS OCEAN

ANCIENT GREECE – PYTHAGOREAN MODEL (420 BC)

OUTER FIRE

FIXED STARS

FIXED STARS

FIXED STARS

JUPITER

MERCURY

THE SUN

VENUS

SATURN

30 DAYS

THE EARTH (SPHERICAL)

THE MOON

24 HOURS

CENTRAL FIRE (Not visible from the known hemisphere)

24 HOURS

365 DAYS

COUNTER-EARTH (ANTICHTHON)

MARS

CHAOS

CHARIOT OF HELIOS

OLYMPUS (ABODE OF THE GODS)

ISLANDS OF THE BLESSED

RIVER ACHERON

RIVER OCEANUS

ELYSIUM (PARADISE)

FIELDS OF ASPHODEL

RIVER STYX

RIVER OCEANUS

TARTARUS

CHAOS

ANCIENT GREECE – MYTHICAL MODEL (700 BC)

All heavenly spheres rotate in unison around the Earth once every 24 hours, and also slowly turn in relation to each other

29.5 YEARS

SPHERE OF THE FIXED STARS

SPHERE OF SATURN

11.9 YEARS

SPHERE OF MARS

1 YEAR

EQUANTS

1 YEAR

1 YEAR

SPHERE OF THE SUN

SPHERE OF JUPITER

SPHERE OF VENUS

11.9 YEARS

SPHERE OF MERCURY

1.9 YEARS

LUNAR SPHERE

EQUANTS

225 DAYS

88 DAYS

1 YEAR

AIR

1 YEAR

STATIONARY EARTH

FIRE

A ET H E R

1 YEAR

30 DAYS

ANCIENT GREECE – PTOLEMAIC MODEL (150 AD)

CELTIC IRELAND (400 AD)

INFINITELY EXTENDING HEAVENS

SATURN (THE EARTH STAR)

MERCURY (THE FIRE STAR)

JUPITER (THE WOOD STAR)

MARS (THE FIRE STAR)

VENUS (THE METAL STAR)

DÌQIÚ (THE EARTH)

ANCIENT CHINA (300 BC)

MAGH MOR (UPPER WORLD)

THE SACRED WORLD OAK

VENUS

QUETZAL (BIRD OF THE HEAVENS)

REALM OF THE GODS

MIDDLE WORLD

SACRED PYRAMID MOUNTAIN

MUIR (SEA)

LAND OF YOUTH

LAND OF WOMEN

LAND OF IMMORTALITY

XIBALBA, 'PLACE OF FEAR' (THE UNDERWORLD)

TIR ANDOMAIN (LOWER WORLD)

ANCIENT MAYA (300 BC)

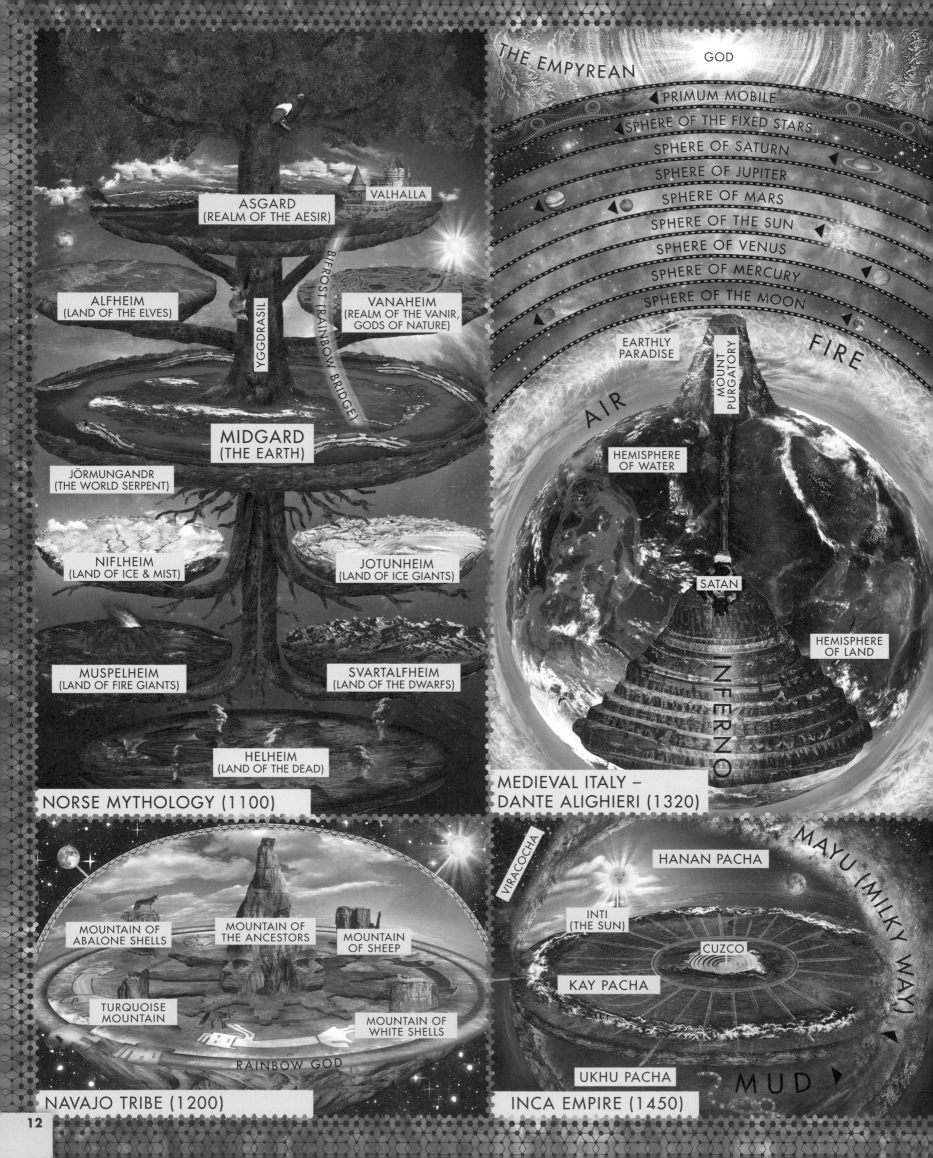

THE EMPYREAN

GOD

PRIMUM MOBILE

SPHERE OF THE FIXED STARS

SPHERE OF SATURN

SPHERE OF JUPITER

SPHERE OF MARS

SPHERE OF THE SUN

SPHERE OF VENUS

SPHERE OF MERCURY

SPHERE OF THE MOON

FIRE

AIR

EARTHLY PARADISE

MOUNT PURGATORY

HEMISPHERE OF WATER

SATAN

HEMISPHERE OF LAND

INFERNO

MEDIEVAL ITALY –
DANTE ALIGHIERI (1320)

ASGARD
(REALM OF THE AESIR)

VALHALLA

ALFHEIM
(LAND OF THE ELVES)

VANAHEIM
(REALM OF THE VANIR,
GODS OF NATURE)

YGGDRASIL

BIFROST (RAINBOW BRIDGE)

MIDGARD
(THE EARTH)

JÖRMUNGANDR
(THE WORLD SERPENT)

NIFLHEIM
(LAND OF ICE & MIST)

JOTUNHEIM
(LAND OF ICE GIANTS)

MUSPELHEIM
(LAND OF FIRE GIANTS)

SVARTALFHEIM
(LAND OF THE DWARFS)

HELHEIM
(LAND OF THE DEAD)

NORSE MYTHOLOGY (1100)

MOUNTAIN OF
ABALONE SHELLS

MOUNTAIN OF
THE ANCESTORS

MOUNTAIN
OF SHEEP

TURQUOISE
MOUNTAIN

MOUNTAIN OF
WHITE SHELLS

RAINBOW GOD

NAVAJO TRIBE (1200)

VIRACOCHA

HANAN PACHA

MAYU (MILKY WAY)

INTI
(THE SUN)

CUZCO

KAY PACHA

UKHU PACHA

MUD

INCA EMPIRE (1450)

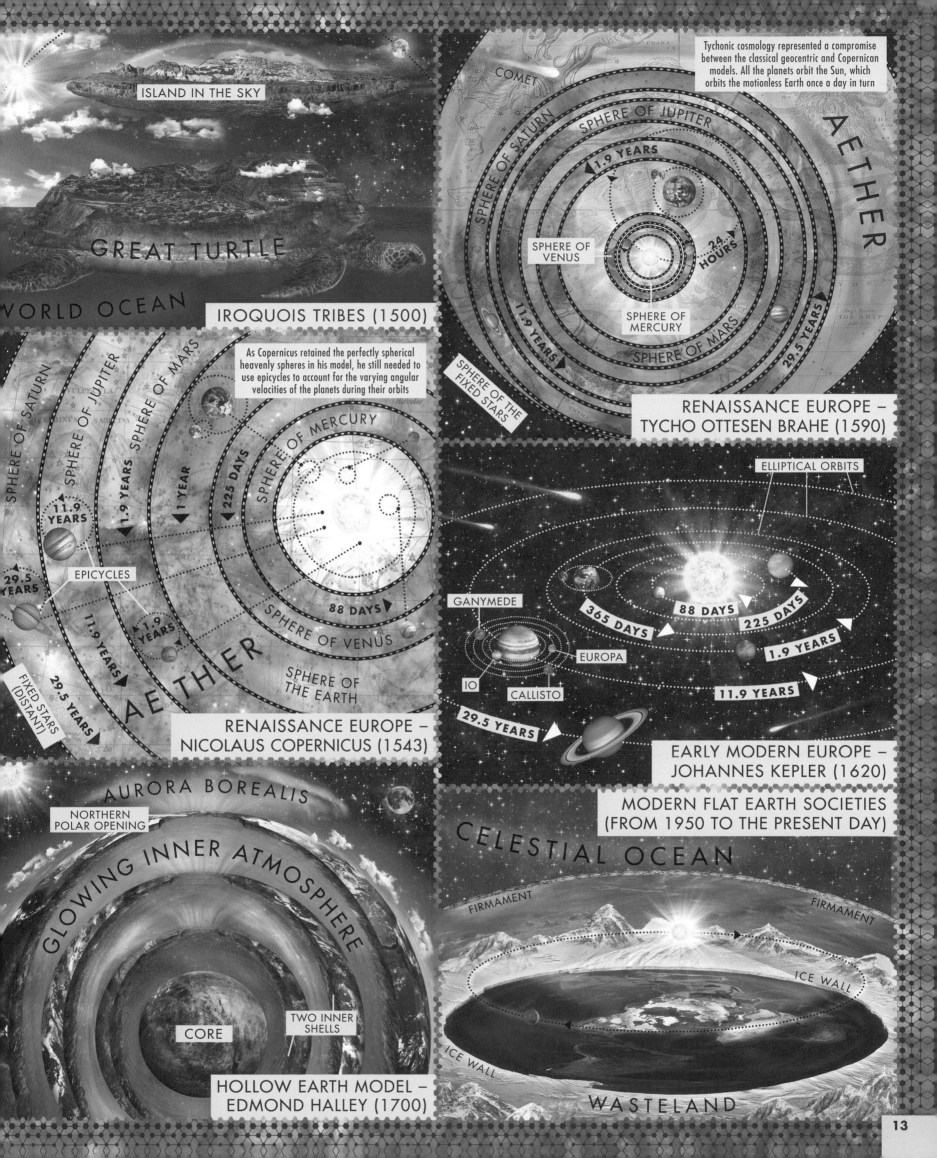

ISLAND IN THE SKY

GREAT TURTLE

WORLD OCEAN

IROQUOIS TRIBES (1500)

Tychonic cosmology represented a compromise between the classical geocentric and Copernican models. All the planets orbit the Sun, which orbits the motionless Earth once a day in turn

COMET

SPHERE OF SATURN
SPHERE OF JUPITER

1.9 YEARS

SPHERE OF VENUS

24 HOURS

SPHERE OF MERCURY

11.9 YEARS

SPHERE OF MARS

29.5 YEARS

AETHER

SPHERE OF THE FIXED STARS

RENAISSANCE EUROPE –
TYCHO OTTESEN BRAHE (1590)

As Copernicus retained the perfectly spherical heavenly spheres in his model, he still needed to use epicycles to account for the varying angular velocities of the planets during their orbits

SPHERE OF SATURN
SPHERE OF JUPITER
SPHERE OF MARS

SPHERE OF MERCURY

11.9 YEARS

1.9 YEARS

1 YEAR

225 DAYS

29.5 YEARS

EPICYCLES

11.9 YEARS

1.9 YEARS

88 DAYS

SPHERE OF VENUS

FIXED STARS (DISTANT)

29.5 YEARS

AETHER

SPHERE OF THE EARTH

RENAISSANCE EUROPE –
NICOLAUS COPERNICUS (1543)

ELLIPTICAL ORBITS

GANYMEDE

365 DAYS

88 DAYS

225 DAYS

EUROPA

1.9 YEARS

IO

CALLISTO

11.9 YEARS

29.5 YEARS

EARLY MODERN EUROPE –
JOHANNES KEPLER (1620)

MODERN FLAT EARTH SOCIETIES
(FROM 1950 TO THE PRESENT DAY)

AURORA BOREALIS

NORTHERN POLAR OPENING

GLOWING INNER ATMOSPHERE

CELESTIAL OCEAN

FIRMAMENT

FIRMAMENT

CORE

TWO INNER SHELLS

ICE WALL

HOLLOW EARTH MODEL –
EDMOND HALLEY (1700)

ICE WALL

WASTELAND

JOURNEY INTO OUTER SPACE

While the observation of celestial bodies in outer space, separated from us by distances we can hardly comprehend, predates recorded history, humanity has always been confined to the surface of the Earth. The heavens were thought of os unreachable, immaterial and impossibly distant, and perhaps that's why our ancestors believed the sky was the domain of immortal gods, who were barely concerned with the petty affairs of the mortals. In many cultures all over the world, birds, with their gift of flight, held a very important place as the messengers between mortals and the gods. It's no coincidence that the Holy Spirit is most often depicted as a dove, that Odin, king of the Norse gods, is associated with ravens, or that Hermes, the divine emissary in Greek mythology, is depicted with winged sandals and a winged helmet.

The earliest stories and legends about humans who acquired the gift of flight and could fly to the heavens date to ancient Greece. In the myth of Daedalus and Icarus, a skilled craftsman makes himself and his son wings from wax and feathers so they could escape their imprisonment, and in another legend, the hero Bellerophon tames the flying horse Pegasus. In both stories, however, there is a moral lesson against hubris and recklessness. After Icarus flies too close to the Sun, his wings melt and he falls into the sea below, and after Bellerophon decides to fly high enough to reach the Olympian gods, he is knocked down from Pegasus by Zeus and becomes a crippled beggar.

Mythical and religious perspectives about the heavens were eventually replaced by philosophical arguments. In the 4th century BC, Aristotle (often considered the father of science) theorized that the world was divided into five elements. He organized them into concentric shells, stating that 'the earth is surrounded by water, just as that is by the sphere of air, and that again by the sphere called that of fire'. The celestial spheres and astronomical bodies located beyond were composed of the element 'aether'. Aristotle published the first comprehensive treatise on meteorology, or 'the study of things high in the air'. In it, he provided some of the earliest accounts and explanations for phenomena such as shooting stars, the Northern Lights, comets, zodiacal light and the Milky Way, believing them to be caused by the interactions of fire and air in the upper atmosphere (as well as describing the winds and the hydrologic cycle in fascinating detail).

These theories were further expanded upon by other natural philosophers in classical antiquity, and in the 2nd century AD, were incorporated by Ptolemy into his model of the solar system. The notion of an 'aether' filling the void of space was eventually adapted by early modern physicists such as Isaac Newton and Christiaan Huygens to explain the propagation of gravitational forces and light, but it was later proven to be unnecessary by James Clerk Maxwell and Albert Einstein.

It took centuries of gradual progress in science and craftsmanship until the first humans were able to leave the confines of the Earth's surface. Building upon the work of the previous generations, we were able to learn more about the layers of the atmosphere and about what lies beyond, to conquer the skies and ultimately reach outer space.

The history of aviation extends for more than 2,000 years, beginning in ancient China with the invention of the kite. They were flat, rectangular sails made of silk, with a lightweight bamboo framework. The Chinese used them for a variety of purposes, such as measuring distances and long-distance communication during

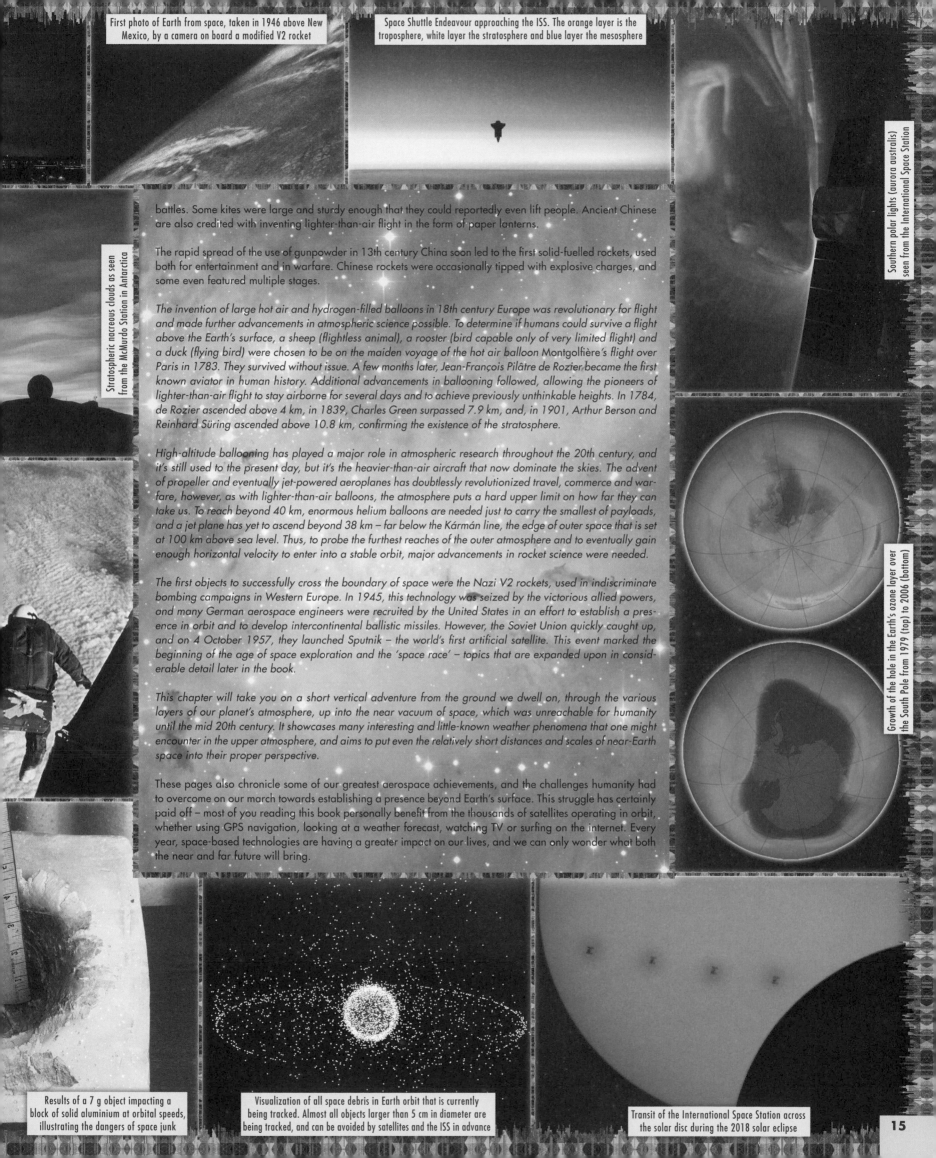

First photo of Earth from space, taken in 1946 above New Mexico, by a camera on board a modified V2 rocket

Space Shuttle Endeavour approaching the ISS. The orange layer is the troposphere, white layer the stratosphere and blue layer the mesosphere

Southern polar lights (aurora australis) seen from the International Space Station

Stratospheric nacreous clouds as seen from the McMurdo Station in Antarctica

battles. Some kites were large and sturdy enough that they could reportedly even lift people. Ancient Chinese are also credited with inventing lighter-than-air flight in the form of paper lanterns.

The rapid spread of the use of gunpowder in 13th century China soon led to the first solid-fuelled rockets, used both for entertainment and in warfare. Chinese rockets were occasionally tipped with explosive charges, and some even featured multiple stages.

The invention of large hot air and hydrogen-filled balloons in 18th century Europe was revolutionary for flight and made further advancements in atmospheric science possible. To determine if humans could survive a flight above the Earth's surface, a sheep (flightless animal), a rooster (bird capable only of very limited flight) and a duck (flying bird) were chosen to be on the maiden voyage of the hot air balloon Montgolfière's flight over Paris in 1783. They survived without issue. A few months later, Jean-François Pilâtre de Rozier became the first known aviator in human history. Additional advancements in ballooning followed, allowing the pioneers of lighter-than-air flight to stay airborne for several days and to achieve previously unthinkable heights. In 1784, de Rozier ascended above 4 km, in 1839, Charles Green surpassed 7.9 km, and, in 1901, Arthur Berson and Reinhard Süring ascended above 10.8 km, confirming the existence of the stratosphere.

High-altitude ballooning has played a major role in atmospheric research throughout the 20th century, and it's still used to the present day, but it's the heavier-than-air aircraft that now dominate the skies. The advent of propeller and eventually jet-powered aeroplanes has doubtlessly revolutionized travel, commerce and warfare, however, as with lighter-than-air balloons, the atmosphere puts a hard upper limit on how far they can take us. To reach beyond 40 km, enormous helium balloons are needed just to carry the smallest of payloads, and a jet plane has yet to ascend beyond 38 km – far below the Kármán line, the edge of outer space that is set at 100 km above sea level. Thus, to probe the furthest reaches of the outer atmosphere and to eventually gain enough horizontal velocity to enter into a stable orbit, major advancements in rocket science were needed.

The first objects to successfully cross the boundary of space were the Nazi V2 rockets, used in indiscriminate bombing campaigns in Western Europe. In 1945, this technology was seized by the victorious allied powers, and many German aerospace engineers were recruited by the United States in an effort to establish a presence in orbit and to develop intercontinental ballistic missiles. However, the Soviet Union quickly caught up, and on 4 October 1957, they launched Sputnik – the world's first artificial satellite. This event marked the beginning of the age of space exploration and the 'space race' – topics that are expanded upon in considerable detail later in the book.

This chapter will take you on a short vertical adventure from the ground we dwell on, through the various layers of our planet's atmosphere, up into the near vacuum of space, which was unreachable for humanity until the mid 20th century. It showcases many interesting and little-known weather phenomena that one might encounter in the upper atmosphere, and aims to put even the relatively short distances and scales of near-Earth space into their proper perspective.

These pages also chronicle some of our greatest aerospace achievements, and the challenges humanity had to overcome on our march towards establishing a presence beyond Earth's surface. This struggle has certainly paid off – most of you reading this book personally benefit from the thousands of satellites operating in orbit, whether using GPS navigation, looking at a weather forecast, watching TV or surfing on the internet. Every year, space-based technologies are having a greater impact on our lives, and we can only wonder what both the near and far future will bring.

Growth of the hole in the Earth's ozone layer over the South Pole from 1979 (top) to 2006 (bottom)

Results of a 7 g object impacting a block of solid aluminium at orbital speeds, illustrating the dangers of space junk

Visualization of all space debris in Earth orbit that is currently being tracked. Almost all objects larger than 5 cm in diameter are being tracked, and can be avoided by satellites and the ISS in advance

Transit of the International Space Station across the solar disc during the 2018 solar eclipse

CUMULONIMBU

50% OF THE TOTAL MASS OF THE
ATMOSPHERE LIES BELOW THIS LINE

ALTOCUMULUS

CUMULUS

NIMBOSTRATUS

ALTOSTRATUS

7 km
Most toy balloons burst

6.4 km
At this pressure,
water boils at 80 °C

4.8 km
Highest altitude
reached by a kite
In 2014, by the Australian
Robert Moore

3.8 km
Standard sport
skydiving altitude
Provides more than
60 seconds of free fall

2.7 km
Altitude reached by the
first hydrogen balloon
In 1783, by Jacques
Alexandre Charles

4.6 km
Highest bungee-jump
In 2002, by Curtis Rivers
from a hot air balloon

1.5 km
Highest altitude
reached by a jetpack

Mount Rainier

Mauna Kea

Mount Fuji

Mount Kilimanjaro

Matterhorn

5.1 km
La Rinconada, Peru
Highest permanently
inhabited human settlement

3.1 km
Altitude reachable by
a rifle bullet fired vertically

1.5 km
Highest altitude reachable
by consumer-grade
model rockets

2.4 km
Altitude sickness sets in
Caused by the low concentration
of oxygen, characterized by
fatigue, headaches and nausea

3.6 km
Average altitude
of La Paz, Bolivia
Highest capital city
in the world

Burj Khalifa

Shanghai Tower

Tokyo Skytree

Lotte World Tower

CN Tower

Taipei 101

Petronas Towers

Empire State Building

Ryugyong Hotel

Eiffel Tower

Woolworth Building

Pyramids of Giza

Uluru (Ayers Rock)

1 km

2 km

3 km

4 km

5 km

6 km

7 km

−20 °C

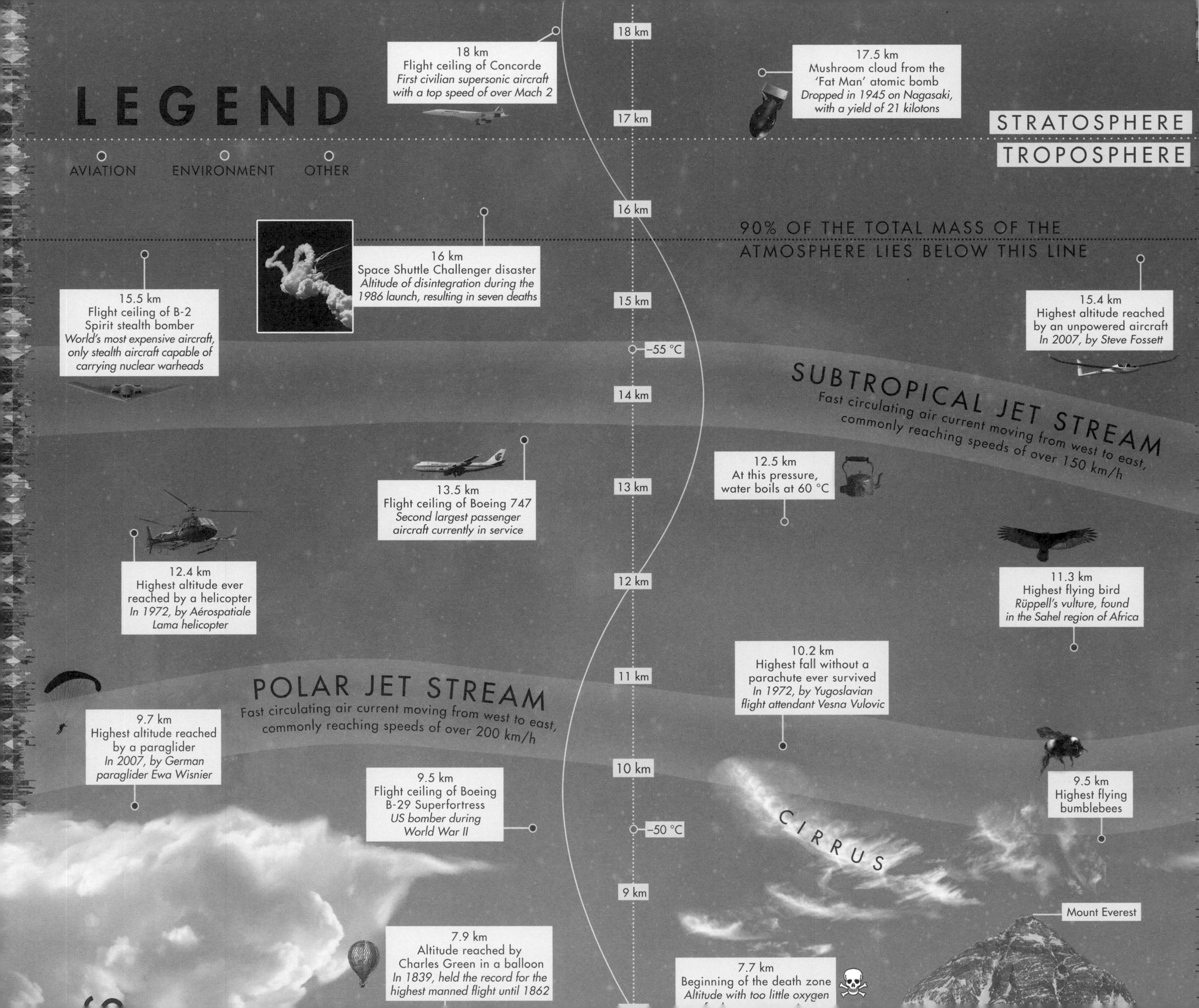

18 km
Flight ceiling of Concorde
First civilian supersonic aircraft
with a top speed of over Mach 2

17.5 km
Mushroom cloud from the
'Fat Man' atomic bomb
Dropped in 1945 on Nagasaki,
with a yield of 21 kilotons

15.5 km
Flight ceiling of B-2
Spirit stealth bomber
World's most expensive aircraft,
only stealth aircraft capable of
carrying nuclear warheads

16 km
Space Shuttle Challenger disaster
Altitude of disintegration during the
1986 launch, resulting in seven deaths

12.4 km
Highest altitude ever
reached by a helicopter
In 1972, by Aérospatiale
Lama helicopter

13.5 km
Flight ceiling of Boeing 747
Second largest passenger
aircraft currently in service

9.7 km
Highest altitude reached
by a paraglider
In 2007, by German
paraglider Ewa Wiśnier

POLAR JET STREAM
Fast circulating air current moving from west to east,
commonly reaching speeds of over 200 km/h

9.5 km
Flight ceiling of Boeing
B-29 Superfortress
US bomber during
World War II

7.9 km
Altitude reached by
Charles Green in a balloon
In 1839, held the record for the
highest manned flight until 1862

**90% OF THE TOTAL MASS OF THE
ATMOSPHERE LIES BELOW THIS LINE**

STRATOSPHERE

TROPOSPHERE

15.4 km
Highest altitude reached
by an unpowered aircraft
In 2007, by Steve Fossett

18 km · **17 km** · **16 km** · **15 km** · **14 km** · **13 km** · **12 km** · **11 km** · **10 km** · **9 km**

−55 °C · −50 °C

12.5 km
At this pressure,
water boils at 60 °C

SUBTROPICAL JET STREAM
Fast circulating air current moving from west to east,
commonly reaching speeds of over 150 km/h

10.2 km
Highest fall without a
parachute ever survived
In 1972, by Yugoslavian
flight attendant Vesna Vulović

11.3 km
Highest flying bird
Rüppell's vulture, found
in the Sahel region of Africa

9.5 km
Highest flying
bumblebees

7.7 km
Beginning of the death zone
Altitude with too little oxygen

C I R R U S

Mount Everest

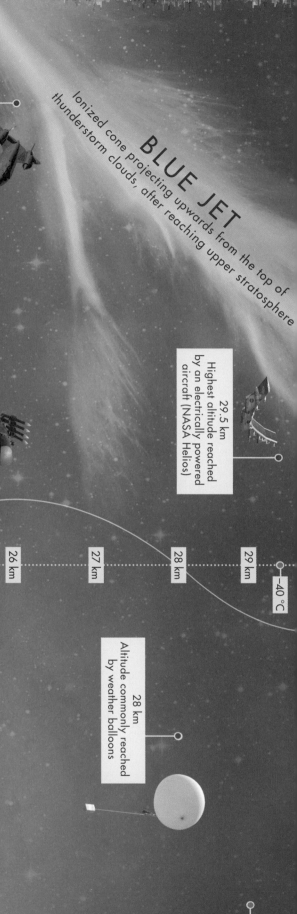

BLUE JET

Ionized cone projecting upwards from the top of thunderstorm clouds, after reaching upper stratosphere

POLAR NACREOUS CLOUDS

OZONE LAYER

22.6 km
Altitude reached by the first supersonic aircraft
Rocket-powered Bell X 1 in 1948, piloted by Chuck Yeager

26 km
Cruising altitude of Lockheed SR-71 Blackbird
World's fastest air-breathing manned aircraft (over Mach 3)

21.5 km
Planned operational altitude of High-Altitude Airships (HAA)
Proposed as a much cheaper alternative to communications and surveillance satellites

25 km
Altitude limit of a Buk surface-to-air missile
Probably used in 2014 to down the Malaysian Airlines Flight 17 over the Ukraine, killing 298 people

29.5 km
Highest altitude reached by an electrically powered aircraft (NASA Helios)

19 km | 20 km | 21 km | 22 km | 23 km | −50 °C | 24 km | 25 km | 26 km | 27 km | 28 km | 29 km | −40 °C

21 km
Highest altitude ever reached by a hot air balloon
In 2005, by Vijaypat Singhania

24 km
Mushroom cloud from a modern strategic nuclear warhead
Deployable at any time using ICBMs by most nuclear powers. Their yield varies from 300 to 500 kilotons

28 km
Altitude commonly reached by weather balloons

19 km
Armstrong Limit
Above this point, humans cannot survive without a pressurized suit. Exposed bodily fluids readily boil

22 km
Airburst of the Chelyabinsk meteorite
Heavier than the Eiffel Tower, exploded over Russia in 2013

25.5 km
At this pressure, water boils at room temperature (23 °C)

29 km
Majority of the sky appears black during daytime

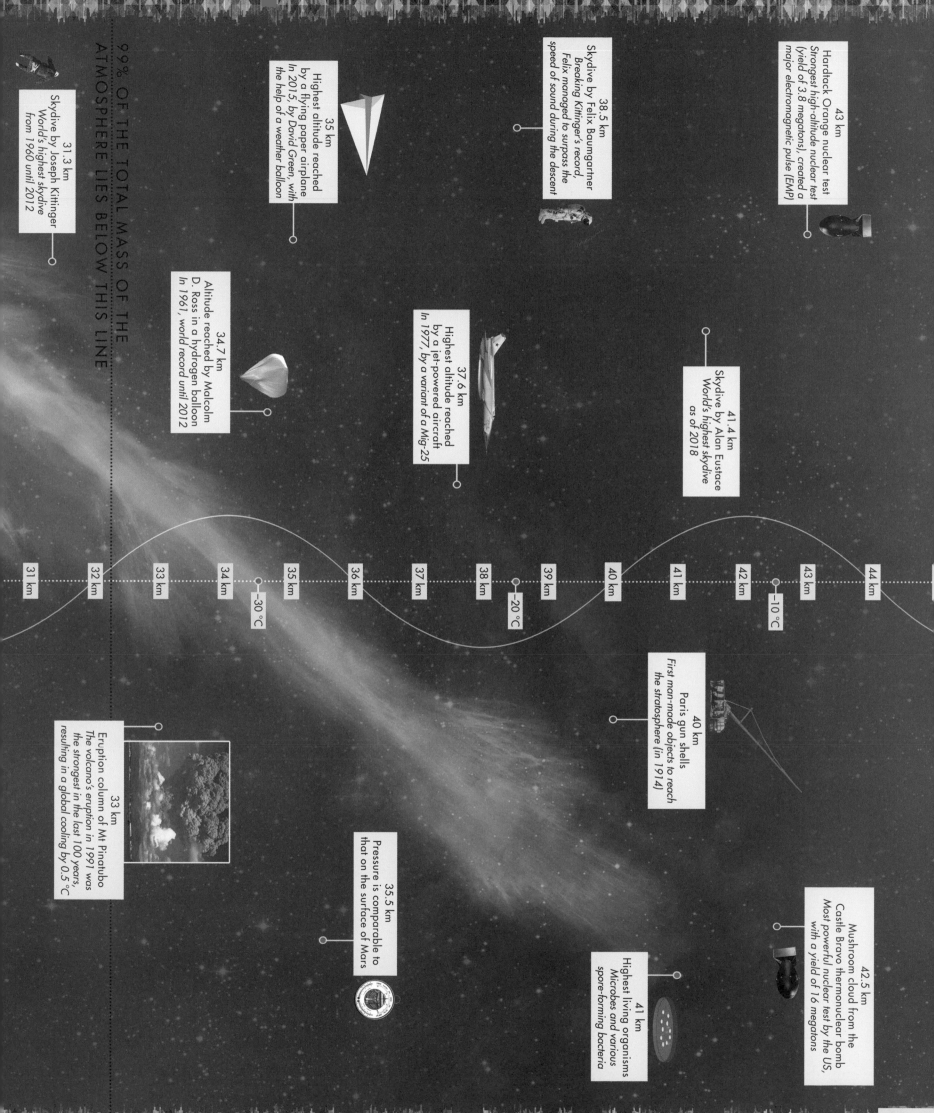

31.3 km
Skydive by Joseph Kittinger
World's highest skydive
from 1960 until 2012

35 km
Highest altitude reached
by a flying paper airplane
In 2015, by David Green, with
the help of a weather balloon

38.5 km
Skydive by Felix Baumgartner
Breaking Kittinger's record,
Felix managed to surpass the
speed of sound during the descent

43 km
Hardtack Orange nuclear test
Strongest high-altitude nuclear test
(yield of 3.8 megatons), created a
major electromagnetic pulse (EMP)

34.7 km
Altitude reached by Malcolm
D. Ross in a hydrogen balloon
In 1961, world record until 2012

37.6 km
Highest altitude reached
by a jet-powered aircraft
In 1977, by a variant of a Mig-25

41.4 km
Skydive by Alan Eustace
World's highest skydive
as of 2018

31 km
32 km
33 km
34 km
35 km
36 km
37 km
38 km
39 km
40 km
41 km
42 km
43 km
44 km

−30 °C
−20 °C
−10 °C

33 km
Eruption column of Mt Pinatubo
The volcano's eruption in 1991 was
the strongest in the last 100 years,
resulting in a global cooling by 0.5 °C

40 km
Paris gun shells
First man-made objects to reach
the stratosphere (in 1914)

35.5 km
Pressure is comparable to
that on the surface of Mars

42.5 km
Mushroom cloud from the
Castle Bravo thermonuclear bomb
Most powerful nuclear test by the US,
with a yield of 16 megatons

41 km
Highest living organisms
Microbes and various
spore-forming bacteria

51 km
Altitude reached by
the Winzen balloon
Balloon altitude world
record from 1972 to 2002

56 km
Mushroom cloud
from the Tsar Bomba
Most powerful nuclear test ever,
with a yield of over 50 megatons

64 km
Altitude reached by Albert
First primate astronaut (USA).
Suffocated during the flight

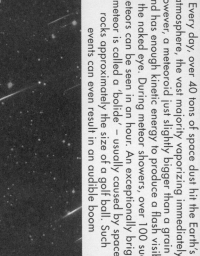

Every day, over 40 tons of space dust hit the Earth's
atmosphere, the vast majority vaporizing immediately.
However, a meteoroid just slightly bigger than a grain of
sand has enough kinetic energy to produce a flash visible
to the naked eye. During meteor showers, over 100 such
meteors can be seen in an hour. An exceptionally bright
meteor is called a 'bolide' – usually caused by space
rocks approximately the size of a golf ball. Such
events can even result in an audible boom

46 km
Earth's gravity is 99% as
strong compared to the surface

53 km
Highest altitude
reached by a balloon
BU60-1, in 2002

60 km
Space Shuttle Columbia disaster
Altitude of disintegration during the
1986 launch, resulting in seven deaths

45 km 46 km 47 km 48 km 49 km 50 km 52 km 54 km 56 km 58 km 60 km 62 km 64 km 66 km 68 km 70 km

−3 °C   −10 °C   −30 °C   −50 °C

50 km
Bottom of the ionosphere
Layer of the atmosphere electrically
charged by solar UV radiation

56 km
Camelopardalids burn up
Slowest meteor shower, with
a velocity of only 16 km/s

62 km
Planned altitude of the artificial meteor
shower proposed for the Tokyo 2020
summer Olympics opening ceremony
Conceived by the startup Sky Canvas, it would
involve a micro-satellite designed to fire up to
1,000 pellets towards the Earth from space

47.5 km
Sutter's Mill meteorite
Exploded over California
in 2012

MESOSPHERE
STRATOSPHERE

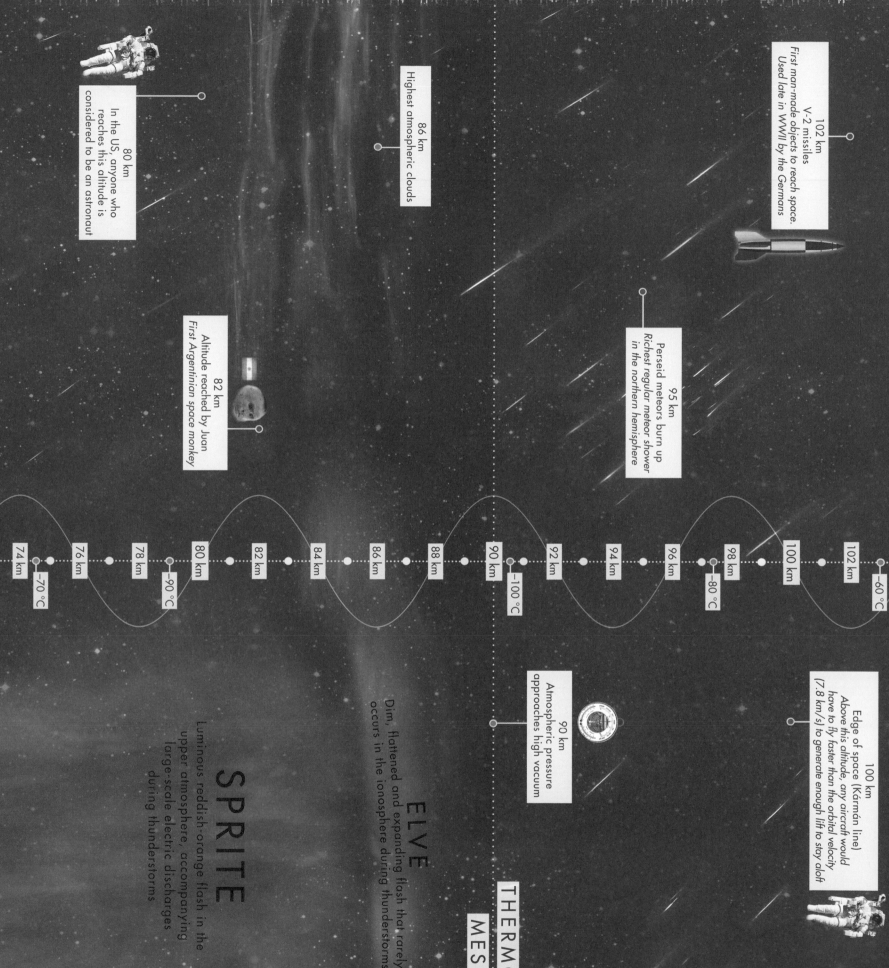

80 km
In the US, anyone who reaches this altitude is considered to be an astronaut

86 km
Highest atmospheric clouds

82 km
Altitude reached by Juan
First Argentinian space monkey

102 km
V-2 missiles
First man-made objects to reach space.
Used late in WWII by the Germans

95 km
Perseid meteors burn up
Richest regular meteor shower in the northern hemisphere

74 km	76 km	78 km	80 km	82 km	84 km	86 km	88 km	90 km	92 km	94 km	96 km	98 km	100 km	102 km

−70 °C   −90 °C   −100 °C   −80 °C   −60 °C

90 km
Atmospheric pressure approaches high vacuum

100 km
Edge of space (Kármán line)
Above this altitude, any aircraft would have to fly faster than the orbital velocity (7.8 km/s) to generate enough lift to stay aloft

SPRITE
Luminous reddish-orange flash in the upper atmosphere, accompanying large-scale electric discharges during thunderstorms

ELVE
Dim, flattened and expanding flash that rarely occurs in the ionosphere during thunderstorms

THERMOSPHERE
MESOSPHERE

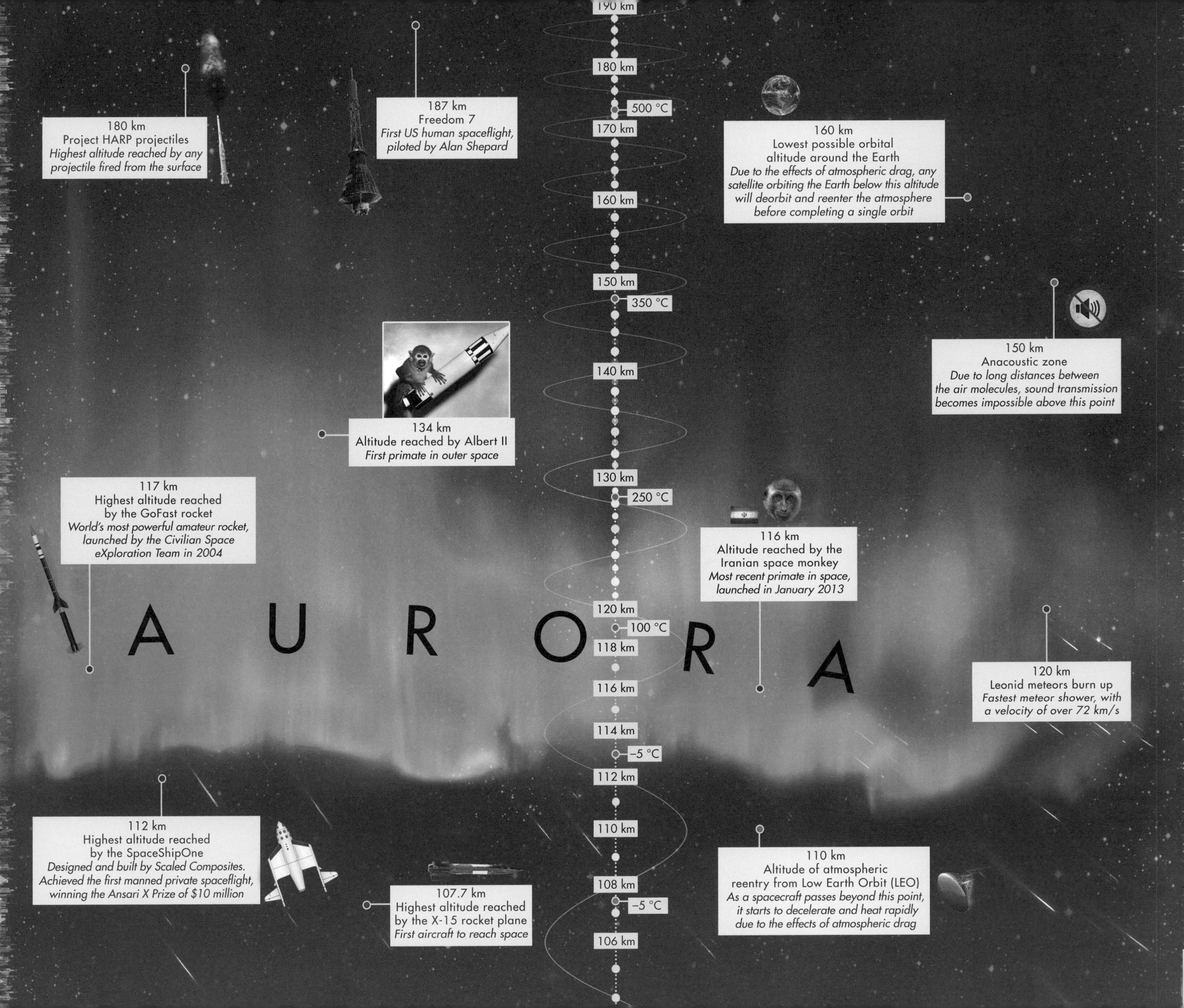

A U R O R A

180 km
Project HARP projectiles
Highest altitude reached by any
projectile fired from the surface

117 km
Highest altitude reached
by the GoFast rocket
World's most powerful amateur rocket,
launched by the Civilian Space
eXploration Team in 2004

112 km
Highest altitude reached
by the SpaceShipOne
Designed and built by Scaled Composites.
Achieved the first manned private spaceflight,
winning the Ansari X Prize of $10 million

187 km
Freedom 7
First US human spaceflight,
piloted by Alan Shepard

134 km
Altitude reached by Albert II
First primate in outer space

107.7 km
Highest altitude reached
by the X-15 rocket plane
First aircraft to reach space

106 km · 108 km · 110 km · 112 km · 114 km · 116 km · 118 km · 120 km · 130 km · 140 km · 150 km · 160 km · 170 km · 180 km · 190 km

−5 °C · −5 °C · 100 °C · 250 °C · 350 °C · 500 °C

116 km
Altitude reached by the
Iranian space monkey
Most recent primate in space,
launched in January 2013

160 km
Lowest possible orbital
altitude around the Earth
Due to the effects of atmospheric drag, any
satellite orbiting the Earth below this altitude
will deorbit and reenter the atmosphere
before completing a single orbit

110 km
Altitude of atmospheric
reentry from Low Earth Orbit (LEO)
As a spacecraft passes beyond this point,
it starts to decelerate and heat rapidly
due to the effects of atmospheric drag

120 km
Leonid meteors burn up
Fastest meteor shower, with
a velocity of over 72 km/s

150 km
Anacoustic zone
Due to long distances between
the air molecules, sound transmission
becomes impossible above this point

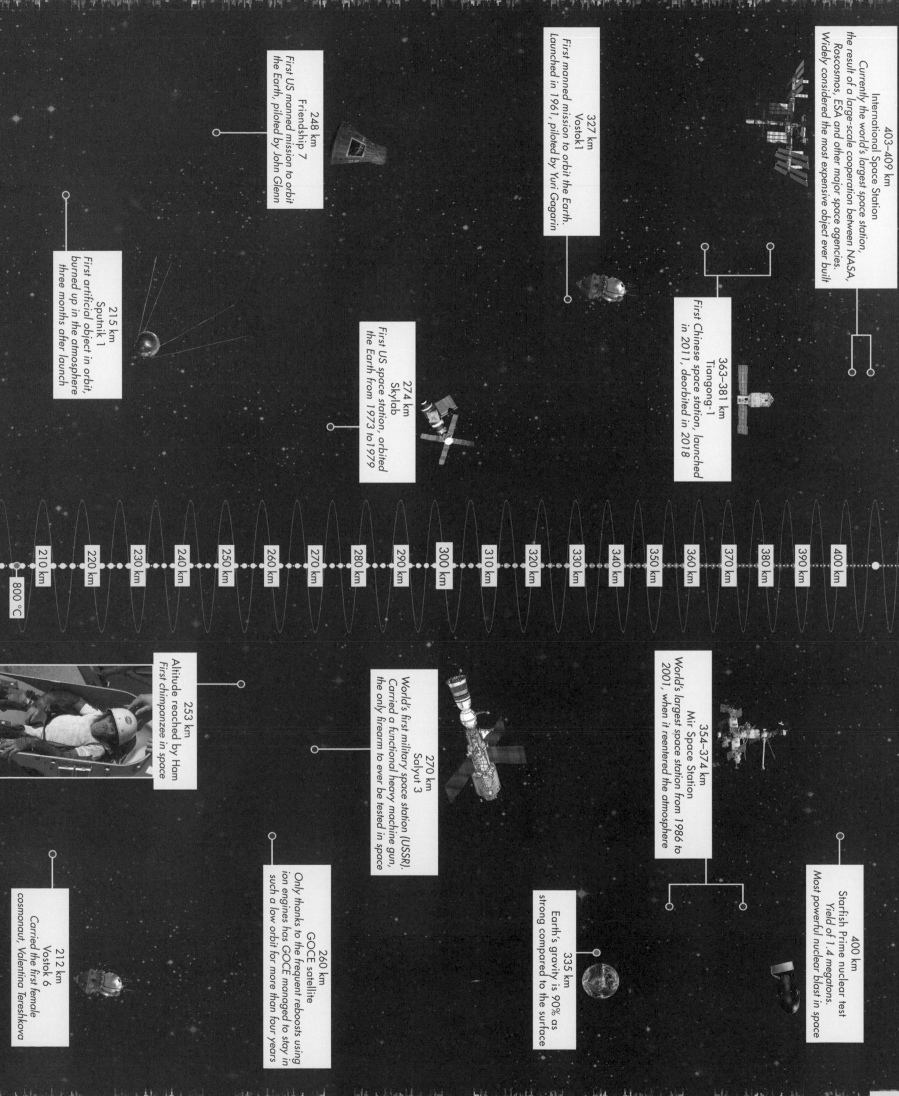

403–409 km
International Space Station
Currently the world's largest space station,
the result of a large-scale cooperation between NASA,
Roscosmos, ESA and other major space agencies.
Widely considered the most expensive object ever built

363–381 km
Tiangong-1
First Chinese space station, launched
in 2011, deorbited in 2018

327 km
Vostok1
First manned mission to orbit the Earth.
Launched in 1961, piloted by Yuri Gagarin

274 km
Skylab
First US space station, orbited
the Earth from 1973 to1979

248 km
Friendship 7
First US manned mission to orbit
the Earth, piloted by John Glenn

215 km
Sputnik 1
First artificial object in orbit,
burned up in the atmosphere
three months after launch

354–374 km
Mir Space Station
World's largest space station from 1986 to
2001, when it reentered the atmosphere

400 km
Starfish Prime nuclear test
Yield of 1.4 megatons.
Most powerful nuclear blast in space

335 km
Earth's gravity is 90% as
strong compared to the surface

270 km
Salyut 3
World's first military space station (USSR).
Carried a functional heavy machine gun,
the only firearm to ever be tested in space

260 km
GOCE satellite
Only thanks to the frequent reboosts using
ion engines has GOCE managed to stay in
such a low orbit for more than four years

253 km
Altitude reached by Ham
First chimpanzee in space

212 km
Vostok 6
Carried the first female
cosmonaut, Valentina Tereshkova

800 °C

200 | 210 km | 220 km | 230 km | 240 km | 250 km | 260 km | 270 km | 280 km | 290 km | 300 km | 310 km | 320 km | 330 km | 340 km | 350 km | 360 km | 370 km | 380 km | 390 km | 400 km

EXOSPHERE

THERMOSPHERE

789 km
2009 satellite accident
Collision of the Iridium 33 and
Kosmos-2251 satellites, producing
a large amount of space debris

615 km
Highest altitude ever reached
by the Space Shuttle
Space Shuttle Discovery in 1990,
during the STS 31 mission that deployed
the Hubble Space Telescope

491 km
WISE
NASA infrared space telescope
that helped to discover thousands
of minor planets and asteroids

781 km
Iridium satellites
Array of 66 communications satellites providing
signal coverage over the Earth's entire surface.
Due to their reflective antennas, they sometimes
produce the iconic bright 'Iridium flares,' that
can be visible even during daylight hours

540 km
Hubble Space Telescope
Most powerful visible light space telescope.
In operation for over 26 years, scheduled
to be succeeded by the much more powerful
James Webb Space Telescope in 2020

830 km
Densest concentration
of space debris

600 km
Above this point, Earth's atmosphere
is mostly composed of helium

1996 Cerise satellite accident
670 km
First verified accidental collision of
a satellite with artificial space debris

480 km
Background radiation levels are nearly
100 times higher than at the surface

530 km
Argus III nuclear test
Yield of 1.7 kilotons
Highest nuclear test ever

700 km
Most off-the-shelf electronic devices
stop working unless shielded heavily.
Due to high levels of cosmic radiation,
all satellites operating beyond this
altitude are equipped with specialized
radiation-hardened integrated circuits

500 km

600 km

700 km

800 km

1,604 km
Echo 1
Metallized inflatable balloon
satellite 31 m in diameter, acting
as a reflector of microwave signals

N.A.S.A.

1,484 km
First British satellite, its launch in 1962
made the UK the third country in the
world to operate a satellite

Apogee of Ariel 1

984 km
Apogee of Badr-1
First Pakistani satellite

890 km
Cosmic Background Explorer
Microwave and infrared observatory
that provided valuable evidence
supporting the Big Bang theory

1,005 km
Alouette 1
First Canadian
satellite

1,400 km
Altitude reachable by the most
powerful 'sounding rockets'
A cheap method to probe the layers of upper
atmosphere unreachable by weather balloons,
testing equipment that will be used in more
expensive and risky orbital spaceflight missions

1,659 km
Altitude reached by Laika
A stray dog sent to space on
board Sputnik 2, as the first
animal to orbit the Earth

900 km • • • • • • • 1,000 km • • • • 1,100 km • • • 1,200 km • • 1,300 km • 1,400 km 1,500 km 1,600 km 1,700 km

985 km
Earth's gravity is 75% as
strong compared to the surface

1,369 km
Apogee of Gemini 11
Highest altitude reached by a manned
spacecraft outside of the Apollo programme

1,000 km
Lower boundary of the inner
Van Allen radiation belt
Layer of energetic protons held in place by the
Earth's magnetic field, extending upwards to
9,500 km. In this region, radiation levels are over
25,000 times higher than at the Earth's surface

1,484 km
SCORE apogee
World's first communications satellite.
Launched by the USA in 1958, it was
incorporated into an Atlas B rocket

865 km
2007 Chinese anti-satellite missile test
Largest recorded creation of space debris
in history, creating more than 2,000
fragments larger than a golf ball

Even though we have managed to reach the outermost layers of the Earth's
atmosphere, continuing at this scale, it would require 13 more pages just to
reach the orbital altitude of GPS satellites, 24 pages to reach the geostationary
orbit, 272 pages to reach the Moon, and 53,700 more pages to reach the Sun

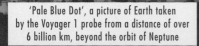
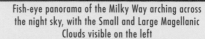
Fish-eye panorama of the Milky Way arching across the night sky, with the Small and Large Magellanic Clouds visible on the left

South pole of the gas giant Jupiter, imaged by the NASA Juno probe

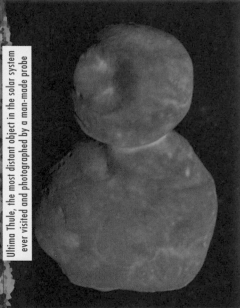
Ultima Thule, the most distant object in the solar system ever visited and photographed by a man-made probe

Dark dust clouds that will eventually collapse to form young stars, located in the Rosette Nebula

SCALE OF THE UNIVERSE

In the previous chapter, we voyaged through the atmosphere into the nearest reaches of outer space – a trip that all astronauts and satellites have to undergo. However, our journey through the universe has only just started.

The astronauts of Apollo 13, that ill-fated mission to the Moon, which fortunately didn't end in tragedy, earned the record of travelling the furthest distance from the Earth's surface, further than any humans before or after them – 400,171 km. Even though this is over ten times the circumference of our planet, it pales in comparison to the distances we encounter in just our own solar system. The average distance from the Earth to the Sun (one astronomical unit, often shortened to AU) is almost 150 million km. It would take over a century of non-stop driving at a typical motorway speed to cover this distance; however, a light ray from the Sun accomplishes it in less than 9 minutes. It takes 43 minutes for sunlight to travel to Jupiter, 1 hour and 20 minutes to Saturn and over 4 hours to Neptune. The Voyager 1 probe, the most distant artificial object, is currently located over 145 AU away, and it takes over 20 hours for its radio signals to cover the distance back to Earth.

The speed of light in a vacuum (299,79,458 m/s) is one of the most important fundamental constants in physics. It's the maximum speed anything in the universe can travel at, it connects space and time in the theory of relativity, and it also appears in the famous mass–energy equivalence $E = mc^2$. When talking about astronomical distances, units based on the distance light covers in a given time (such as light minutes, light days and light years) become the most convenient.

A light year is approximately equal to 63,241 AU, or 9,460,000,000,000,000 km. Another unit of length commonly used in astronomy is the parsec. One parsec is equal to about 3.26 light years, and is defined as the distance at which an object with a diameter of 1 AU (such as a red giant star) covers an angle of 1 arcsecond (1/3600 of a degree, 1/1800 of the apparent size of the lunar disc) in the sky.

Using these units of length, the closest star to the Sun, Proxima Centauri, lies approximately 4.244 light years (1.3 parsecs) away. Most of the few thousand stars visible to the naked eye in the night sky are less than 1,000 light years away; however, that's just a tiny fraction compared to the size of our own galaxy, the Milky Way, which is over 150,000 light years in diameter, and contains upwards of 1 trillion stars. Then the nearest large galaxy, Andromeda, is roughly 2 million light years away, while the diameter of our observable universe has been calculated as 93 billion light years.

101955 Bennu, a potentially dangerous Earth-crossing asteroid 500 m in diameter, photographed by the OSIRIS-REx probe

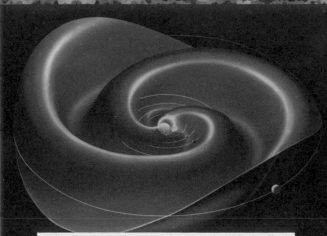
Heliospheric current sheet, a spiral structure shaped by the effects of the Sun's rotating magnetic field on the interplanetary medium

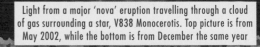
Light from a major 'nova' eruption travelling through a cloud of gas surrounding a star, V838 Monocerotis. Top picture is from May 2002, while the bottom is from December the same year

Just the raw numbers make it very difficult for a person to grasp how mind-boggling the distances we're talking about are. The sheer scale of the universe can't be imagined and easily put into perspective by a human mind.

This chapter aims to help readers to understand the sheer scale and variety of the macrocosmic realms by comparing the relative sizes of different objects in our universe in a series of panels, each on a larger scale than the one preceding it. It's a journey from tiny satellites and the smallest of asteroids to galaxies and, ultimately, to galactic superclusters – the largest known structures in the universe. Special attention is given to exoplanets – extraterrestrial worlds, one of the most rapidly advancing fields of modern astronomy, which will undoubtedly provide many surprises in the future.

Keeping in mind the sheer scale of the universe and our place in it, our troubles and preoccupations that once seemed important evaporate away. However, this should not depress us – you can stand in awe at the inconceivably large cosmos, and be motivated to value and cherish what and who you have in your own life even more deeply.

JAMES WEBB SPACE TELESCOPE
(NASA visible and infrared light telescope, planned to be launched in 2021)

NEW HORIZONS
(NASA outer solar system probe that provided the first high-resolution pictures of Pluto)

SPITZER SPACE TELESCOPE
(NASA infrared telescope)

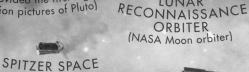

LUNAR RECONNAISSANCE ORBITER
(NASA Moon orbiter)

GPS SATELLITE
(One of the 31 satellites of the Global Positioning System operated by the US Air Force)

CASSINI
(Internationally operated Saturn orbiter)

MAVEN
(NASA Mars orbiter)

EARTH RADIATION BUDGET SATELLITE
(NASA scientific satellite)

EC...
(Passi...
micro...

INTERNATIONAL SPACE STATION

HUBBLE SPACE TELESCOPE
(NASA visible light telescope)

UARS
(NASA orbital observatory studying the Earth's atmosphere)

COMPTON GAMMA RAY OBSERVATORY
(NASA gamma ray observatory)

MESSENGER
(NASA Mercury orbiter)

KEPLER SPACE OBSERVATORY
(NASA space observatory dedicated to discovering Earth-sized exoplanets)

TERRA
(NASA climate and environment monitoring satellite)

EXOMARS ORBITER
(ESA Mars telecommunications orbiter and atmospheric gas analyser)

CHANDRA
(NASA X-ray observatory)

SALYUT 7
(Soviet space station)

GEMINI SPACECRAFT
(Used during the Gemini program to develop and test space travel techniques and orbital manoeuvres necessary for the Apollo programme)

MARS RECONNAISSANCE ORBITER
(NASA Mars orbiter)

OAO-2
(First orbital observatory)

COBE
(NASA observatory studying the cosmic microwave background)

EXTREME ULTRAVIOLET EXPLORER
(NASA ultraviolet observatory)

SPUTNIK
(Soviet satellite, first artificial object in orbit)

SWIFT
(NASA gamma ray observatory)

CARBON OBSERVATORY 2
(NASA climate and environment monitoring satellite)

HUMANS

JUNO
(NASA Jupiter orbiter)

ROSETTA
(ESA interplanetary probe sent to the comet Churyumov-Gerasimenko)

OSIRIS-ReX
(Nasa probe sent towards the asteroid Bennu)

ELON MUSK'S ROADSTER
(Strapped on top of a Falcon Heavy upper stage, orbiting the Sun between the Earth's orbit and the asteroid belt)

SKYLAB
(US space station)

SPUTNIK 2
(Soviet satellite that carried Laika, first dog in orbit)

MIR
(Russian space station)

NuSTAR
(Nasa orbital X-ray observatory)

10 M

ORION
(NASA spacecraft currently under development, intended to carry a crew of four beyond low Earth orbit)

IRAS
(NASA satellite, world's first infrared space telescope)

VENERA 7
(Soviet probe, first object to land softly on Venus)

B2100
(Inflatable space habitat designed by Bigelow Aerospace. Fully expanded, it would have twice the volume of the ISS)

CURIOSITY
(NASA Mars rover in operation since 2011)

DAWN
(NASA interplanetary probe, first spacecraft to fly by Ceres)

ECHO 1
(Passive reflector of microwave signals)

SPACE EXPLORATION VEHICLE
(Planned NASA crewed rover to be used for long-distance Moon and Mars surface travel)

MAGELLAN
(NASA Venus orbiter)

PIONEER 10
(NASA interplanetary probe, first artificial object to fly by Jupiter)

B330
(Inflatable space habitat being privately developed by Bigelow Aerospace)

HO 2
flector of ave signals)

PASCOMSAT
(Passive reflector of microwave signals)

EMU SUIT
(NASA spacesuit used during ISS spacewalks)

OPPORTUNITY
(NASA Mars rover that operated for a record 14 years)

Although it is quite difficult to observe, all light that shines on a surface acts with a tiny amount of pressure. Satellites powered by solar sails take an advantage of this effect – facing the Sun, the light slowly accelerates them in the desired direction, allowing them to slowly, but very efficiently, travel across the solar system without a need for any fuel

IKAROS
(Japanese probe powered by a solar sail)

VOYAGER 1
NASA exploratory probe hat visited Jupiter, Saturn Uranus and Neptune)

APOLLO COMMAND & SERVICE MODULE

GALILEO
(NASA Jupiter orbiter)

APOLLO LUNAR MODULE
(Used for the six successful Moon landings of the Apollo programme)

In 1908, a massive explosion occurred in Central Siberia, near the Tunguska river. It was equivalent to over 15 megatons of TNT, and flattened over 2,000 km² of forest. The most likely culprit was a rocky asteroid roughly 100 m in diameter, that exploded after entering the Earth's atmosphere, not leaving any known crater. Most of the small to medium-sized impact craters, such as the Meteor Crater in Arizona, have been caused by much denser iron asteroids, that can penetrate the atmosphere far more easily

CHELYABINSK ASTEROID
(Exploded over the Russian city of Chelyabinsk in 2014)

MARINER 4
(NASA probe, first spacecraft to fly by Mars)

BLACK HOLE OF ONE JUPITER MASS

SPACE SHUTTLE ORBITER
(Reusable part of the Space Shuttle, to deliver the Hubble Telescope and the majority of the ISS components into orbit)

SUTTER'S MILL ASTEROID
(Exploded over California in 2012)

TUNGUSKA ASTEROID

DRAGON 2
(Crewed orbital spacecraft designed and built by SpaceX, made its first unmanned flight in 2019)

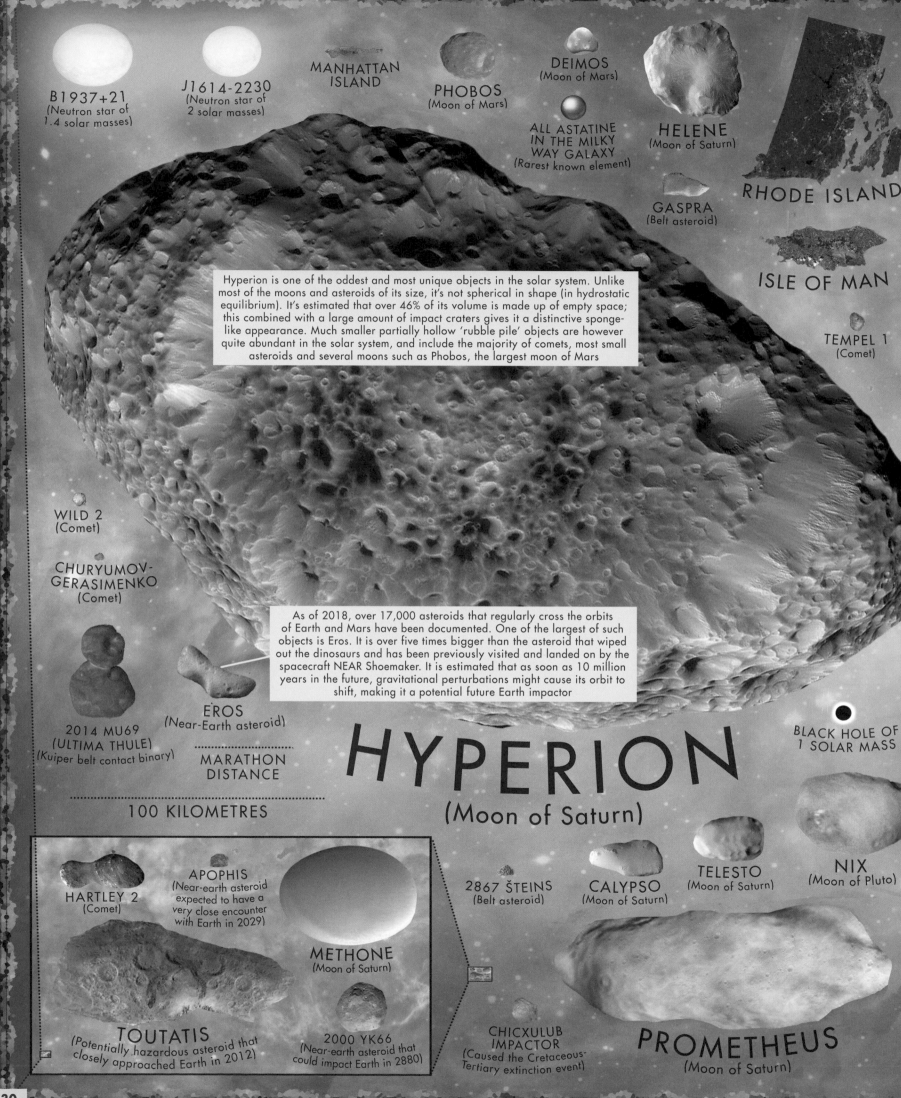

B1937+21
(Neutron star of
1.4 solar masses)

J1614-2230
(Neutron star of
2 solar masses)

MANHATTAN
ISLAND

PHOBOS
(Moon of Mars)

DEIMOS
(Moon of Mars)

ALL ASTATINE
IN THE MILKY
WAY GALAXY
(Rarest known element)

HELENE
(Moon of Saturn)

GASPRA
(Belt asteroid)

RHODE ISLAND

ISLE OF MAN

TEMPEL 1
(Comet)

Hyperion is one of the oddest and most unique objects in the solar system. Unlike most of the moons and asteroids of its size, it's not spherical in shape (in hydrostatic equilibrium). It's estimated that over 46% of its volume is made up of empty space; this combined with a large amount of impact craters gives it a distinctive sponge-like appearance. Much smaller partially hollow 'rubble pile' objects are however quite abundant in the solar system, and include the majority of comets, most small asteroids and several moons such as Phobos, the largest moon of Mars

WILD 2
(Comet)

CHURYUMOV-
GERASIMENKO
(Comet)

As of 2018, over 17,000 asteroids that regularly cross the orbits of Earth and Mars have been documented. One of the largest of such objects is Eros. It is over five times bigger than the asteroid that wiped out the dinosaurs and has been previously visited and landed on by the spacecraft NEAR Shoemaker. It is estimated that as soon as 10 million years in the future, gravitational perturbations might cause its orbit to shift, making it a potential future Earth impactor

2014 MU69
(ULTIMA THULE)
(Kuiper belt contact binary)

EROS
(Near-Earth asteroid)

MARATHON
DISTANCE

100 KILOMETRES

BLACK HOLE OF
1 SOLAR MASS

HYPERION
(Moon of Saturn)

HARTLEY 2
(Comet)

APOPHIS
(Near-earth asteroid
expected to have a
very close encounter
with Earth in 2029)

METHONE
(Moon of Saturn)

TOUTATIS
(Potentially hazardous asteroid that
closely approached Earth in 2012)

2000 YK66
(Near-earth asteroid that
could impact Earth in 2880)

2867 ŠTEINS
(Belt asteroid)

CALYPSO
(Moon of Saturn)

TELESTO
(Moon of Saturn)

NIX
(Moon of Pluto)

CHICXULUB
IMPACTOR
(Caused the Cretaceous-
Tertiary extinction event)

PROMETHEUS
(Moon of Saturn)

NIOBE
(Belt asteroid)

243 IDA
(Belt asteroid)

Planets and dwarf planets are not the only objects in the solar system that have their own moons. 243 Ida, a belt asteroid that was visited in 1993 by the Galileo spacecraft, is one of them. Its moon, Dactyl, is 1.2 km in diameter, and orbits Ida at a distance of 90 km. Dactyl's orbital velocity is just 10 m/s – comparable to the peak running speed of a trained sprinter

VOLCANIC EJECTA FROM THE HUCKLEBERRY RIDGE YELLOWSTONE ERUPTION
(2.1 million years ago)

VOLCANIC EJECTA FROM THE 1815 MOUNT TAMBORA ERUPTION

PANDORA
(Moon of Saturn)

MALTA

ALL LIVING MASS ON EARTH

JANUS
(Moon of Saturn)

VREDEFORT IMPACTOR
(Created the largest known meteor crater on Earth, located in modern-day South Africa

HALLEY'S COMET

BEATRIX
(Belt asteroid)

KERBEROS
(Moon of Pluto)

3200 PHAETHON
(Inner solar system asteroid of the Geminids meteor shower)

STYX
(Moon of Pluto)

EPIMETHEUS
(Moon of Saturn)

LUTETIA
(Belt asteroid)

MATHILDE
(Belt asteroid)

HAWAII ISLAND

TRITON
(Moon of Neptune)

HAUMEA
(Dwarf planet)

ARIEL
(Moon of Uranus)

HI'IAKA
(Moon of Haumea)

TETHYS
(Moon of Saturn)

UMBRIEL
(Moon of Uranus)

RHEA
(Moon of Saturn)

IAPETUS
(Moon of Saturn)

CHARON
(Moon of Pluto)

DIONE
(Moon of Saturn)

OBERON
(Moon of Uranus)

MIMAS
(Moon of Saturn)

TITANIA
(Moon of Uranus)

PLUTO

2007 OR10
(Kuiper Belt object)

MAKEMAKE
(Dwarf planet)

ENCELADUS
(Moon of Saturn)

DYSNOMIA
(Kuiper Belt object)

KEPLER-3b
(Smallest known exoplanet)

MIRANDA
(Moon of Uranus)

TITAN
(Moon of Saturn)

ERIS
(Dwarf planet)

EARTH'S INNER
CORE

NEREID
(Moon of
Neptune)

PUCK
(Moon of
Uranus)

LARISSA
(Moon of Neptune)

2002 MS4
(Black hole of
23 solar masses)

2002 MS4
(Kuiper belt object)

PROTEUS
(Moon of
Neptune)

IXION
(Kuiper Belt
object)

EUROPA
(Moon of Jupiter)

BRITISH ISLES

Due to tidal friction caused by Jupiter and the moons
Europa and Ganymede, Io is by far the most volcanically
active object in the solar system. Along with the Earth's
Moon, it is the only moon in the solar system thought to
have a partially molten rocky interior, however Io's surface
is relatively young, covered by a thick layer of sulphur

IO
(Moon of Jupiter)

SALACIA
(Kuiper Belt object)

ALL WATER
ON EARTH

KLEOPATRA
(Belt asteroid)

EARTH

SEDNA
(Detached object)

EARTH'S MOON

VARDA
(Kuiper Belt object)

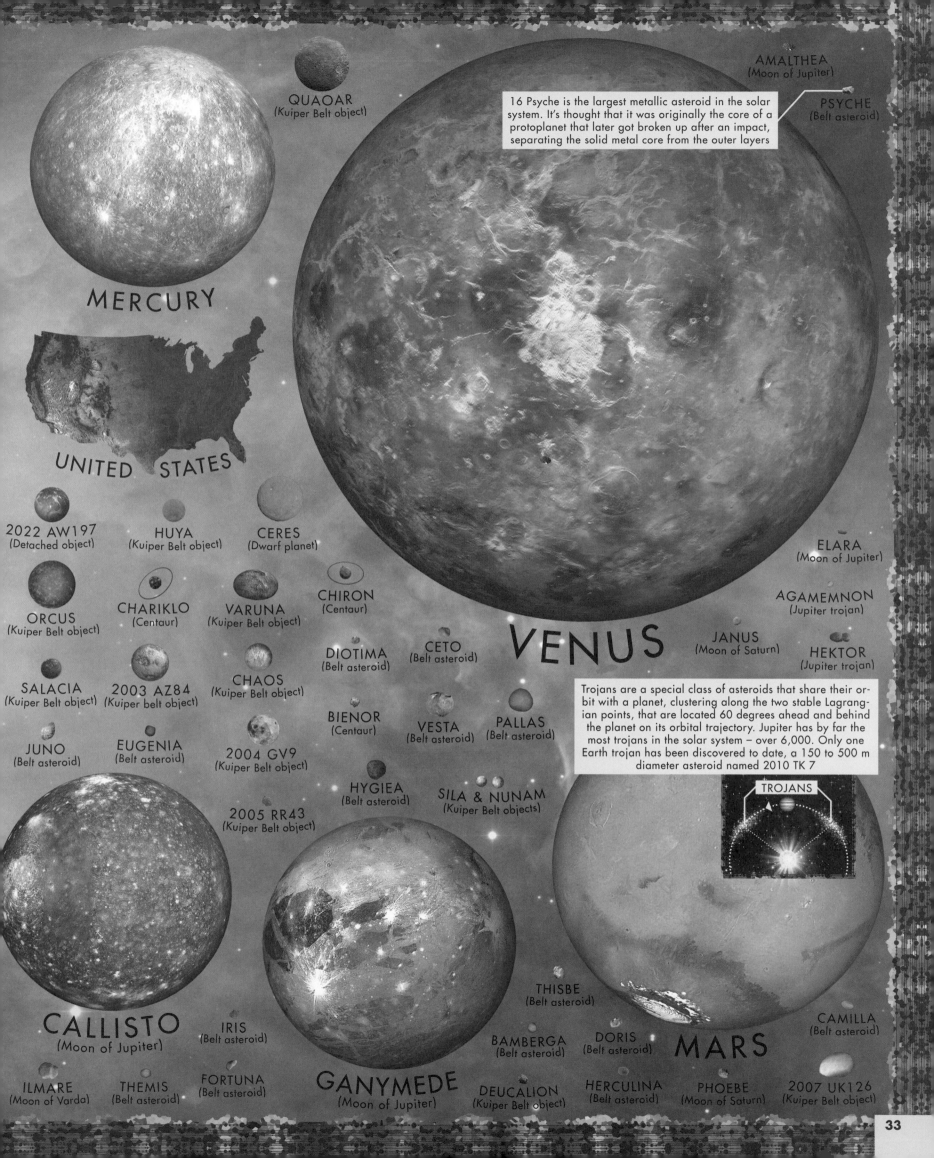

MERCURY

QUAOAR
(Kuiper Belt object)

AMALTHEA
(Moon of Jupiter)

16 Psyche is the largest metallic asteroid in the solar system. It's thought that it was originally the core of a protoplanet that later got broken up after an impact, separating the solid metal core from the outer layers

PSYCHE
(Belt asteroid)

UNITED STATES

VENUS

2022 AW197
(Detached object)

HUYA
(Kuiper Belt object)

CERES
(Dwarf planet)

ELARA
(Moon of Jupiter)

ORCUS
(Kuiper Belt object)

CHARIKLO
(Centaur)

VARUNA
(Kuiper Belt object)

CHIRON
(Centaur)

AGAMEMNON
(Jupiter trojan)

SALACIA
(Kuiper Belt object)

2003 AZ84
(Kuiper belt object)

DIOTIMA
(Belt asteroid)

CETO
(Belt asteroid)

CHAOS
(Kuiper Belt object)

JANUS
(Moon of Saturn)

HEKTOR
(Jupiter trojan)

JUNO
(Belt asteroid)

EUGENIA
(Belt asteroid)

2004 GV9
(Kuiper Belt object)

BIENOR
(Centaur)

VESTA
(Belt asteroid)

PALLAS
(Belt asteroid)

Trojans are a special class of asteroids that share their orbit with a planet, clustering along the two stable Lagrangian points, that are located 60 degrees ahead and behind the planet on its orbital trajectory. Jupiter has by far the most trojans in the solar system – over 6,000. Only one Earth trojan has been discovered to date, a 150 to 500 m diameter asteroid named 2010 TK 7

HYGIEA
(Belt asteroid)

SILA & NUNAM
(Kuiper Belt objects)

2005 RR43
(Kuiper Belt object)

TROJANS

CALLISTO
(Moon of Jupiter)

IRIS
(Belt asteroid)

THISBE
(Belt asteroid)

CAMILLA
(Belt asteroid)

ILMARE
(Moon of Varda)

THEMIS
(Belt asteroid)

FORTUNA
(Belt asteroid)

GANYMEDE
(Moon of Jupiter)

BAMBERGA
(Belt asteroid)

DEUCALION
(Kuiper Belt object)

DORIS
(Belt asteroid)

HERCULINA
(Belt asteroid)

MARS

PHOEBE
(Moon of Saturn)

2007 UK126
(Kuiper Belt object)

MOA-2011-BLG-028L b

OGLE-2005-169L-b

MOA-2007-BLG-192L b

METHUSELAH
(PSR B 1620-26 b)

OGLE-2008-BLG-092L b

URANUS

HYPOTHETICAL PLANET NINE

NEPTUNE

NY VIR c

OGLE-2013-BLG-0341LBb

GJ 229 Ab

OGLE-2016-BLG-1195L b

GJ 649 b

PSR B1620-26 b is the oldest exoplanet ever discovered, with its age estimated at 12.7 billion years. It orbits around a binary system composed of a white dwarf and a millisecond pulsar, inside a dense stellar cluster Messier 4. The planet probably did not form where it is found today, but was instead captured by the pulsar along with its mother star (which later became the white dwarf)

OGLE-2005-BLG-169Lb

SATURN

GLIESE 433 c

KEPLER-1630 B

KEPLER-167 e

BD-06 1339 d

GJ 667 C g

OGLE-2007-BLG-349(AB)b

KIC 12454613 b

HD 219134 h

GJ 221 BD-06 1339 d

GJ 667 C d

HD 219

AEGIR
(EPSILON ERIDANI E

KEPLER-1540 b

OGLE-2005-BLG-390Lb

HD 99492 c

Planet Nine is a theorized large planet located far beyond the orbit of Neptune. Its gravitational effects are thought to be the best explanation for the unusually similar orbits of numerous trans-Neptunian objects. According to models, it would take the planet over 15,000 years to complete just one orbit around the Sun

A large amount of currently known exoplanets belong to the class of 'ice giants' – large planets that lack a solid surface and have a composition similar to that of Uranus and Neptune. In contrast to gas giants, ice giants are not majority hydrogen and helium by mass, but are mostly composed of water, ammonia and light hydrocarbons

HD 113538 b

GJ 180 c

A PLETHORA OF
EXOPLANETS

Since the discovery of the first planet orbiting another world in the year 1991, we have managed to discover over 3,500 exoplanets using various diverse methods. This chapter aims to showcase the vast and incredible variety of other worlds discovered to date, visualizing almost 1,000 different planets in our galaxy. For better clarity, the different planets are arranged horizontally according to their equilibrium temperature levels. Continuing with the 'Scale of the Universe', the last page of this chapter is dedicated to small stars, brown dwarfs and stellar remnants.

Fomalhaut B, located around 25 light years away, is the first exoplanet that was observed directly, and currently the faintest extrasolar object ever imaged

KAPTEYN C

DAGON
(FOMALHAUT B)

HD 219134 e

HD 204941 b

HD

TEMPERATURE ▶   −175 °C   −150 °C

34

GJ 163 d

GJ 3138 d

TRAPPIST-1 h

GLIESE 785 c

KEP.-1600 b

GJ 682 c

HD 82943 d

HD 215456 c

KIC 6436029 b

SANCHO
(MU ARAE e)

KEPLER-186 f

KEPLER-48 e

EPIC 201598502 b

HD 141399 e

The visual appearance of gas giants (planets without a solid surface composed mostly of hydrogen and helium) is highly influenced by the chemical composition of their cloud layers. As different elements and compounds condense into clouds at different temperatures, we can predict how gas giants of various temperatures are going to look. At temperatures below −150 °C, ammonia and ammonium hydrosulphide cloud decks are expected to be dominant – resulting in a reddish-yellow appearance

HD 150433 b

KEPLER-1536 b

KEPLER-174 d

JUPITER

HD 134060 c

HD 141399 e

GJ 3323 c

HIP 57050 b

GJ 682 b

GJ 687 b

876 e

HD 10180 h

HD 126525 b

KEPLER-16 (AB) b

HD 85390 b

LIPPERHEY
(55 CANCRI d)

04149 b

In 1992, a pulsar (rapidly rotating neutron star), PSR 1257+12, became the first extrasolar object proven to host a planetary system of its own. This was unexpected, as pulsars are only formed when a sufficiently massive star explodes in a violent supernova. Due to the massive amount of ionizing radiation emitted by the stellar remnant, pulsar planets are highly unlikely to be habitable

HD 4208 b

41 b

HD 134987 c

PHOBETOR
(PSR 1257+12 d)

HIP 79431 b

KEPLER-1097 b

TRAPPIST-1 g

KEP.-441 b

KEP.-421 b

−125 °C

−100 °C

KEPLER-459 B

TAU CETI f

HD 40307 g

KEPLER-296 f

TRAPPIST-1 f

KEPLER-455 b

LHS 1140 b

GJ 667 C f

KEP.-150 f

GJ 3293 c

POLTERGEIST (PSR B1257+12 c)

DOPPLER SPECTROSCOPY

In any planetary system, the star and its planets orbit a shared centre of mass called a barycentre. Often, the barycentre is located within the star, but it doesn't have to be – in fact, in the case of our own solar system, it is currently located outside the Sun itself.

As a planet orbits along and tugs on its home star, the star slightly wobbles back and forth in relation to the solar system. Because of the Doppler effect, these changes in radial velocity then cause detectable cyclical changes in its observed spectral lines.

Most of the early exoplanet discoveries were accomplished using this method. It's especially suitable for discovering massive gas giants orbiting low-mass stars, but useless detecting terrestrial planets around Sun-sized stars.

Potential exoplanet

Blue-shifted light

Barycentre

Red-shifted light

TRANSIT PHOTOMETRY

When an exoplanet transits between its host star and the Earth, it blocks a small percentage of the star's light and causes a slight dimming. The transit method then makes it possible to determine the planet's radius and orbit size.

Unfortunately, it's only possible to detect planets if their orbit is in a suitable alignment – that is the case for less than 1% of exoplanets located within the habitable zone. The rate of false positives is considerable as well, so additional methods often need to be used to confirm an exoplanet discovery.

Despite its limitations, over 75% of known exoplanets have been discovered using transit photometry. Several major ground and space observatories (most notably Kepler) use this method, and are continuously monitoring tens of thousands of stars for any changes in their brightness.

Additionally, the transit method occasionally allows astronomers to observe the emission spectrum of the planet's atmosphere and detect specific elements and compounds present there.

KEPLER-1229 b

KEPLER-1634 b

KIC 9662267 b

GJ 832 c

KIC 6372194 b

HARRIOT (55 CANCRI f)

HD 218566 b

55 CANCRI f

KEPLER-443 b

KEPLER-68 d

KEPLER-712 c

KEPL

HF 150433 b

KEPLER-34 (AB) b

KOI-682 b

HD 38858 b

PROXIMA CENTAURI b

In 2017, a system of seven terrestrial planets was discovered around the red dwarf star TRAPPIST-1, three of them being within the habitable zone. Tidal locking and frequent fluctuations of star brightness are a detriment to the habitability of red dwarf planets. However, as red dwarf lifespans can extend to several trillion years, any potential lifeforms are going to have far more time to evolve and thrive than the life on Earth

KEPLER 1647 b

TAPHAO KAEW (47 UMA C)

BD-08 2823 c

MARS

TRAPPIST-1 e

HD 65216 b

KEPLER-553 c

KEPLER-1086 c

KOI-4427.01

KOI-5929 b

HD 10180 g

GJ 3293 d

KEPLER-1341 b

KEPLER-1318 b

KIC 5094412 b

KEPLER-1318 b

KEPL

−75 °C

HABITABLE ZONE

−50 °C

Over 1,000 of the known exoplanets belong to the class of so-called 'Super-Earths' – planets made mostly out of silicate rock, but considerably more massive than the Earth. The largest super-Earths are estimated to have a radius around 2.2 times that of the Earth. Due to their mass, they are expected to have an atmosphere considerably denser than Earth's

HD 85512 b

TRAPPIST -1 d

KEPLER-1606 b

KEPLER-439 b

KEPLER-397 c

KEPLER-458 b

HD 51608 c

GJ 667 C c

KEPLER-90 h

KEPLER-169 f

HD 159868 c

GJ 625 b

KEPLER-452 b

KEPLER-1632 b

KEPLER-86 b

KEPLER-309 c

HD 102365 b

HD 93083 b

GJ 3323 b

KEPLER-1632 b

KIC 9663113 b

KEPLER-1410 b

KEPLER-453 (AB) b

KEPLER-298 d

GLIESE 876 c

ROSS 128 b

Discovered in November 2017, Ross 128 b is the second closest habitable exoplanet to the Earth. It is probably tidally locked to its star, and expected to be a so-called 'eyeball planet', with liquid water only covering the side permanently facing the Sun

KIC 10024862 b

KOI-620.02

HD 164509 b

KEPLER-283 c

KIC 8012732 b

KEPLER-62 e

KEPLER-351 d

GJ 422 b

KEPLER-560 b

KEPLER-296 f

3 b

HD 31527 d

GLIESE 581 g

In 1996, the first potentially habitable planet was discovered orbiting a red dwarf, Gliese 681 g. As the majority of the star's light is in the infrared wavelengths, it is theorized that in order to absorb the light more effectively, any potential plant-like lifeforms would evolve pitch-black leaves

KEP.-458 b

KEPLER-953 b

KEPLER-1544 b

KEPLER-549 b

EARTH

HIP 68468 c

HID 215456 b

KIC 5522786 b

KEPLER-1636 b

KEPLER-947 c

KEPLER-436 b

KEPLER-1038 b

KEP.-505 b

HD 20794 e

Current data suggests that 1/5 of sun-like stars are expected to have a roughly Earth-sized planet within their habitable zone. If red dwarf systems are included, there might be over 40 billion potentially habitable terrestrial planets just in the Milky Way galaxy

KEPLER-1567 b

MAJRITI (UPSILON ANDROMEDAE d)

KEPLER-111 c

DRAUGR (PSR B1257+12 b)

TADMOR (GAMMA CEPHEI A b)

KEPLER-62 f

KEPLER-442 b

KIC 5010054 b

HD 210277 b

KEPLER-991 b

KEPLER-235 e

KEPLER-22 b

KEPLER-61 b

TRAPPIST-1c

7 c

−25 °C

0 °C

KEPLER-155 c

KEPLER-610 c

KEPLER-331 d

KEPLER-236 c

KEPLER-55 b

KEPLER-482 b

KEPLER-630 b

KEPLER-616 c

KEPLER-225 c

KEPLER-460 c

KEPLER-83 c

KEPLER-51 c

KEPLER-1388 e

KEPLER-568 b

KEPLER-401

KEPLER-108 c

Kepler-138 d is a so-called 'gas dwarf', a planet with a small rocky core that has accumulated a very thick atmosphere of hydrogen, helium and other gases

KEPLER-138 d

KEPLER-603 c

KEPLER-540 b

K2-3 d

KEPLER-1455 b

EPIC 201465501 b

KEPLER-438 b

KEPLER-440 b

KOI-372 b

KEPLER-1455 d

KEPLER-737 b

KEPLER-289 d

KEPLER-267 d

KEPLER-186 f

KEPLER-231 c

KEP.-399 d

KEPLER-1503 b

K2-26 b

KEPLER-241 c

KEPLER-26 e

KEPLER-49 e

KEPLER-1020 b

YZ CET c

KEPLER-807

It's expected that a substantial fraction of super-Earths are ocean worlds – planets entirely covered by a global ocean. It's likely that during the later stages of their star's lifespan, those planets will develop very thick steamy atmospheres and undergo a runaway greenhouse effect, leaving them with surface conditions similar to those on Venus

KEP.-344 c

KEPLER-251 e

KEPLER-201 c

KEPLER-841 b

KEPLER-782 b

KEP.-1333 b

KEPLER-19 d

KEPLER-1389 b

KEPLER-239 c

KEPLER-159

KEPLER-35 b

KEPLER-985 b

KEPLER-910 b

At temperatures between −50 and 80 °C, the appearance of most gas giants is expected to be light blue, dominated by bands of reflective ice and water clouds. Although the moons of such planets would be located within the habitable zone, tidal heating and the radiation belts produced by the strong magnetic fields of some gas giants (such as Jupiter) might make them inhospitable

KEPLER-849 b

KEPLER-149 d

KEPLER-976 b

KEPLER-1036 b

KEPLER-1185 b

KEPLER-90 g

KEP.-867 b

KEP.-799 b

KEPL

Carbon planets are a theoretical class of planets that contain more carbon than oxygen. Their mantle would probably be composed of silicon carbide and diamond, which could be carried to the surface by volcanic eruptions. Such planets would be devoid of water, but feature oceans of liquid hydrocarbons

KEPLER-1359 b

KEPLER-951 b

KEPLER-1040 b

KEPLER-139 c

KEPLER-296 e

KEPLER-1450 d

KEPLER-1459 b

K2-3 c

KEPLER-395 c

KEPLER-266

KEP.-54 d

KEP.-437 b

KEPLER-30 d

KEPLER-52 d

QUIJOTE
(MU ARAE b)

KEPLER-1090 b

KEPLER-148 d

KEPLER-1638 b

KEPLER-952

25 °C

HABITABLE ZONE

50 °C

7-

At slightly over 16 masses and with a radius 2.2 times that of the Earth, BD+20594 is thought to be at the upper mass limit for rocky super-Earth exoplanets. Any bigger, and it would either contract under its own gravity, or attract enough gases and volatiles during its formation to become an ice giant similar to Neptune

The temperature of exoplanets stated in this chart is only a rough estimate – depending on the planet's reflectivity (albedo) and atmospheric greenhouse effect, it might vary considerably. For instance, if Venus had an atmosphere similar to that of Earth, its average surface temperature would only be about 70 °C instead of the current 460 °C

100 °C   125 °C   150 °C

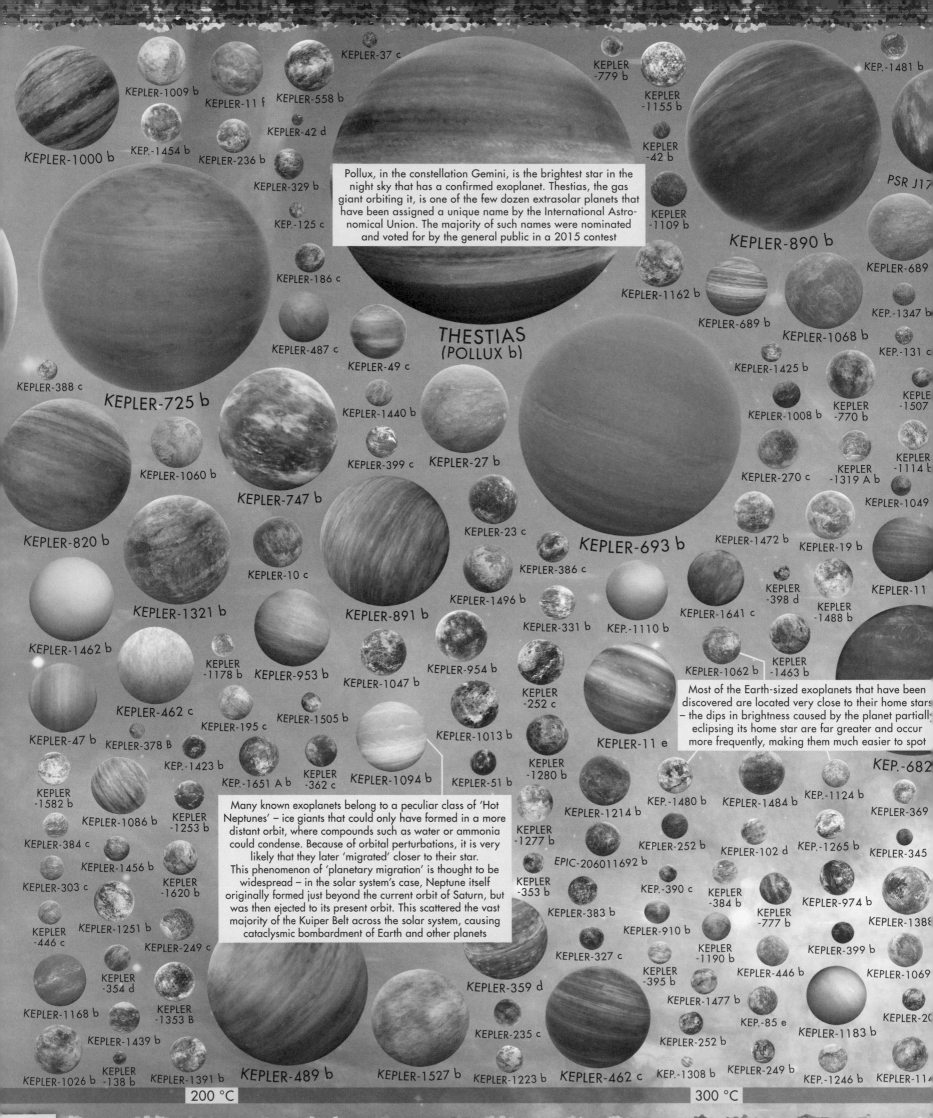

Pollux, in the constellation Gemini, is the brightest star in the night sky that has a confirmed exoplanet. Thestias, the gas giant orbiting it, is one of the few dozen extrasolar planets that have been assigned a unique name by the International Astronomical Union. The majority of such names were nominated and voted for by the general public in a 2015 contest

Most of the Earth-sized exoplanets that have been discovered are located very close to their home stars – the dips in brightness caused by the planet partially eclipsing its home star are far greater and occur more frequently, making them much easier to spot

Many known exoplanets belong to a peculiar class of 'Hot Neptunes' – ice giants that could only have formed in a more distant orbit, where compounds such as water or ammonia could condense. Because of orbital perturbations, it is very likely that they later 'migrated' closer to their star. This phenomenon of 'planetary migration' is thought to be widespread – in the solar system's case, Neptune itself originally formed just beyond the current orbit of Saturn, but was then ejected to its present orbit. This scattered the vast majority of the Kuiper Belt across the solar system, causing cataclysmic bombardment of Earth and other planets

200 °C

300 °C

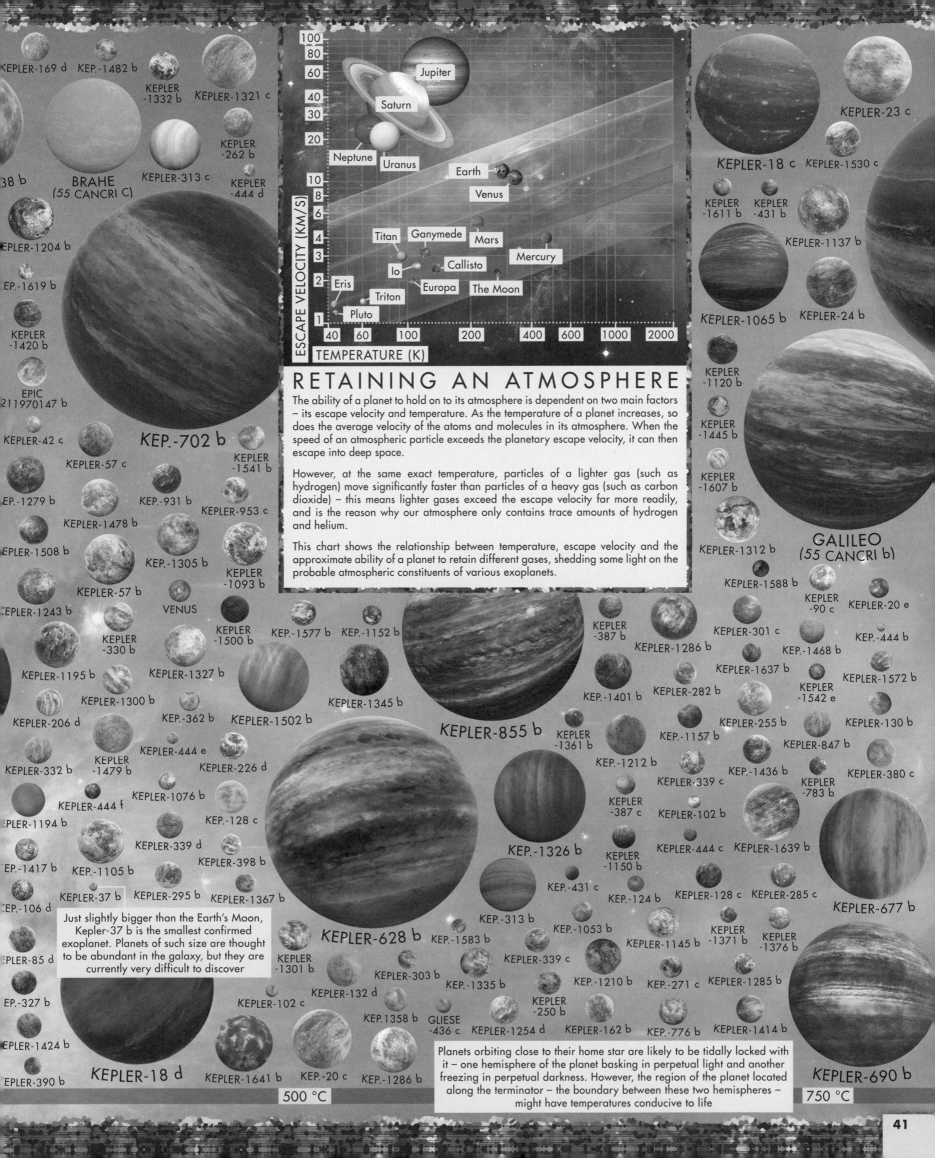

RETAINING AN ATMOSPHERE

The ability of a planet to hold on to its atmosphere is dependent on two main factors – its escape velocity and temperature. As the temperature of a planet increases, so does the average velocity of the atoms and molecules in its atmosphere. When the speed of an atmospheric particle exceeds the planetary escape velocity, it can then escape into deep space.

However, at the same exact temperature, particles of a lighter gas (such as hydrogen) move significantly faster than particles of a heavy gas (such as carbon dioxide) – this means lighter gases exceed the escape velocity far more readily, and is the reason why our atmosphere only contains trace amounts of hydrogen and helium.

This chart shows the relationship between temperature, escape velocity and the approximate ability of a planet to retain different gases, shedding some light on the probable atmospheric constituents of various exoplanets.

Just slightly bigger than the Earth's Moon, Kepler-37 b is the smallest confirmed exoplanet. Planets of such size are thought to be abundant in the galaxy, but they are currently very difficult to discover

Planets orbiting close to their home star are likely to be tidally locked with it – one hemisphere of the planet basking in perpetual light and another freezing in perpetual darkness. However, the region of the planet located along the terminator – the boundary between these two hemispheres – might have temperatures conducive to life

500 °C

750 °C

KEP.-188 b
KEPLER-1342 b
KEPLER-20 b
KEPLER-696 b
KEPLER-18 b
KEP.-1344 b
KEPLER-1154 b
KEPLER-65 c
KEPLER-1409 b
KEP.-89 b
KEPLER-255 d
KEPLER-1352 b

KEPLER-1597 b
KEPLER-1559 b
KEPLER-1153 b
KEPLER-997 b

The vast majority of known extrasolar gas giants belong to the class of 'hot Jupiters' – by far the easiest type of planet to detect using most methods. They are likely to be dark, dominated by clouds of sodium and potassium, which strongly absorb all wave- lengths of visible light. Most of them are thought to be tidally locked, although atmospheric circulation would make the temperature across both hemispheres similar

ARKAS
(41 LYNCIS b)
KEPLER-101 c
ALPHA CENTAURI B c
KEPLER-374 c
KEPLER-1436 b
KEPLER-762 b
KOI-2700 b
EPIC 22881391 b
COROT-2 b
KEPLER-1028 b
KEPLER-605 c
KEPLER-1412 b
KEPLER-756 b
KEPLER-1525 b
KEPLER-1055 b
KEPLER-1398 b
KEPLER-1061 b
KEPLER-664
KEPLER-106 b
KEP.-821 b
KEPLER-1138 b
KEPLER-1158 b
KEPLER-969 c
KEPLER-376 b
KEPLER-880 b
KEPLER-999 b
KEPLER-1320 b
KEPLER-207 b
KEPLER-1027 b
KEP.-1143 b
KEPLER-119 c
KEP.-1225 b
KEPLER-1403 b
KEPLER-402 b
KEPLER-1592 b
KEPLER-1099 b
KEPLER-1004 b
KEPLER-826 b
KEP.-823 b
KEPLER-23 c
KEP.-1336 c
KEPLER-1293 b
KEPLER-1002 b
KEPLER-1560 b
KEPLER-1070 b
KEP.-1240 b
KEPLER-1313 b
KEPLER-1616 b
KEPLER-1148 b
KEPLER-1298 b
KEPLER-271 b
KEPLER-1292 b
KEPLER-1487 b
KEP.-1443 b
KEP.-1115 b
KEPLER-956 b
KEP.-450 d
KEPLER 201713348 b
KEPLER-703 b
KEP.-1578 b
KEPLER-996 b
KOI-1843 b
KEP.-1265 b
KEPLER-1598 b
KEPLER-1601 b
KEPLER-372 b
KEPLER-1016 b
KEPLER-994 b
KEP.-392 b
KEPLER-228 b
KEP.-90 b
KEPLER-301 b
KEPLER-880 b

In 1995, 51 Pegasi b became the first exoplanet discovered around a main-sequence star (using the radial velocity method). Even though it is tidally locked to its home star, it completes one orbit (and thus one rotation around its axis) in only four days, making the Coriolis effect more than sufficient to produce the signature cloud-band patterns that most gas giants possess

KEPLER-322 b
KEPLER-1066 b
KEP.-1219 b
KEPLER-793 b
KEPLER-132 c
KEPLER-292 b
KEPLER-1438 b
KEPLER-1537 b
KEPLER-1576 b
KEPLER-1612 b
KEPLER-1328 b
KEPLER-1387 b
KEPLER-380 b
KEPLER-1563 b
KEPLER-406 c
KEP.-1123 b
KEPLER-1343 b
KEPLER-197 b
KEPLER-746 b
KEPLER-50 b
KEPLER-1153 b
DIMIDIUM
(51 PEGASI b)
KEPLER-381 b
KEPLER-686 b
KEPLER-132 b
KEP.-1357 b
KEPLER-1351 b
KEPLER-1629 b
KEP.-1163 c
KEPLER-1393 b
KEPLER-1402 b
KEPLER-1579 b
KEPLER-887 c
KEP.-403 b
KEPLER-1542 b
KEPLER-1222 b
KEP.-1464 c
KEPLER-1421 b
KEPLER-669 b
KEPLER-932 b
KEPLER-625 c
KEPLER-1278 b
KEPLER-1434 b

1,000 °C

KEPLER -1415 b

KEPLER -207 b

KEPLER -342 e

KEP.-1310 b

KEPLER-1547 b

KEPLER -1368 b

COROT-7 b

TrES-2 b has the lowest albedo of any known exoplanet, reflecting less than 0.1% of any light that hits it. That makes it intrinsically darker than coal or asphalt. However, due to its very high temperature, it glows brightly in the visible spectrum

COROT-3 b

KEPLER -529 b

KEPLER-1173 b

IP 68468 b

KEPLER-1603 b

KEPLER-1372 b

EPLER-1339 b

KEPLER-1349 b

KEP.-1194 b

KEP.-1446 b

KEPLER-1238 b

WASP-12 b

KEPLER-1589 b

KEPLER -1531 b

KEPLER-1340 b

TrES-2 b

KEPLER-70 b

KEPLER-70 c

KEPLER-142 b

KEPLER-1264 b

KEP.-1082 b

KEPLER-65 b

KEPLER -1031 b

KEPLER-1067 b

KEPLER-1258 b

KEPLER-828 b

KEPLER-1284 b

KEP.-1139 b

KEPLER -1394 b

Kepler 70 b is the hottest known exoplanet, with a surface temperature on its day-side approaching 7,000 K. It orbits a post-red giant star that is currently in the process of contraction into a white dwarf. It's estimated that 17 million years ago, it was a medium-sized gas giant orbiting inside the gaseous envelope of its own star, while most of its original mass was stripped away. During its orbit, it passes just 240,000 km from its planetary neighbour, Kepler-70 c – less than the distance between Earth and the Moon

Hot Jupiter Kepler-7 b is the first and currently only planet to have its cloud cover crudely mapped. It is also a so-called 'puffy planet' – even though it is over two times less massive than Jupiter, due to the heating from its star, it has expanded to a considerably bigger size. The largest puffy planets can reach a size of up to two Jupiter radii

KEPLER-1072 b

KEPLER-1427 b

KEPLER -653 c

The hottest of 'hot Jupiters' often have cloud decks consisting of truly unexpected and exotic compounds such as molten silicate rock or iron droplets. Hat-P-7 b is one of the most unique in this regard – its clouds are made of liquid corundum. On the night side of the planet, they condense into crystalline rubies and sapphires, which then continuously rain to the deeper layers of the atmosphere and melt again

EPIC 201637175 b

KEPLER-808 b

KEP.-1284 b

KOI-1843.03

KEPLER-1379 b

KEP.-1106 b

KEPLER -1416 b

JANSSEN (55 CANCRI e)

KEP.-1561 b

KEPLER-1107 b

KEPLER-1244 b

KEP.-1323 b

HAT-P-7 b

KEPLER-990 c

KEP.-1340 b

EPIC 220674823 b

KEPLER-312 b

KEPLER -407 b

KEPLER-7 b

KEPLER-1377 b

PLER-770 c

KEPLER-1297 b

KEPLER-1047 c

KEPLER-1167 b

P.-775 b

KEPLER -1375 b

KEPLER -607 b

PLER 08 b

KEP.-1315 b

KEP.-9 d

KEPLER-1317 b

CVSO-30 b

KEPLER-1373 b

KEPLER -1228 b

SMERTRIOS (HD 149026 b)

P.-336 b

KEP.-115 b

Due to the strong tidal forces and intense irradiance, most terrestrial planets orbiting close to their star are partially or entirely covered with molten lava, with atmospheres consisting mostly of oxygen and sodium

1,000 °C

HD 219828 b

KEP.-908 b

KEPLER -10 b

KEP.-757 b

FORTITUDO (XI AQUILAE b)

KEP.-1356 b

KEP.-217 d

Pulsar planet PSR J1719-1438 b is probably the most extreme and unique exoplanet ever discovered. Over 75% of the planet is made of pure diamond and exotic crystalline forms of carbon. It weighs roughly as much as Jupiter, but gravitational contraction forces have drastically compressed it and increased its density to over 23 g/cm³, more than gold or even osmium. It's unknown how this planet formed, but many scientists speculate that it was originally a brown dwarf or even a light-weight star, that had all its outer layers siphoned off by its neutron star companion, leaving just the carbon–oxygen interior

PSR J1719-1438 b

G 99-47
(W. dwarf)

After stars smaller than 8–10 solar masses pass through the red giant stage, exhausting all of their available fuel, the outer layers of the star are ejected, forming a planetary nebula. The remaining carbon–oxygen-rich core then gradually contracts into an extremely dense planet-sized body composed of so-called 'degenerate matter' – a white dwarf. This object is so dense that a single teaspoon of white dwarf matter can weigh more than an automobile

WZ SAGITTAE
(White dwarf)

GLIESE 381
(White dwarf)

EX HYDRAE a
(White dwarf)

TRAPPIST-1
(Red dwarf star)

AR SCORPII
(White dwarf)

WD 1145+017
(White dwarf)

2M1207
(Brown dwarf)

If a forming protostar doesn't gather enough mass to start hydrogen fusion in its core, it becomes a brown dwarf. The majority of such objects weigh between 15 and 75 Jupiter masses, and their temperature varies from just −50 °C to over 2,000 °C. Such objects are abundant in the universe, with the closest brown dwarf, Luhman 16, being just 6.2 light years away. Some brown dwarfs even have planets of their own, although it's unlikely that they are habitable

Smaller than Saturn and only at 85 Jupiter masses, EBLM J0555-57 Ab is the smallest known star still capable of fusing hydrogen. If it were only slightly smaller, it would be classified as a brown dwarf

THETA HYDRAE
(White dwarf)

TEID
(Brow

EBLM J0555-57 Ab
(Red dwarf star)

DENIS 1058-15
(Brown dwarf)

HD 74389
(White dwarf)

DENIS 0823-49
(Brown dwarf)

All objects slowly lose their orbital energy via the emission of gravitational waves. Although negligible in the case of planets, as objects get more massive and orbit closer together, this effect becomes vastly more prominent. In the case of HM Cancri, a system of two white dwarfs who orbit each other in just five minutes, it will only take 340,000 years until their orbit decays completely, and they violently merge, forming a new red giant star roughly as massive as the Sun

HM CANCRI a & b
(Binary white dwarf)

EPSILON INDI Ba
(Brown dwarf)

GD 40
(White dwarf)

OGLE-TR-122
(Red dwarf star)

FEIGE 55
(White dwarf)

PROXIMA CENTAURI
(Red dwarf, closest star to the Sun)

WOLF 489
(White dwarf)

2MASS 0939-2448 A
(Brown dwarf)

NN SERPENTIS
(White dwarf)

Most currently known white dwarfs are still hot enough to glow mainly in white, blue or ultraviolet light. However, the oldest ones known have already cooled below the solar surface temperature. PSR J2222-0137 in particular approaches a record low of 2,700 °C, and it has probably already solidified. Over 100 billion years in the future, first white dwarfs are predicted to cool down enough to stop glowing in the visible spectrum, becoming 'black dwarfs' – the final stage in the life of a star

PSR J2222-0137
(White dwarf)

EGGR-91 B
(White dwarf)

PSR B1620-26 b
(White dwarf)

T CORONAE
BOREALIS
(White dwarf)

G 240-72
(White dwarf)

WD 0346+246
(White dwarf)

40 ERIDANI B
(White dwarf)

ZETA
CAPRICORNI
(White dwarf)

WD 2459-434
(White dwarf)

KOI-74
(White dwarf)

TVLM 513-46546
(Red dwarf star)

L 97-12
(White dwarf)

ZETA CYGNI b
(White dwarf)

PG 1159-035
(Young white dwarf)

V803 CENTAURI
(White dwarf)

More than 97% of all the stars are bound to become white dwarfs after their fuel is exhausted. Most currently known white dwarfs are the remnants of large stars with shorter lifespans, compressed to the size of the Earth. However, white dwarfs formed from stars smaller and lighter than our Sun can exceed the size of Neptune due to weaker gravitational contraction

HD 49798 b
(White dwarf)

IK Peg B
(White dwarf)

VAN MAANEN 2
(White dwarf)

There is a limit to how massive white dwarfs can get. If a white dwarf ever reaches 1.44 solar masses, so-called 'Chandrasekhar mass' (usually by siphoning the outer layers of another star), its core will get hot enough to allow carbon fusion. This then results in a violent runaway reaction, releasing a tremendous amount of energy and destroying the white dwarf in a violent explosion – a type 1a supernova. Because of this white dwarf mass limit, all type 1a supernovae are roughly equally luminous, making them indispensible in measuring the distances of distant galaxies. This is expected to also be the final fate of the white dwarf IK Peg B, after its stella companion becomes a red giant 1 billion years in the future

Z CHAMAELEONTIS
(White dwarf)

LUHMAN 16
(Brown dwarf)

:-1
(...warf)

KOI-81
(White dwarf)

PSR J0348+0432 B
(Least massive white dwarf known)

GLIESE 229 B
(Brown dwarf)

KEPLER-70
(Contracting post-giant star)

PROCYON B
(White dwarf)

RR CAELI
(White dwarf)

BLACK HOLE
OF 10,000
SOLAR MASSES

Sirius B
(White dwarf)

GLIESE 758 b
(Brown dwarf)

WD 0806-661
(White dwarf)

AVERAGE DISTANCE
BETWEEN THE MOON
AND THE EARTH

AD LEONIS
(Red dwarf star)

WOLF 359
(Red dwarf star)

PROCYON
(White main sequence star)

In the year 2007, the faint coma (nebulous envelope of gas and dust that forms around the nucleus of a comet) of the comet 17P/Holmes rapidly expanded as it approached the inner solar system. Its diameter briefly surpassed that of the Sun, and was visible to the naked eye from Earth at half the size of the Moon

VEGA
(White main sequence star)

2M1207
(Brown dwarf)

ROSS 154
(Red dwarf star)

ALL URANIUM IN THE
MILKY WAY GALAXY

HD 85512
(Orange main sequence star)

DUST COMA OF
COMET HOLMES

LUYTEN 726-8 A
(Red dwarf star)

STRUVE 2398
(Red dwarf star)

ROSS 614
(Red dwarf star)

TW PISCIS AUSTRINI
(Orange main sequence star)

LACAILLE 8760
(Red dwarf star)

SOLAR CORE

KAPTEYN'S STAR
(Red dwarf star)

BLACK HOLE OF
10,000 SOLAR MASSES

GLIESE 832
(Red dwarf star)

BARNARD'S STAR
(Red dwarf star)

THE SUN
(Yellow main sequence star)

ALPHA CENTAURI B
(Orange main sequence star)

ALPHA CENTAURI A
(Yellow main sequence star)

LALANDE 21185
(Red dwarf star)

COROT-3 b
(Brown dwarf)

BLUE DWARF
(Predicted class of star, expected to
develop from very old red dwarfs)

As the temperature of a star's surface changes, so does the spectrum
of the light it emits. Red dwarfs (small stars colder than 4,000 °C)
are the most common type of star in the universe, and shine mostly in
infrared and red light. As stars get hotter, the apparent colour of the
star shifts to orange, yellow, white and blue, while the hottest known
stars emit up to 99% of their light as ionizing UV radiation

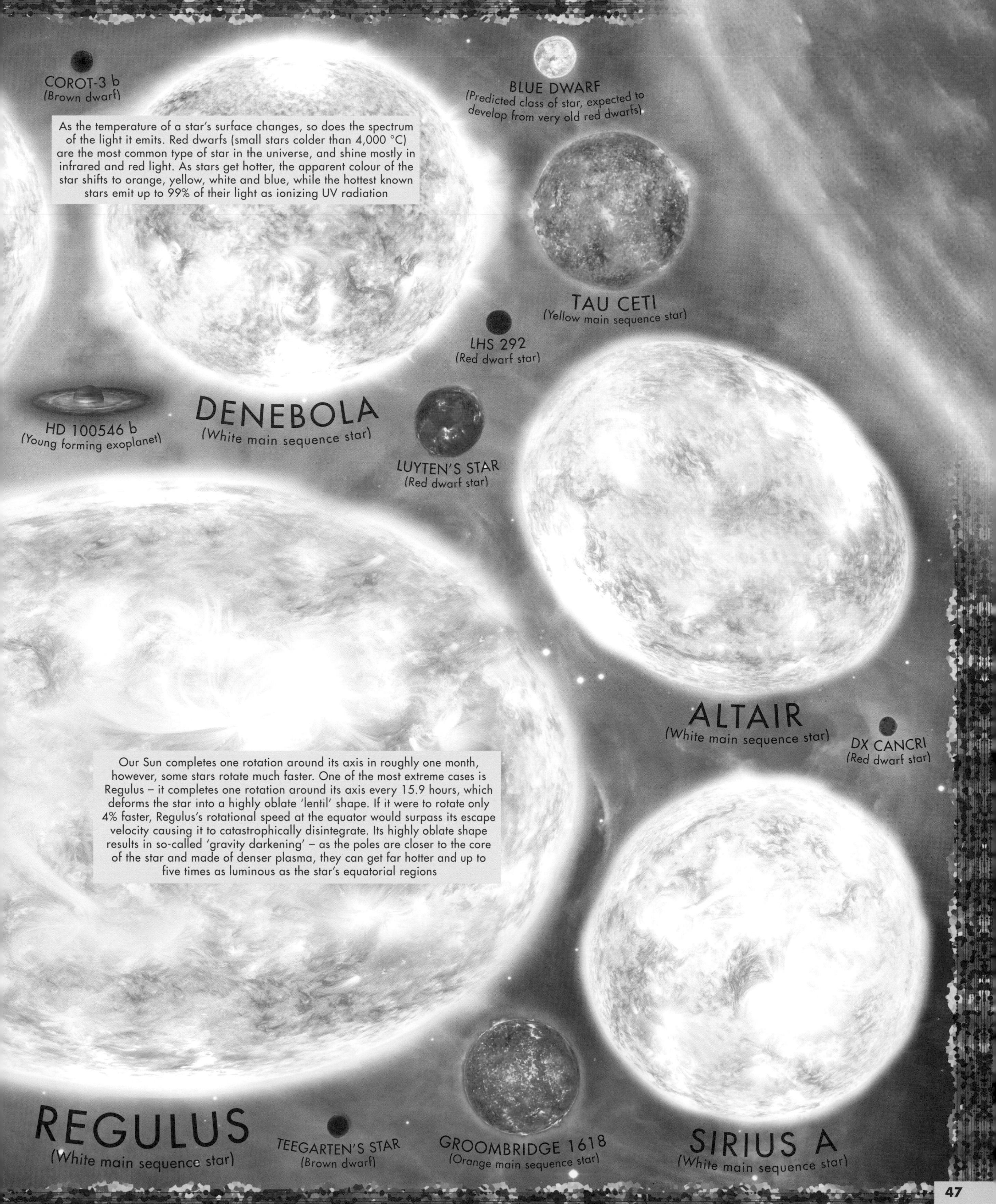

TAU CETI
(Yellow main sequence star)

LHS 292
(Red dwarf star)

HD 100546 b
(Young forming exoplanet)

DENEBOLA
(White main sequence star)

LUYTEN'S STAR
(Red dwarf star)

ALTAIR
(White main sequence star)

DX CANCRI
(Red dwarf star)

Our Sun completes one rotation around its axis in roughly one month,
however, some stars rotate much faster. One of the most extreme cases is
Regulus – it completes one rotation around its axis every 15.9 hours, which
deforms the star into a highly oblate 'lentil' shape. If it were to rotate only
4% faster, Regulus's rotational speed at the equator would surpass its escape
velocity causing it to catastrophically disintegrate. Its highly oblate shape
results in so-called 'gravity darkening' – as the poles are closer to the core
of the star and made of denser plasma, they can get far hotter and up to
five times as luminous as the star's equatorial regions

REGULUS
(White main sequence star)

TEEGARTEN'S STAR
(Brown dwarf)

GROOMBRIDGE 1618
(Orange main sequence star)

SIRIUS A
(White main sequence star)

ALKAID
(Blue main
sequence star)

ALPHERATZ
(Blue main
sequence star)

BETA CASSIOPEAE
(White main sequence star)

ETA OPHIUCHI
(White main sequence sta

CANOPUS
(White supergiant star)

SPICA
(Blue subgiant star)

ALNAIR
(White subgiant star)

KOCHAB
(Orange giant star)

ALL LEAD IN THE
MILKY WAY GALAXY

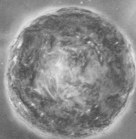

EPSILON SCORPII
(Orange giant star)

ALDEBARAN
(Orange giant star)

BELLATRIX
(White giant star)

CAPELLA
(Orange giant star)

POLLUX
(Orange giant star)

PEACOCK STAR
(Blue main sequence star)

MY Camelopardalis is a binary star system consisting of two massive and extremely bright blue stars orbiting so close that they actually touch each other, and are currently the only so-called 'contact binary' system known. It's expected that the system will eventually merge and transform into a single massive star, however some astrophysicists predict that the system will explode as a supernova instead

ALL LITHIUM IN THE
MILKY WAY GALAXY

ARCTURUS
(Orange giant star)

ALL GOLD IN THE
MILKY WAY GALAXY

KOCHAB
(Yellow supergiant star)

MY CAMELOPARDALIS
(Binary blue star system)

ALL PLATINUM IN THE
MILKY WAY GALAXY

At this scale, the Earth is smaller
than an average human cell

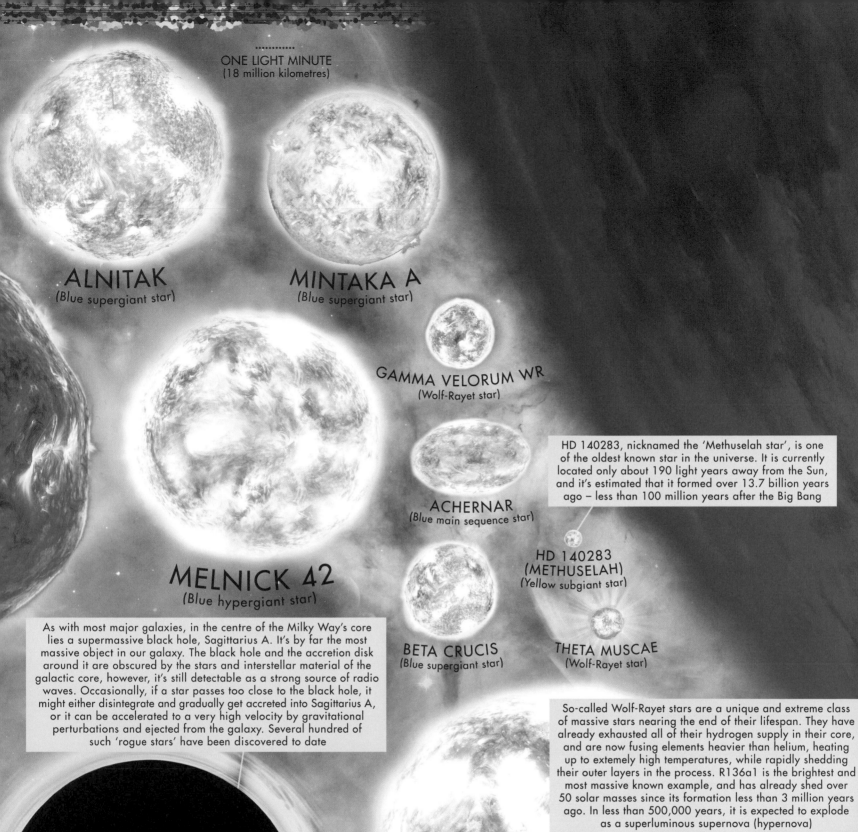

ONE LIGHT MINUTE
(18 million kilometres)

ALNITAK
(Blue supergiant star)

MINTAKA A
(Blue supergiant star)

GAMMA VELORUM WR
(Wolf-Rayet star)

ACHERNAR
(Blue main sequence star)

HD 140283, nicknamed the 'Methuselah star', is one of the oldest known star in the universe. It is currently located only about 190 light years away from the Sun, and it's estimated that it formed over 13.7 billion years ago – less than 100 million years after the Big Bang

MELNICK 42
(Blue hypergiant star)

HD 140283
(METHUSELAH)
(Yellow subgiant star)

BETA CRUCIS
(Blue supergiant star)

THETA MUSCAE
(Wolf-Rayet star)

As with most major galaxies, in the centre of the Milky Way's core lies a supermassive black hole, Sagittarius A. It's by far the most massive object in our galaxy. The black hole and the accretion disk around it are obscured by the stars and interstellar material of the galactic core, however, it's still detectable as a strong source of radio waves. Occasionally, if a star passes too close to the black hole, it might either disintegrate and gradually get accreted into Sagittarius A, or it can be accelerated to a very high velocity by gravitational perturbations and ejected from the galaxy. Several hundred of such 'rogue stars' have been discovered to date

So-called Wolf-Rayet stars are a unique and extreme class of massive stars nearing the end of their lifespan. They have already exhausted all of their hydrogen supply in their core, and are now fusing elements heavier than helium, heating up to extemely high temperatures, while rapidly shedding their outer layers in the process. R136a1 is the brightest and most massive known example, and has already shed over 50 solar masses since its formation less than 3 million years ago. In less than 500,000 years, it is expected to explode as a superluminous supernova (hypernova)

MERAK
(White main sequence star)

SAGITTARIUS A
(Supermassive black hole of 4.4 million solar masses)

SMC WR8
(Wolf-Rayet star)

R136A1
(Wolf-Rayet star)

KOCHAB
(Orange main sequence star)

MENKENT
(Orange giant star)

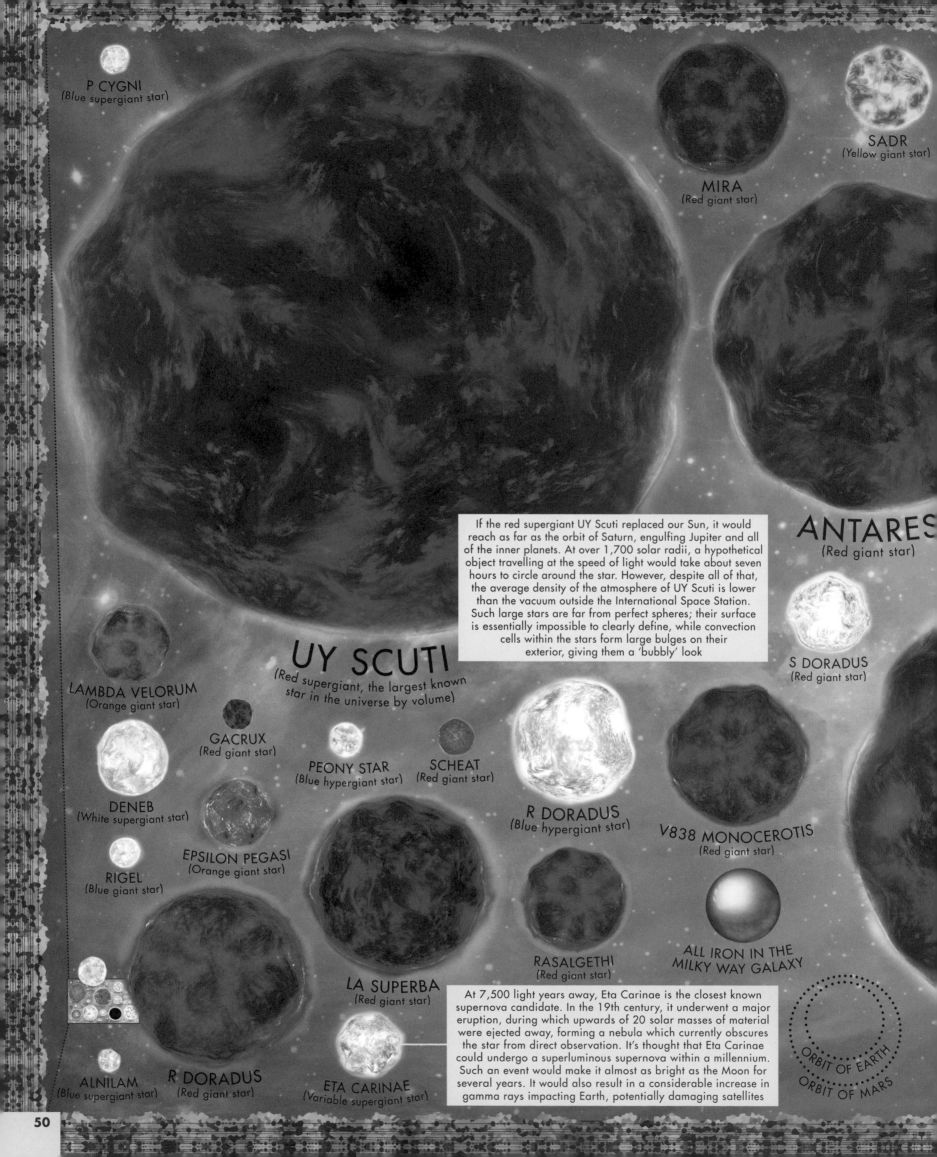

P CYGNI
(Blue supergiant star)

MIRA
(Red giant star)

SADR
(Yellow giant star)

ANTARES
(Red giant star)

If the red supergiant UY Scuti replaced our Sun, it would reach as far as the orbit of Saturn, engulfing Jupiter and all of the inner planets. At over 1,700 solar radii, a hypothetical object travelling at the speed of light would take about seven hours to circle around the star. However, despite all of that, the average density of the atmosphere of UY Scuti is lower than the vacuum outside the International Space Station. Such large stars are far from perfect spheres; their surface is essentially impossible to clearly define, while convection cells within the stars form large bulges on their exterior, giving them a 'bubbly' look

S DORADUS
(Red giant star)

UY SCUTI
(Red supergiant, the largest known star in the universe by volume)

LAMBDA VELORUM
(Orange giant star)

GACRUX
(Red giant star)

PEONY STAR
(Blue hypergiant star)

SCHEAT
(Red giant star)

DENEB
(White supergiant star)

R DORADUS
(Blue hypergiant star)

V838 MONOCEROTIS
(Red giant star)

EPSILON PEGASI
(Orange giant star)

RIGEL
(Blue giant star)

RASALGETHI
(Red giant star)

ALL IRON IN THE
MILKY WAY GALAXY

ALNILAM
(Blue supergiant star)

R DORADUS
(Red giant star)

LA SUPERBA
(Red giant star)

ETA CARINAE
(Variable supergiant star)

At 7,500 light years away, Eta Carinae is the closest known supernova candidate. In the 19th century, it underwent a major eruption, during which upwards of 20 solar masses of material were ejected away, forming a nebula which currently obscures the star from direct observation. It's thought that Eta Carinae could undergo a superluminous supernova within a millennium. Such an event would make it almost as bright as the Moon for several years. It would also result in a considerable increase in gamma rays impacting Earth, potentially damaging satellites

ORBIT OF EARTH
ORBIT OF MARS

ALL SULPHUR IN THE
MILKY WAY GALAXY

ALL ALUMINIUM IN THE
MILKY WAY GALAXY

ALL TITANIUM IN THE
MILKY WAY GALAXY

WEZEN
(Yellow giant star)

HD 50064
(Blue supergiant star)

R99
(Blue hypergiant star)

ALL SILICON IN THE
MILKY WAY GALAXY

ALL CARBON IN THE
MILKY WAY GALAXY

CENTRAL BLACK HOLE
OF THE BODE GALAXY

ETA AQUILAE
(Cepheid variable)

BETA CRUCIS
(Cepheid variable)

CENTRAL BLACK HOLE OF
THE ANDROMEDA GALAXY

AG CARINAE
(Variable supergiant star)

BETELGEUSE
(Red hypergiant star)

Most stars maintain a stable size, brightness and temperature
for very long periods of time; however, this is not the case with the
so-called variable stars. One of their most well-known and understood classes are the classical Cepheid variables, which pulsate
in size and brightness in very regular periods, usually on the order
of days to months. The fact that the length of this period is directly
determined by the star's luminosity allows Cepheids to be used for
calculating distances to nearby galaxies with great precision

THETA SCORPII
(Blue main sequence star)

MU CEPHEI
(Red hypergiant star)

ETA CANIS
MAJORIS
(Blue supergiant star)

BETA CRUCIS
(Wolf-Rayet star)

VY CANIS MAJORIS
(Red supergiant star)

POLARIS
(Cepheid variable)

DISTANCE FROM
NEW HORIZONS
TO THE EARTH

DISTANCE FROM
VOYAGER 2
TO THE EARTH

DISTANCE FROM
PIONEER 10
TO THE EARTH

DISTANCE FROM
VOYAGER 1
TO THE EARTH

TW HYDRAE
(Protoplanetary disc)

GOMEZ'S HAMBURGER
(Protoplanetary disc)

Quasars are by far the most luminous and energetic objects found in the universe. They are the result of supermassive black holes accreting a large amount of gas, which forms into a disk of massive proportions. As matter approaches the black hole, it accelerates to extreme velocities and emits a tremendous amount of energy in the form of light. The brightest known quasars consume up to 1,000 solar masses every year, and can become thousands of times more luminous than the entire Milky Way galaxy. As the supply of gas is limited, though, most quasars eventually become much less active, and revert into ordinary, calm supermassive black holes, similar to the one in the core of our own galaxy

V883 ORIONIS
(Protoplanetary disc)

3C 273
(First quasar ever to be identified, of 550 million solar masses)

IRAS 04248+2612
(Protoplanetary disc)

As the matter in the accretion disc of a quasar approaches the event horizon of the central black hole (the dark boundary where the laws of physics as we know them break down, and from beyond which not even light can escape), it gets accelerated to speeds approaching the speed of light. The material which isn't consumed by the black hole is ejected in the form of extremely energetic 'relativistic jets' from the poles of the black hole. When the jets of a quasar are oriented towards the solar system, such an object is called a 'blazar'. Despite being billions of light years away, they can still be observed by amateur astronomers

ULAS J1120+0641
(Quasar of 2 billion solar masses)

NGC 4889
(Most massive known black hole at 21 billion solar masses)

ORBIT OF SEDNA

ORBIT OF THE COMET HALE-BOPP

PREDICTED ORBIT OF PLANET NINE

VOLUME OF ALL STARS IN THE MILKY WAY

At this scale, the Earth is smaller than an average-sized virus

3C 390.3 QUASAR
(Quasar of 300 million solar masses)

KUIPER BELT

HELIOSPHERE
(Extent of the solar wind)

When a protostar forms, it is still surrounded by a large rotating disc of gas and dust, the so-called 'protoplanetary disc', from which all the planets eventually form. Originally, the disc is mostly uniform in its composition; however, as the star begins to radiate more and more heat, the particles within the disc start to stratify. Metals, silicates and other elements and minerals with a high melting point generally clump together and stay close to the star, while gases and other volatiles, such as water, get pushed outwards by the star's radiation. In the case of our solar system, this explains the differing composition of the inner 'terrestrial' planets and the outer gas and ice giant

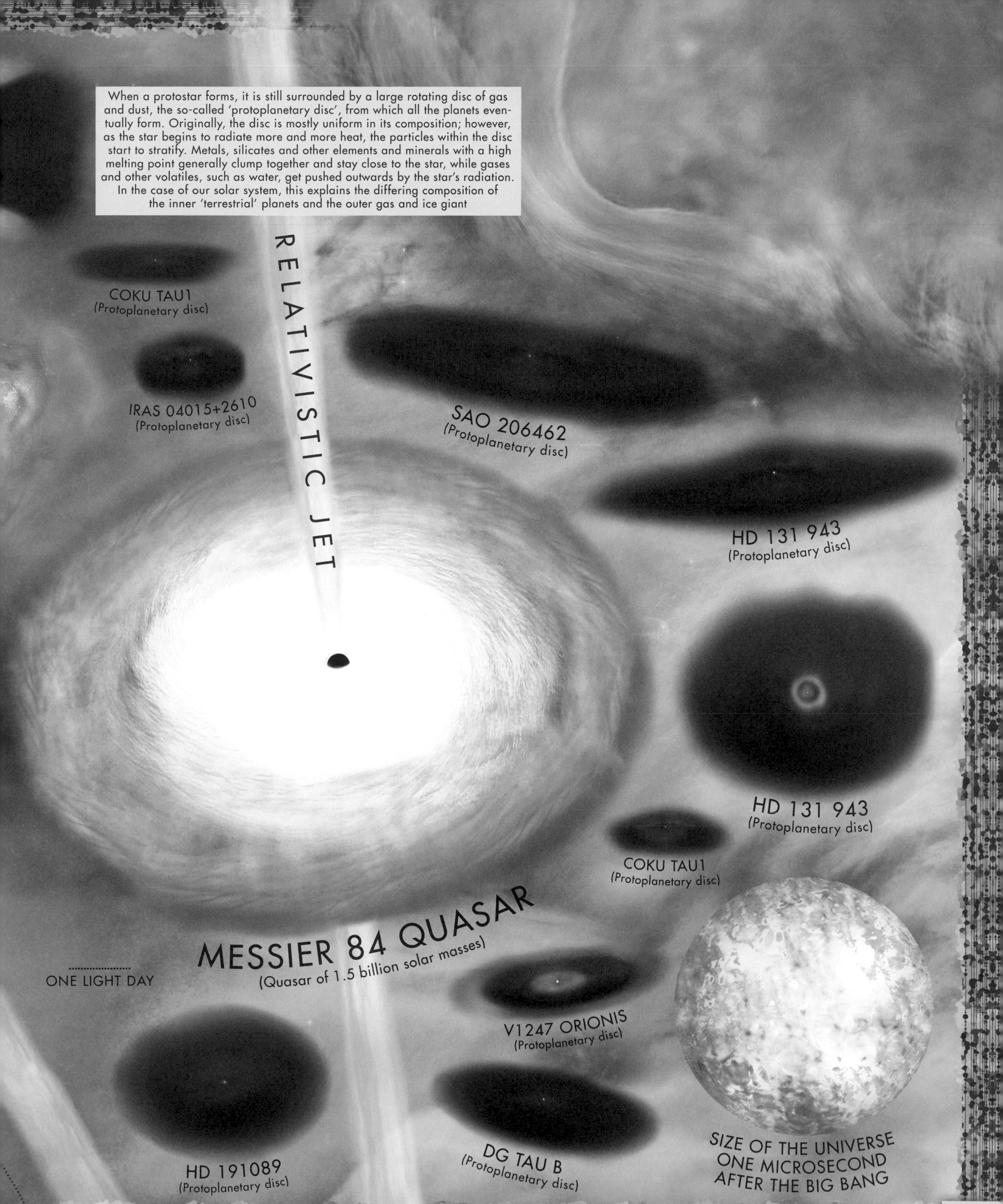

COKU TAU1
(Protoplanetary disc)

RELATIVISTIC JET

IRAS 04015+2610
(Protoplanetary disc)

SAO 206462
(Protoplanetary disc)

HD 131 943
(Protoplanetary disc)

HD 131 943
(Protoplanetary disc)

COKU TAU1
(Protoplanetary disc)

MESSIER 84 QUASAR
(Quasar of 1.5 billion solar masses)

ONE LIGHT DAY

V1247 ORIONIS
(Protoplanetary disc)

SIZE OF THE UNIVERSE
ONE MICROSECOND
AFTER THE BIG BANG

DG TAU B
(Protoplanetary disc)

HD 191089
(Protoplanetary disc)

53

ONE LIGHT YEAR
(Distance a photon travels in 1 year)

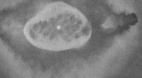

NGC 6826
(Planetary nebula)

NGC 5307
(Planetary nebula)

NGC 3918
(Planetary nebula)

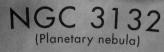

NGC 3132
(Planetary nebula)

HEN 2-47
(Planetary nebula)

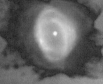

NGC 6741
(or 'Phantom Streak Nebula')

MINKOWSKI 92
(Planetory nebula)

LITTLE GEM NEBULA
(Planetary nebula)

After a star of 0.8 to 8 solar masses reaches the end of its lifespan as a red giant, the strong stellar winds eject its ouyer layers, and the star sheds the majority of its mass, leaving just its hot core – the white dwarf. The expanding shell of gas becomes a 'planetary nebula'. Ultraviolet light from the young white dwarf continuously ionizes and energizes the nebulous gas, causing it to glow with varied bright colours. Hydrogen tends to glow red to violet, and oxygen is greenish-blue. Large concentrations of dust particles move the colour towards dark orange

NGC 6886
(Planetary nebula)

GHOST OF JUPITER NEBULA
(Planetary nebula)

CAT'S EYE NEBULA
(Planetary nebula)

NGC 2440
(Planetary nebula)

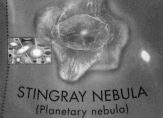

STINGRAY NEBULA
(Planetary nebula)

NGC 7027
(Planetary nebula)

ESKIMO NEBULA
(Planetary nebula)

NGC 2022
(Planetary nebula)

NGC 6210
(Planetary nebula)

ANT NEBULA
(Planetary nebula)

Planetary nebulae are a relatively short-lived phenonon, as it only takes a few thousand years for the expanding gas of the nebula to disperse into interstellar space. This is one of the main mechanisms by which heavy elements are dispersed throughout the galaxy, eventually finding their way into the next generation of stars and planets

BUTTERFLY NEBULA
(Planetary nebula)

HOMUNCULUS NEBULA
(Planetary nebula)

RED RECTANGLE NEBULA
(Planetary nebula)

SATURN NEBULA
(Planetary nebula)

RETINA NEBULA
(Planetary nebula)

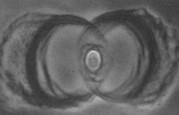

SPIROGRAPH NEBULA
(Planetary nebula)

HOURGLASS NEBULA
(Planetary nebula)

SHAPLEY 1
(Planetary nebula)

NGC 2818
(Planetary nebula)

KEPLER'S SUPERNOVA
(Supernova remnant)

SOUTHERN OWL
NEBULA
(Planetary nebula)

RING NEBULA
(Planetary nebula)

TYCHO'S SUPERNOVA
(Supernova remnant)

After white dwarfs or massive stars explode as supernovae, they eject the majority of their mass to interstellar space at very high speeds – often at 1–10% the speed of light. Astronomers have observed exceptional supernovae for centuries – Tycho's supernova in 1572 reportedly exceeded even the glow of Venus! With the help of historical records, we can now find a number of such supernova remnants, such as Tycho and Kepler's supernovae. It's theorized that even the biblical star of Bethlehem was in fact a bright supernova – however, no potential remnant of it has been discovered yet

CASSIOPEIA a
(Supernova remnant)

EYE OF SAURON
NEBULA
(Planetary nebula)

DISTANCE TO
BARNARD'S STAR

DISTANCE TO
ALPHA CENTAURI

COTTON CANDY
NEBULA
(Pre-planetary nebula)

DISTANCE TO SIRIUS

OORT CLOUD
(Outermost part of
the solar system)

NGC 2261
(Planetary nebula)

DISTANCE TO
EPSILON ERIDANI

DISTANCE TO PROCYON

FOX FUR NEBULA
(Diffuse nebula)

BUG NEBULA
(Blue giant star)

PENCIL NEBULA
(Supernova remnant)

NGC 6357
(Diffuse nebula)

FLAME NEBULA
(Emission nebula)

HORSEHEAD
NEBULA
(Absorption nebula)

NGC 5148
(Planetary nebula)

Pillars of Creation, a portion of the Eagle Nebula, feature in one of the most iconic pictures ever taken by the Hubble Space Telecope. The pillars are an active star-forming region, where the interstellar dust and gas gravitationally contract and start forming a protoplanetary disc. Recently, a hot cloud, thought to be a result of a recent supernova, was discovered in the vicinity of the Eagle Nebula, revealing that the pillars were in fact blown off by the shockwave 6,000 years ago. However, as the Pillars of Creation are over 7,000 light years away, it will take over 1,000 years for their destruction to become visible from the Earth

FROSTY LEO NEBULA
(Pre-planetary nebula)

HELIX NEBULA
(Planetary nebula)

At this scale, the Earth is smaller than a single helium atom

PILLARS OF CREATION
(Part of the Eagle Nebula)

BOOMERANG NEBULA
(Pre-planetary nebula)

BUBBLE NEBULA
(Planetary nebula)

CALIFORNIA NEBULA
(Emission nebula)

DISTANCE TO VEGA

NGC 7129
(Reflection nebula)

CRAB NEBULA
(Supernova remnant)

Unlike Tycho's and Kepler's supernovae, which are the remnants of exploding white dwarfs, the 'Crab Nebula' formed when a massive star went supernova at the end of its lifespan. This is the reason for its unusual, irregular shape. It formed in 1054 after a supernova that was independently observed both in medieval China and Europe. In the centre of the nebula lies a rapidly spinning neutron star, a so-called 'pulsar'. Just 30 km in diameter, it completes 30 rotations every second

THE UNIVERSE ONE SECOND AFTER THE BIG BANG

ORION NEBULA
(Diffuse nebula)

CONE NEBULA
(Hybrid nebula)

DUMBBELL NEBULA
(Planetary nebula)

G2920+1.8
(Supernova remnant)

RCW 103
(Supernova remnant)

SOUL NEBULA
(Emission nebula)

PACMAN
NEBULA
(Diffuse nebula)

RCW 103
(Supernova remnant)

HEART NEBULA
(Emission nebula)

SN 185
(Supernova remnant)

SIMEIS 147
(Supernova remnant)

N49
(Supernova remnant)

NORTH AMERICA
NEBULA
(Emission nebula)

CRESCENT
NEBULA
(Emission nebula)

NGC 346
(Diffuse nebula)

PUPPIS A
(Supernova remnant)

BARNARD'S LOOP
(Emission nebula)

LAGOON NEBULA
(Emission nebula)

MESSIER 10
(Globular cluster)

ROSETTE NEBULA
(Diffuse nebula)

·········
100 LIGHT YEARS

IC 2118
(Diffuse nebula)

EAGLE NEBULA
(Diffuse nebula)

OMEGA CENTAURI
(Globular cluster)

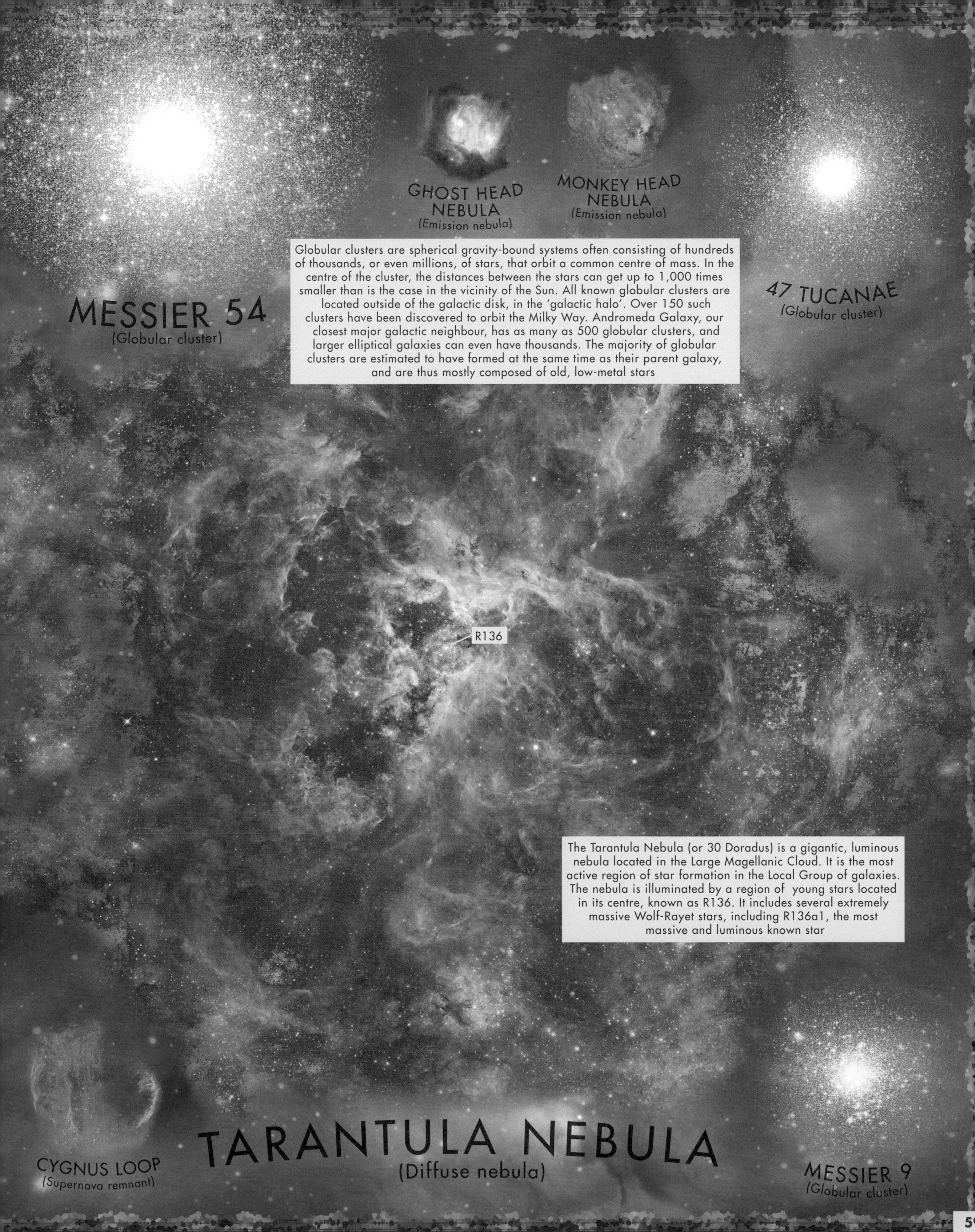

GHOST HEAD
NEBULA
(Emission nebula)

MONKEY HEAD
NEBULA
(Emission nebula)

47 TUCANAE
(Globular cluster)

MESSIER 54
(Globular cluster)

Globular clusters are spherical gravity-bound systems often consisting of hundreds of thousands, or even millions, of stars, that orbit a common centre of mass. In the centre of the cluster, the distances between the stars can get up to 1,000 times smaller than is the case in the vicinity of the Sun. All known globular clusters are located outside of the galactic disk, in the 'galactic halo'. Over 150 such clusters have been discovered to orbit the Milky Way. Andromeda Galaxy, our closest major galactic neighbour, has as many as 500 globular clusters, and larger elliptical galaxies can even have thousands. The majority of globular clusters are estimated to have formed at the same time as their parent galaxy, and are thus mostly composed of old, low-metal stars

R136

The Tarantula Nebula (or 30 Doradus) is a gigantic, luminous nebula located in the Large Magellanic Cloud. It is the most active region of star formation in the Local Group of galaxies. The nebula is illuminated by a region of young stars located in its centre, known as R136. It includes several extremely massive Wolf-Rayet stars, including R136a1, the most massive and luminous known star

TARANTULA NEBULA
(Diffuse nebula)

CYGNUS LOOP
(Supernova remnant)

MESSIER 9
(Globular cluster)

THE UNIVERSE
10 MINUTES
AFTER THE BIG BANG

URSA MAJOR II GALAXY
(Irregular dwarf galaxy)

FORNAX DWARF
(Irregular dwarf galaxy)

PGC 51017
(Dwarf galaxy)

1,000 LIGHT YEARS

WLM GALAXY
(Irregular dwarf galaxy)

NGC 2419
(Globular cluster)

CARINA NEBULA
(Diffuse nebula)

IC-10 is the only known 'starburst galaxy' in the
Local Group, undergoing an unusually high rate
of star formation. Despite its small size, it's just as
luminous as the entire Small Magellanic Cloud. The
galaxy produces stars at the rate of 0.04–0.08 solar
masses every year. Taking into account its small size,
that makes it over 100 times more active than the
Milky Way. The dwarf galaxy still has a sufficient supply of gas to maintain this for several billion years

LEO II DWARF
(Irregular dwarf galaxy)

NGC 1569
(Irregular dwarf galaxy)

IC-10 has an exceptionally high amount of massive,
blue stars, including many extreme Wolf-Rayet stars.
The UV light emanating from those stars continuously ionizes the clouds of gas that stretch across the
galaxy, causing them to glow in vibrant colours.
It is also home to the most massive known stellar
black hole, IC 10 X-1, of over 32 solar masses

TUCANA DWARF
(Irregular dwarf galaxy)

HENIZE 2-10
(Irregular dwarf galaxy)

MGC1
(Globular cluster)

ESO 540
(Dwarf galaxy)

SCULPTOR DWARF
(Irregular dwarf galaxy)

I ZWICKY 18
(Irregular dwarf galaxy)

IC-10
(Irregular dwarf galaxy)

NGC 604
(Diffuse nebula)

LEO A
(Irregular dwarf galaxy)

CARINA DWARF
(Dwarf galaxy)

ANTLIA DWARF
(Dwarf galaxy)

UGC 4879
(Dwarf galaxy)

DISTANCE TRAVELLED BY THE
SUN AROUND THE GALACTIC
CORE IN 5 MILLION YEARS

SOMBRERO GALAXY
(Spiral galaxy)

RADIO BUBBLE
(Extent of human radio signals)

In addition to numerous globular clusters, the Milky Way is orbited
by 20 known dwarf galaxies – the largest of which being the Small
and Large Magellanic Clouds. They both contain several billion
stars, however over 90% of their visible mass is made up of interstellar gas and dust. The two Magellanic Clouds are connected by
a bridge of gas – suggesting that they tidally interact. The Large
Magellanic Cloud shows clear signs of a bar structure – it's
believed that it was a dwarf spiral galaxy, before its spiral arms
were disrupted by the Milky Way. Both dwarf galaxies show a
very high intensity of star formation – they especially feature a
large amount of young, massive, bright Wolf-Rayet stars

ESO 489-056
(Irregular dwarf galaxy)

SMALL MAGELLANIC CLOUD
(Irregular dwarf galaxy)

Centaurus A is the closest 'active' galaxy to the Milky Way. Its centre contains a black hole of 55 million solar masses, which is currently in the process of accreting and consuming massive amounts of matter. This causes two powerful X-ray jets to be generated at the black hole's poles, ejecting mass from the galaxy. Particles at the centre of this jet are moving at nearly half the speed of light. Centaurus A is a star-burst galaxy, with a very high rate of new star formation. This is probably the result of a recent galactic merger, which destabilized the orbits of a number of stars, and perturbed interstellar gas clouds

NGC 55
(Spiral galaxy)

SILVERADO GALAXY
(Spiral galaxy)

ANTENNAE GALAXIES
(Interacting galaxies)

NGC 3982
(Spiral galaxy)

CENTAURUS A
(Lenticular active galaxy)

The solar system is located within one of the spiral arms of a mid-sized galaxy, the Milky Way. Although a large portion of it is obscured from our view by the gas, and stars, it has been calculated that the Milky Way is composed of 100–400 billion stars. According to the data gathered by the Kepler telescope there could be as many as 11 billion Earth-sized planets orbiting within the habitable zones of Sun-like stars. All stars, including our Sun, orbit the galactic core at a speed of approximately 220 km per second, regardless of the distance. That's due to the vast majority of the galaxy's mass being located in the dark-matter halo surrounding it

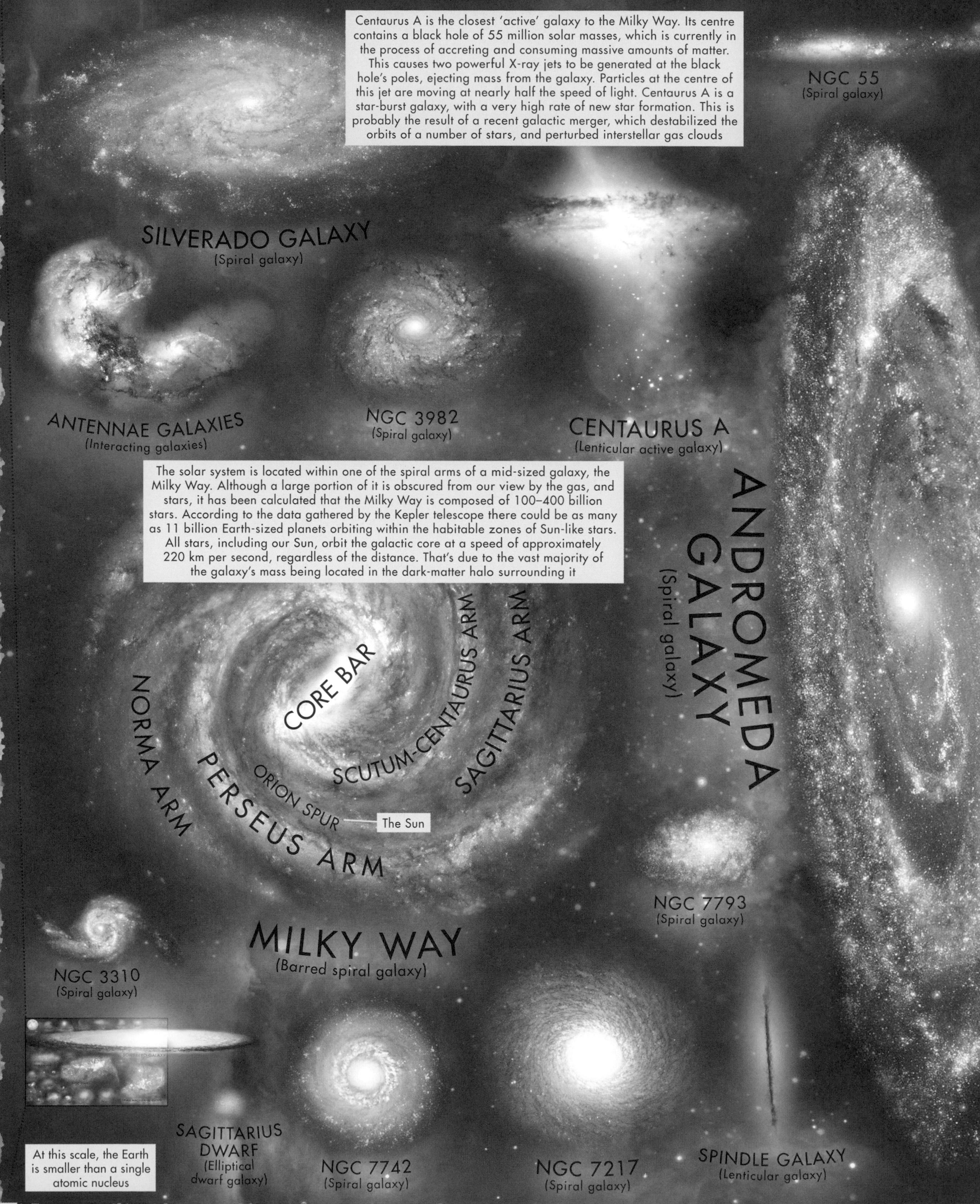

ANDROMEDA GALAXY
(Spiral galaxy)

CORE BAR

SCUTUM-CENTAURUS ARM

SAGITTARIUS ARM

NORMA ARM

PERSEUS ARM

ORION SPUR

The Sun

NGC 7793
(Spiral galaxy)

MILKY WAY
(Barred spiral galaxy)

NGC 3310
(Spiral galaxy)

SAGITTARIUS DWARF
(Elliptical dwarf galaxy)

NGC 7742
(Spiral galaxy)

NGC 7217
(Spiral galaxy)

SPINDLE GALAXY
(Lenticular galaxy)

At this scale, the Earth is smaller than a single atomic nucleus

NGC 4945
(Spiral galaxy)

NGC 2976
(Lenticular galaxy)

NGC 2207+IC 2163
(Interacting galaxies)

TRIANGULUM GALAXY
(Spiral galaxy)

The majority of galaxies as we see them are a result of a dynamic and complex process of galactic evolution. Since the first galaxies formed, they have been interacting and merging with each other. When any two galaxies collide, they don't coalesce immediately, but become tidally disrupted as they move past each other. Surprisingly, it's extremely unlikely that any two stars from the merging galaxies collide – distances between any two star systems are simply too large. Galactic collisions are fairly common – even the Milky Way is on a collision course with its largest neighbour, Andromeda Galaxy. However, the merger is only expected to happen over 4.5 billion years in the future

MESSIER 96
(Lenticular galaxy)

NGC 2841
(Spiral galaxy)

LARGE
MAGELLANIC CLOUD
(Disrupted dwarf spiral galaxy)

NGC 1672
(Spiral galaxy)

MAFFEI 2
(Spiral galaxy)

NGC 1512
(Spiral galaxy)

CIGAR GALAXY
(Starburst galaxy)

NGC 6745
(Irregular galaxy)

NGC 1300
(Barred spiral galaxy)

BLACK EYE GALAXY
(Spiral galaxy)

NGC 1409 & 1410
(Interacting spiral galaxies)

AM 0644-741
(Lenticular galaxy)

HCG 90 GROUP
(Galaxy group)

NGC 1448
(Spiral galaxy)

NGC 6240
(Irregular galaxy)

MESSIER 49
(Elliptical galaxy)

MESSIER 86
(Elliptical galaxy)

NGC 1566
(Spiral galaxy)

ABELL 2199
(Galaxy cluster)

VIRGO A
(Giant elliptical galaxy)

SEYFERT'S SEXTET
(Galaxy group)

DISTANCE FROM THE MILKY WAY
TO THE ANDROMEDA GALAXY

In several billion years, roughly at the same time the Sun will reach the final stage of its lifespan and transform into a red giant, the Milky Way is set to collide with its largest neighbour – Andromeda Galaxy. The process is expected to take several hundred million years. The supermassive black holes at the cores of both galaxies will merge, and the collisions of gas clouds during the merger will trigger a massive wave of star formation all over the combined galaxy. The final result will be a significantly more massive, elliptical galaxy with no defined spiral arms. Over the course of the collision, the majority of the stars are expected to have their orbits significantly disturbed – many might end up too close to the supermassive black holes, or, more likely, get ejected into intergalactic space

NGC 4696
(Spiral galaxy)

DARK MATTER HALO
OF THE MILKY WAY

SUNFLOWER
GALAXY
(Spiral galaxy)

NGC 1232
(Spiral galaxy)

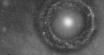

ARP 256
(Interacting spiral galaxies)

PINWHEEL
GALAXY
(Spiral galaxy)

HOAG'S OBJECT
(Lenticular galaxy)

NGC 6744
(Spiral galaxy)

THE UNIVERSE
50,000 YEARS AFTER
THE BIG BANG

500,000 LIGHT YEARS

NGC 3314
(Interacting spiral galaxies)

NEEDLE GALAXY
(Spiral galaxy)

CONDOR GALAXY
(Barred spiral galaxy)

NGC 5257 & 5258
(Interacting spiral galaxies)

CARTWHEEL
GALAXY
(Lenticular galaxy)

NGC 4921
(Barred spiral galaxy)

The supermassive IC 1101 is by far the largest known galaxy in the universe. It's located at the centre of a dense galaxy cluster, Abell 2029, over 1 billion light years away. The galaxy is home to roughly 100 trillion stars, making it roughly 500 times larger than the Milky Way. In the IC 1101's centre lies a supermassive black hole of almost 100 billion solar masses – equivalent to the mass of an entire galaxy

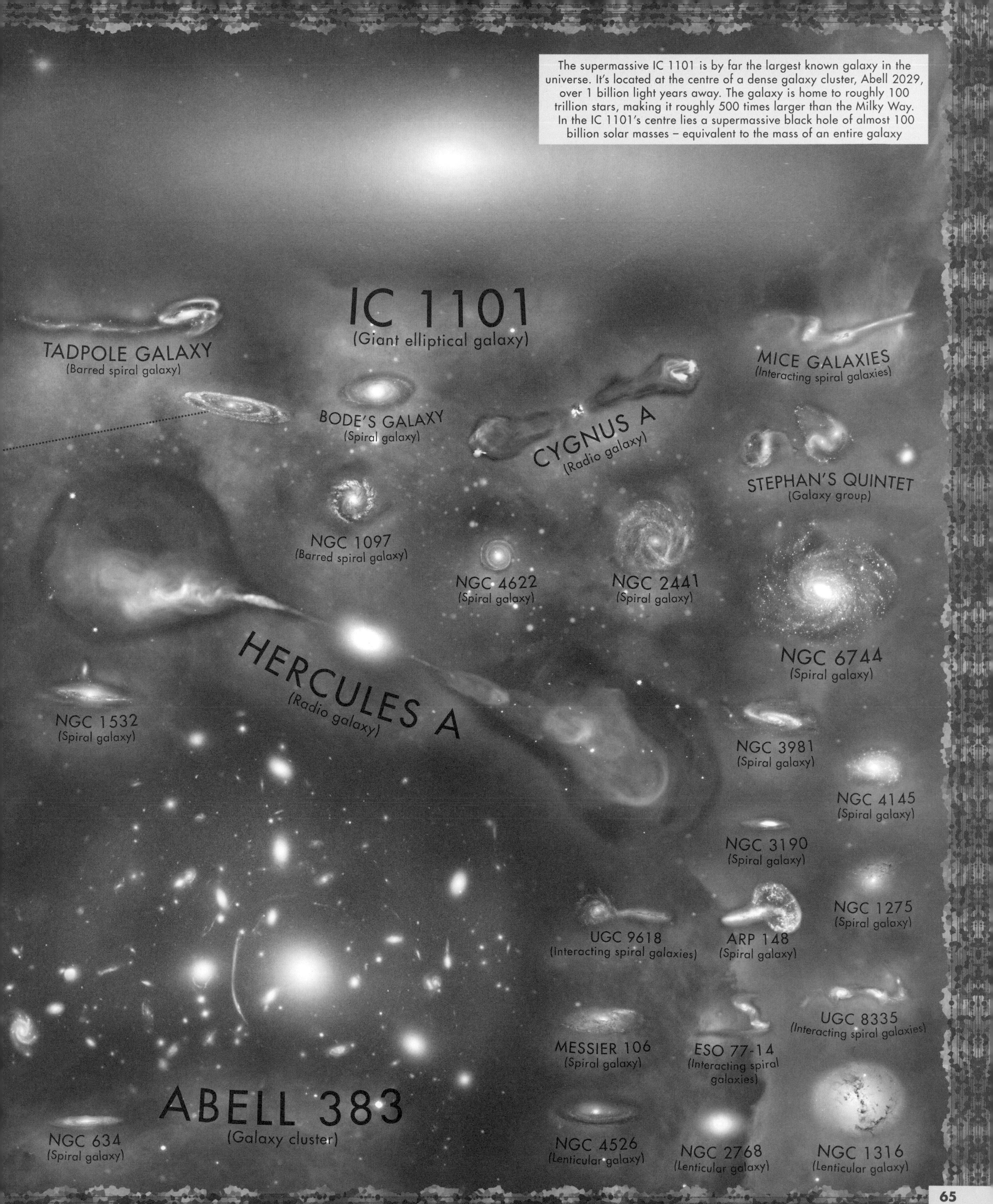

IC 1101
(Giant elliptical galaxy)

TADPOLE GALAXY
(Barred spiral galaxy)

BODE'S GALAXY
(Spiral galaxy)

MICE GALAXIES
(Interacting spiral galaxies)

CYGNUS A
(Radio galaxy)

STEPHAN'S QUINTET
(Galaxy group)

NGC 1097
(Barred spiral galaxy)

NGC 4622
(Spiral galaxy)

NGC 2441
(Spiral galaxy)

NGC 6744
(Spiral galaxy)

HERCULES A
(Radio galaxy)

NGC 1532
(Spiral galaxy)

NGC 3981
(Spiral galaxy)

NGC 4145
(Spiral galaxy)

NGC 3190
(Spiral galaxy)

NGC 1275
(Spiral galaxy)

UGC 9618
(Interacting spiral galaxies)

ARP 148
(Spiral galaxy)

UGC 8335
(Interacting spiral galaxies)

MESSIER 106
(Spiral galaxy)

ESO 77-14
(Interacting spiral galaxies)

ABELL 383
(Galaxy cluster)

NGC 634
(Spiral galaxy)

NGC 4526
(Lenticular galaxy)

NGC 2768
(Lenticular galaxy)

NGC 1316
(Lenticular galaxy)

The actual size of the entire universe is unknown – some believe that it is infinite, while many think that it curves into itself at some higher dimension, being continuous and borderless in 3-dimensional space, but finite in volume. The size of the spherical bubble of our observable universe, however, has been calculated to be 94 billion light years in diameter, and includes over 2 trillion galaxies

NGC 7582 GROUP

NGC 6744 GROUP

Milky Way

Andromeda Galaxy

LOCA GROU

NGC 7090

NGC 7793

SCULPTOR GROUP

MAFFE GROU

Recently, it was discovered that the Local Supercluster is itself a part of an even larger structure – the Laniakea Supercluster. Beyond this point, the entirety of our universe consists of a largely homogeneous, interlaced web of galactic supercluster filaments, separated by massive, empty voids

NGC 1023 GROUP

NGC 1566

DORADO CLUSTER

NGC 1399

NGC 1097

FORNAX CLUSTER

NGC 1316

NGC 1448

ERIDANUS CLUSTER

NGC 1300

NGC 1232

At this scale, the Earth is over 100 times smaller than a single proton

10 MILLION LIGHT YEARS

LOCAL SUP

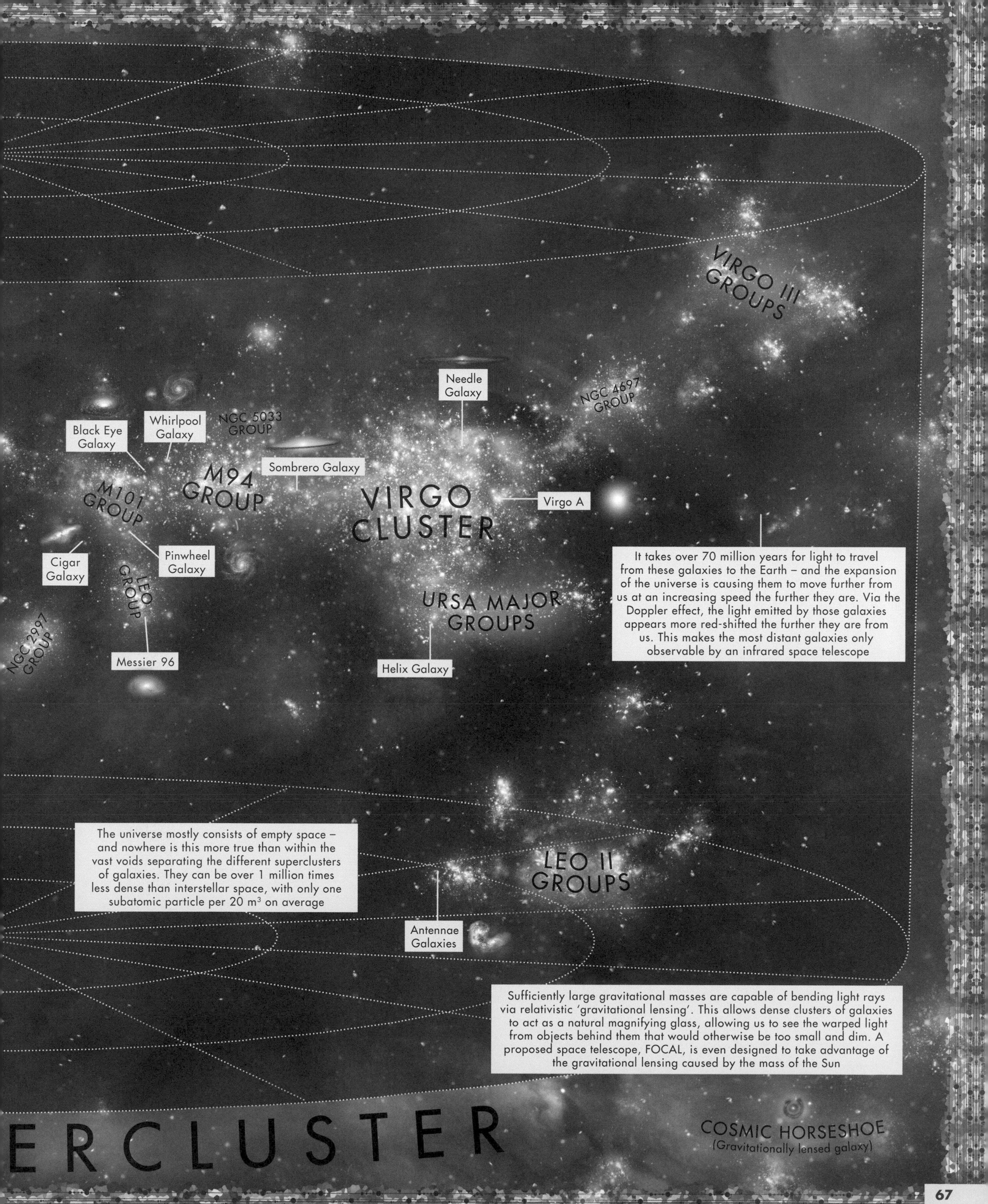

Needle
Galaxy

NGC 4697
GROUP

Black Eye
Galaxy

Whirlpool
Galaxy

NGC 5033
GROUP

M94
GROUP

Sombrero Galaxy

M101
GROUP

VIRGO
CLUSTER

Virgo A

Cigar
Galaxy

Pinwheel
Galaxy

LEO
GROUP

It takes over 70 million years for light to travel
from these galaxies to the Earth – and the expansion
of the universe is causing them to move further from
us at an increasing speed the further they are. Via the
Doppler effect, the light emitted by those galaxies
appears more red-shifted the further they are from
us. This makes the most distant galaxies only
observable by an infrared space telescope

URSA MAJOR
GROUPS

NGC 2997
GROUP

Messier 96

Helix Galaxy

The universe mostly consists of empty space –
and nowhere is this more true than within the
vast voids separating the different superclusters
of galaxies. They can be over 1 million times
less dense than interstellar space, with only one
subatomic particle per 20 m³ on average

LEO II
GROUPS

Antennae
Galaxies

Sufficiently large gravitational masses are capable of bending light rays
via relativistic 'gravitational lensing'. This allows dense clusters of galaxies
to act as a natural magnifying glass, allowing us to see the warped light
from objects behind them that would otherwise be too small and dim. A
proposed space telescope, FOCAL, is even designed to take advantage of
the gravitational lensing caused by the mass of the Sun

ERCLUSTER

COSMIC HORSESHOE
(Gravitationally lensed galaxy)

THE NIGHT SKY

Since antiquity, people have examined the night sky in exceptional detail. Creative human minds quickly came up with various patterns in the randomly scattered stars, grouped them into constellations, and spread legends about their supposed origins. This is a graphical representation of the night sky, as seen during optimal conditions, without the light pollution that makes it impossible for most people on Earth to see the sky in all its splendour. It's separated into two hemispheres, showing how the sky looks from the North and South pole respectively. Stars are relatively scaled to each other based on their apparent magnitude and are shown in their true colour, while some of the brightest galaxies, nebulae and star clusters are represented as well.

In the 2nd century BC, ancient Greek astronomer Hipparchus divided the stars in the night sky into six 'magnitudes' according to their brightness – the brightest were of magnitude 1, while the faintest still visible to the naked eye were magnitude 6. In the 19th century, this system was formalized into a descending logarithmic scale, with every magnitude being 2.51 times brighter than the one preceding it. A star with an apparent magnitude of 0 will thus be roughly 100 times fainter than a star of magnitude 5. As the human eye only perceives certain wavelengths, and is especially sensitive to green light, this scale, while still helpful, is inherently flawed. While sun-like stars glow mostly in visible wavelengths, red supergiants such as Betelgeuse and hot blue stars such as Hadar radiate most of their energy in invisible infrared and high-energy UV wavelengths respectively.

In the night sky, there are approximately 14,000 stars brighter than magnitude 7
...4,800 stars brighter than magnitude 6
...1,602 stars brighter than magnitude 5
...513 stars brighter than magnitude 4
...171 stars brighter than magnitude 3
...48 stars brighter than magnitude 2

LEGEND

Star of a magnitude:
-1 0 1 2 3 4 5 6

Globular star cluster
Open star cluster
Planetary nebula
Diffuse nebula
Close galaxy

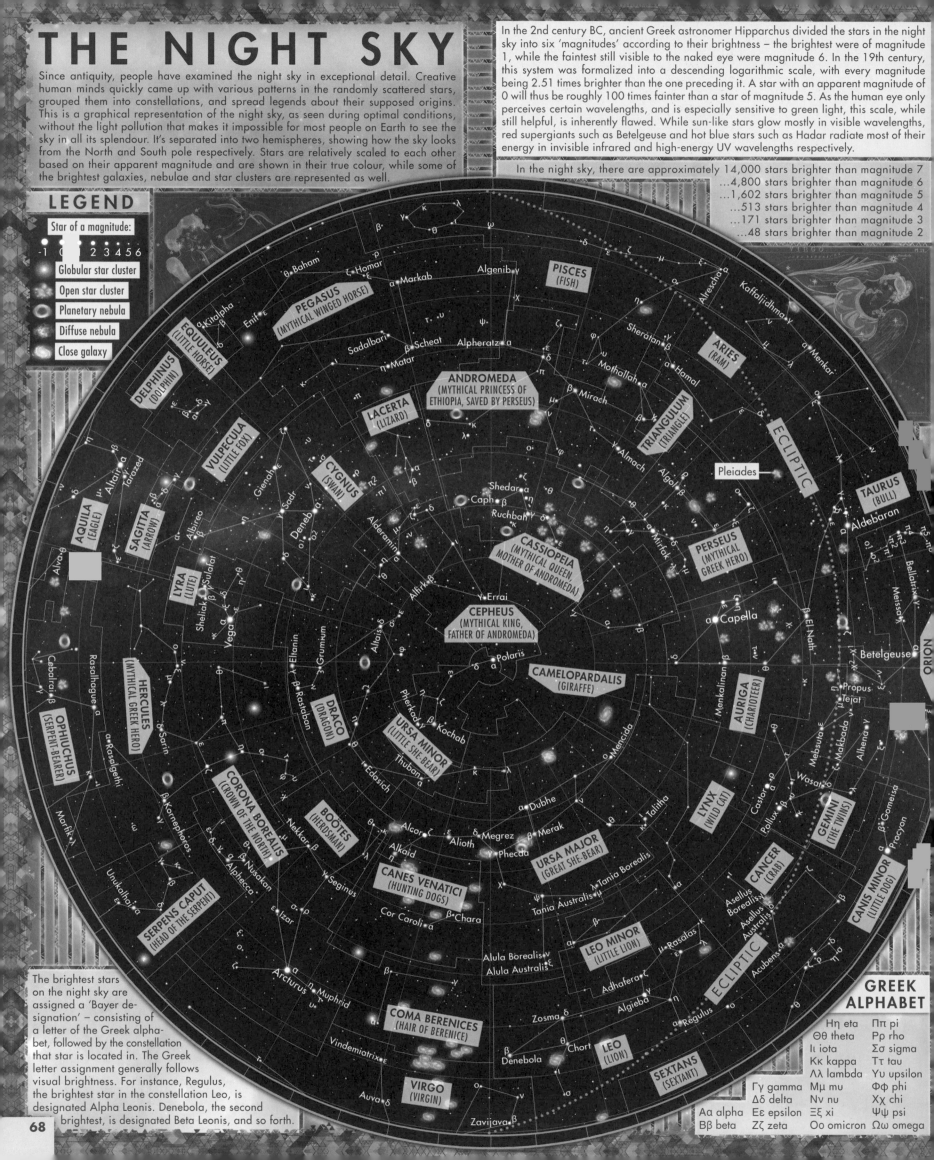

The brightest stars on the night sky are assigned a 'Bayer designation' – consisting of a letter of the Greek alphabet, followed by the constellation that star is located in. The Greek letter assignment generally follows visual brightness. For instance, Regulus, the brightest star in the constellation Leo, is designated Alpha Leonis. Denebola, the second brightest, is designated Beta Leonis, and so forth.

GREEK ALPHABET

Ηη eta	Ππ pi
Θθ theta	Ρρ rho
Ιι iota	Σσ sigma
Κκ kappa	Ττ tau
Λλ lambda	Υυ upsilon
Γγ gamma	Μμ mu	Φφ phi
Δδ delta	Νν nu	Χχ chi
	Ξξ xi	Ψψ psi
Αα alpha	Εε epsilon	Οο omicron	Ωω omega
Ββ beta	Ζζ zeta		

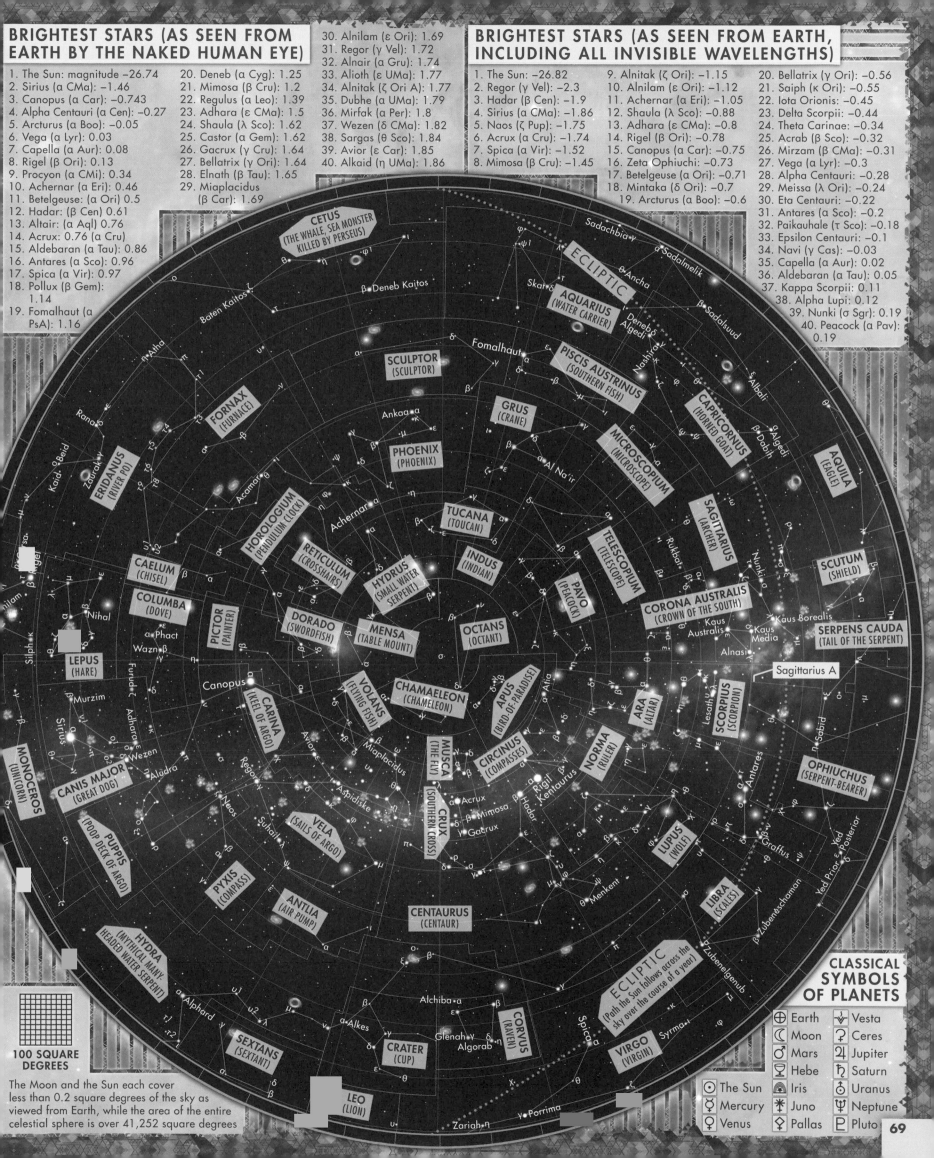

BRIGHTEST STARS (AS SEEN FROM EARTH BY THE NAKED HUMAN EYE)

1. The Sun: magnitude −26.74
2. Sirius (α CMa): −1.46
3. Canopus (α Car): −0.743
4. Alpha Centauri (α Cen): −0.27
5. Arcturus (α Boo): −0.05
6. Vega (α Lyr): 0.03
7. Capella (α Aur): 0.08
8. Rigel (β Ori): 0.13
9. Procyon (α CMi): 0.34
10. Achernar (α Eri): 0.46
11. Betelgeuse: (α Ori) 0.5
12. Hadar: (β Cen) 0.61
13. Altair: (α Aql) 0.76
14. Acrux: 0.76 (α Cru)
15. Aldebaran (α Tau): 0.86
16. Antares (α Sco): 0.96
17. Spica (α Vir): 0.97
18. Pollux (β Gem): 1.14
19. Fomalhaut (α PsA): 1.16
20. Deneb (α Cyg): 1.25
21. Mimosa (β Cru): 1.2
22. Regulus (α Leo): 1.39
23. Adhara (ε CMa): 1.5
24. Shaula (λ Sco): 1.62
25. Castor (α Gem): 1.62
26. Gacrux (γ Cru): 1.64
27. Bellatrix (γ Ori): 1.64
28. Elnath (β Tau): 1.65
29. Miaplacidus (β Car): 1.69
30. Alnilam (ε Ori): 1.69
31. Regor (γ Vel): 1.72
32. Alnair (α Gru): 1.74
33. Alioth (ε UMa): 1.77
34. Alnitak (ζ Ori A): 1.77
35. Dubhe (α UMa): 1.79
36. Mirfak (α Per): 1.8
37. Wezen (δ CMa): 1.82
38. Sargas (θ Sco): 1.84
39. Avior (ε Car): 1.85
40. Alkaid (η UMa): 1.86

BRIGHTEST STARS (AS SEEN FROM EARTH, INCLUDING ALL INVISIBLE WAVELENGTHS)

1. The Sun: −26.82
2. Regor (γ Vel): −2.3
3. Hadar (β Cen): −1.9
4. Sirius (α CMa): −1.86
5. Naos (ζ Pup): −1.75
6. Acrux (α Cru): −1.74
7. Spica (α Vir): −1.52
8. Mimosa (β Cru): −1.45
9. Alnitak (ζ Ori): −1.15
10. Alnilam (ε Ori): −1.12
11. Achernar (α Eri): −1.05
12. Shaula (λ Sco): −0.88
13. Adhara (ε CMa): −0.8
14. Rigel (β Ori): −0.78
15. Canopus (α Car): −0.75
16. Zeta Ophiuchi: −0.73
17. Betelgeuse (α Ori): −0.71
18. Mintaka (δ Ori): −0.7
19. Arcturus (α Boo): −0.6
20. Bellatrix (γ Ori): −0.56
21. Saiph (κ Ori): −0.55
22. Iota Orionis: −0.45
23. Delta Scorpii: −0.44
24. Theta Carinae: −0.34
25. Acrab (β Sco): −0.32
26. Mirzam (β CMa): −0.31
27. Vega (α Lyr): −0.3
28. Alpha Centauri: −0.28
29. Meissa (λ Ori): −0.24
30. Eta Centauri: −0.22
31. Antares (α Sco): −0.2
32. Paikauhale (τ Sco): −0.18
33. Epsilon Centauri: −0.1
34. Navi (γ Cas): −0.03
35. Capella (α Aur): 0.02
36. Aldebaran (α Tau): 0.05
37. Kappa Scorpii: 0.11
38. Alpha Lupi: 0.12
39. Nunki (σ Sgr): 0.19
40. Peacock (α Pav): 0.19

100 SQUARE DEGREES

The Moon and the Sun each cover less than 0.2 square degrees of the sky as viewed from Earth, while the area of the entire celestial sphere is over 41,252 square degrees

CLASSICAL SYMBOLS OF PLANETS

⊕ Earth	⚶ Vesta	☉ The Sun	⚴ Iris
☽ Moon	⚳ Ceres	☿ Mercury	⚵ Juno
♂ Mars	♃ Jupiter	♀ Venus	♇ Pluto
⚱ Hebe	♄ Saturn		⚸ Pallas
	♅ Uranus		
	♆ Neptune		

Soyuz, the world's most used rocket, hauling a crew up to the International Space Station

Space Shuttle Atlantis preparing to land after delivering the Columbus science module to the ISS

SpaceX's Falcon Heavy rocket lifts off from the Kennedy Space Center

Launch of RTV-G-4 Bumper, the world's first multistage space rocket. It could reach an altitude of nearly 400 km

Cosmonaut Sergey Volkov working during a space-walk outside the International Space Station

American rocket pioneer Robert Goddard stands next to his 1926 invention – the world's first liquid-fuelled rocket

EXPLORING THE COSMOS

Publication of Galileo Galilei's telescopic observations in the year 1610 sparked a widespread interest in astronomy all over early modern Europe. The idea that the Moon, Sun and the planets were large physical bodies subject to the same laws as the Earth suggested that they could be inhabited by beings much like us. In 1638, Francis Goodwin published the book *The Man in the Moone*, about an adventurer who built a flying machine that took him to the Moon, discovering that it was home to a utopian civilization. Many people consider this to be the earliest example of literary science fiction.

Over the decades, this work gradually faded into obscurity. Only in the 19th century, at the time of revolutionary innovation and progress in all fields of science and technology, did the notion of space travel start to gather literary prominence once again. In the famous science fiction novels *From the Earth to the Moon* and *Around the Moon* by Jules Verne, a massive cannon, Columbiad, constructed in Florida, is used to launch a capsule with a crew of three travellers on a journey into outer space. They travel around the Moon's far side, and successfully return back to Earth, splashing into the ocean. This story shares many striking similarities with the real-world Apollo missions that took place nearly a century later.

Inspired by the works of Jules Verne, Russian rocket scientist Konstantin Tsiolkovsky helped to lay the foundations of modern spaceflight, and is widely considered to be one of the founders of modern rocket science. Tsiolkovsky's work in the late 19th and early 20th centuries provided a much-needed theoretical basis for modern rocketry, and included treatises on rocket engines, optimal ascent and descent trajectories for spacecraft, and designs of multi-stage rockets fuelled by liquid oxygen and liquid hydrogen. He had a major influence on Sergey Korolev, the mastermind behind the Soviet space programme.

Meanwhile, in America, the physicist and engineer Robert Goddard had a more practical approach to rocket science. For a long time, conventional rockets converted only about 2% of the thermal energy in their fuel into thrust. After building and testing many diverse designs of solid-fuelled rockets, he discovered that using hourglass-shaped de Laval nozzles increased conversion efficiencies upwards of 60%, making practical high-altitude rocketry possible. In 1926, Goddard successfully launched the world's first liquid-fuelled rocket, using gasoline and liquid oxygen. Before the beginning of the Second World War, he improved rocket technology further, incorporating a gyroscopically controlled guiding system, fuel turbopumps, and regeneratively cooled engine combustion chambers – key features of most rockets still used in the present day.

The trio of founding fathers of rocketry is completed by a German scientist, Hermann Oberth. He mentored Wernher von Braun and, during the war, greatly contributed to the development of the V2 – the first space-worthy rocket. After the end of the war, the United States recruited von Braun as part of Operation Paperclip, while the Soviets kicked their space programme into full gear as well.

Throughout the first decade of the Cold War, both superpowers worked on their aerospace programmes in strict secrecy. However, in 1957, the Soviet Union surprised the entire world by launching a satellite, Sputnik 1, into orbit. The United States hurried to catch up with a satellite of its own, Explorer 1, initiating the 'space race', a decades-long struggle between the United States and the Soviet Union over technological and (indirectly) ideological superiority.

Illustrations from Jules Verne's novel *From the Earth to the Moon*, where a giant cannon is used to launch the first humans into space

Konstantin Tsiolkovsky, the father of theoretical rocketry, and some of his designs for liquid fuel and compressed-air powered rockets

Main mirror of the James Webb Space Telescope being assembled in the Goddard Space Flight Center

Soyuz spacecraft carrying three crew members to the International Space Station

Mars Pathfinder and the Sojourner rover being folded into their launch position

Galileo Galilei presenting the telescope used in his groundbreaking observations to the Doge of Venice

The space race was responsible for tremendous progress in aerospace technology in a very short time. After the Soviet Union launched the first cosmonaut, Yuri Gagarin, into space, the United States sensed that they were falling behind. In 1962, President John F. Kennedy rallied America towards a milestone far exceeding anything that had been done before – to land a man on the Moon, *'not because it is easy, but because it is hard; because that goal will serve to organize and measure the best of our energies and skills, because that challenge is one that we are willing to accept, one we are unwilling to postpone, and one we intend to win'*. On 21 July 1969, America did what it promised, when Neil Armstrong became the first man to set foot on the Moon. It was a small step for man, but a giant leap for all of mankind.

After the Soviets failed in their own lunar programme, they eventually conceded the space race to the United States, and in 1975, the superpowers conducted the first joint spaceflight, Apollo-Soyuz. Since then, space exploration has transitioned into a far more cooperative and international matter. The European Space Agency, founded in 1974, unites twenty-two European countries in a common purpose, to conduct manned and unmanned spaceflight and to learn more about the Earth and the solar system. The ESA has a long history of joint missions with NASA, JAXA (Japan Aerospace Exploration Agency) and numerous other space agencies worldwide. But, nothing embodies the spirit of international space cooperation more than the International Space Station (ISS) – a marvel of modern-day engineering, jointly constructed by the US, Russia, Europe, Japan and Canada.

After the end of the space race, the focus of NASA has mostly shifted from human spaceflight towards unmanned solar system probes and the operation of space observatories. However, this is about to change, as it aims to push the boundaries of human spaceflight forwards to the Moon once again, and then perhaps on to Mars. The advent of reusable rocketry, pioneered by Elon Musk and his private company SpaceX, is pushing the launch costs of orbital payloads down into new territory, and the launch of Blue Origin by Jeff Bezos will also contribute to the shift from the rigid and expensive government-run space programmes of yesteryear towards the more ambitious and flexible ventures of today, funded by private investors willing to take a risk for the chance of higher rewards.

We can take all of this as a sign of great times to come. Hopefully, many of you will live long enough to see a future in which space travel will be a lot more widespread, far safer and more open and affordable, opening up new frontiers and opportunities for progress and greater prosperity.

This chapter is dedicated to the achievements in exploration of both near and deep space over the last seventy years, and is split into three distinct sections.

The first part, Timeline of Space Exploration, chronicles the most iconic and important missions in the history of spaceflight, and provides a sneak peek at what we can expect in the near future. It includes the most important milestones of human spaceflight during the space age, but is uniquely focused on the missions aiming to explore, visit and learn more about the planets, asteroids and other bodies in our cosmic neighbourhood.

The second part focuses on rockets, currently the only proven way to deliver payloads into orbit. It includes over a hundred rockets from different countries, comparing their size and lifting capabilities. These pages display many historical classes of rockets used both during and after the space race, rockets that are still in active use, and also ones that are currently under development and are expected to complete their maiden flight in the near future.

Finally, the last part is dedicated to a comparison of the light-gathering mirrors of some of the largest observatories and telescopes in the world. These intricate instruments are indispensable if we want to learn more about the distant stars, nebulae and galaxies – objects that we can't ever hope to physically reach, but that can still tell us a tremendous amount about the inner workings of our universe.

Astronaut Buzz Aldrin standing on the surface of the Moon next to the lunar module

'Leviathan of Parsonstown', a massive 1.8-m diameter telescope used by William Parsons to discover the spiral form of several galaxies

Hooker Telescope at Mount Wilson observatory, used by Edwin Hubble to prove that the universe is expanding (Credit: Doc Searls)

Parkes Radio Telescope as photographed in 1969, while receiving signals from the Apollo 11 Moon landing (Credit: CSIRO)

Arecibo Radio Telescope in Puerto Rico, the second largest in the world

Very Large Telescope (VLT) in Chile, featuring advanced adaptive optics that enhance its resolving power (Credit: ESO)

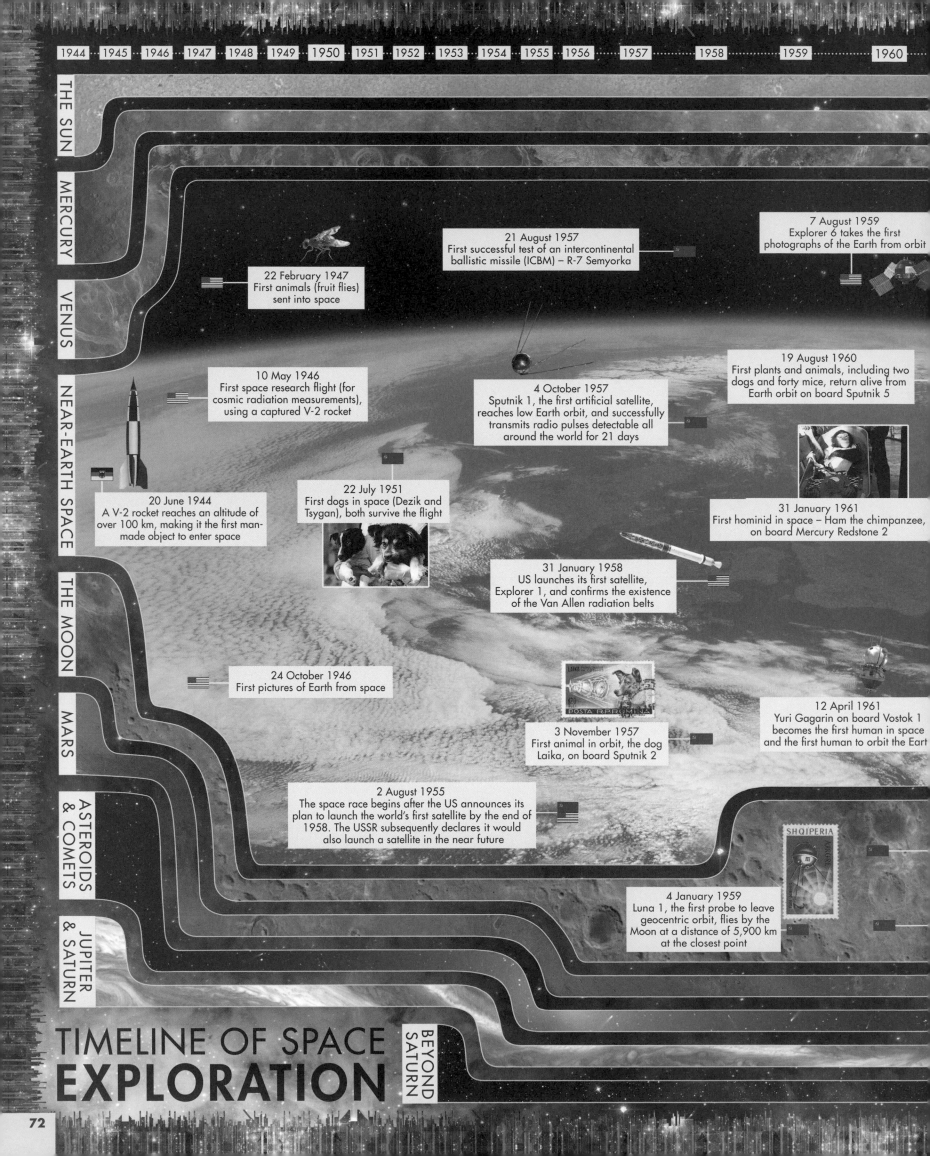

THE SUN

MERCURY

VENUS

NEAR-EARTH SPACE

THE MOON

MARS

ASTEROIDS & COMETS

JUPITER & SATURN

BEYOND SATURN

7 August 1959
Explorer 6 takes the first photographs of the Earth from orbit

21 August 1957
First successful test of an intercontinental ballistic missile (ICBM) – R-7 Semyorka

22 February 1947
First animals (fruit flies) sent into space

19 August 1960
First plants and animals, including two dogs and forty mice, return alive from Earth orbit on board Sputnik 5

10 May 1946
First space research flight (for cosmic radiation measurements), using a captured V-2 rocket

4 October 1957
Sputnik 1, the first artificial satellite, reaches low Earth orbit, and successfully transmits radio pulses detectable all around the world for 21 days

31 January 1961
First hominid in space – Ham the chimpanzee, on board Mercury Redstone 2

20 June 1944
A V-2 rocket reaches an altitude of over 100 km, making it the first man-made object to enter space

22 July 1951
First dogs in space (Dezik and Tsygan), both survive the flight

31 January 1958
US launches its first satellite, Explorer 1, and confirms the existence of the Van Allen radiation belts

24 October 1946
First pictures of Earth from space

12 April 1961
Yuri Gagarin on board Vostok 1 becomes the first human in space and the first human to orbit the Earth

3 November 1957
First animal in orbit, the dog Laika, on board Sputnik 2

2 August 1955
The space race begins after the US announces its plan to launch the world's first satellite by the end of 1958. The USSR subsequently declares it would also launch a satellite in the near future

4 January 1959
Luna 1, the first probe to leave geocentric orbit, flies by the Moon at a distance of 5,900 km at the closest point

TIMELINE OF SPACE
EXPLORATION

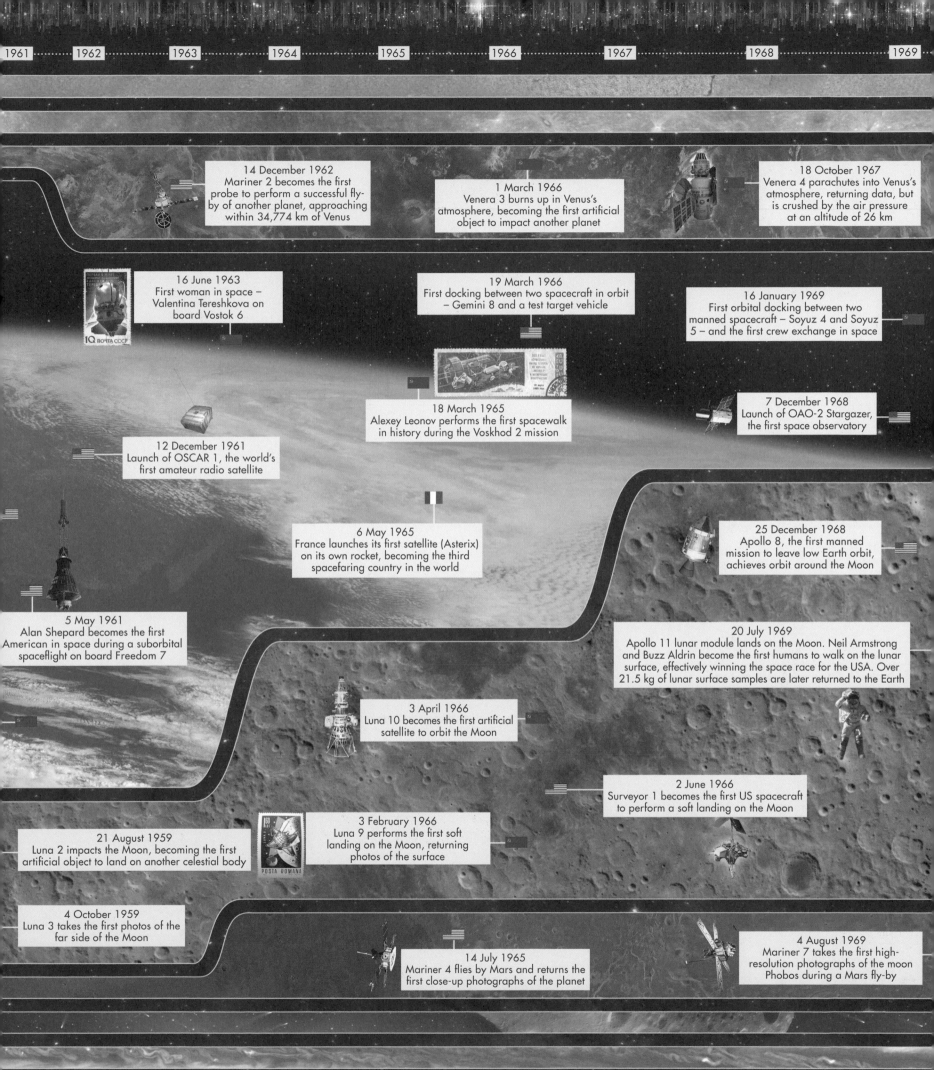

14 December 1962
Mariner 2 becomes the first probe to perform a successful fly-by of another planet, approaching within 34,774 km of Venus

1 March 1966
Venera 3 burns up in Venus's atmosphere, becoming the first artificial object to impact another planet

18 October 1967
Venera 4 parachutes into Venus's atmosphere, returning data, but is crushed by the air pressure at an altitude of 26 km

16 June 1963
First woman in space – Valentina Tereshkova on board Vostok 6

19 March 1966
First docking between two spacecraft in orbit – Gemini 8 and a test target vehicle

16 January 1969
First orbital docking between two manned spacecraft – Soyuz 4 and Soyuz 5 – and the first crew exchange in space

18 March 1965
Alexey Leonov performs the first spacewalk in history during the Voskhod 2 mission

7 December 1968
Launch of OAO-2 Stargazer, the first space observatory

12 December 1961
Launch of OSCAR 1, the world's first amateur radio satellite

6 May 1965
France launches its first satellite (Asterix) on its own rocket, becoming the third spacefaring country in the world

25 December 1968
Apollo 8, the first manned mission to leave low Earth orbit, achieves orbit around the Moon

5 May 1961
Alan Shepard becomes the first American in space during a suborbital spaceflight on board Freedom 7

20 July 1969
Apollo 11 lunar module lands on the Moon. Neil Armstrong and Buzz Aldrin become the first humans to walk on the lunar surface, effectively winning the space race for the USA. Over 21.5 kg of lunar surface samples are later returned to the Earth

3 April 1966
Luna 10 becomes the first artificial satellite to orbit the Moon

2 June 1966
Surveyor 1 becomes the first US spacecraft to perform a soft landing on the Moon

21 August 1959
Luna 2 impacts the Moon, becoming the first artificial object to land on another celestial body

3 February 1966
Luna 9 performs the first soft landing on the Moon, returning photos of the surface

4 October 1959
Luna 3 takes the first photos of the far side of the Moon

14 July 1965
Mariner 4 flies by Mars and returns the first close-up photographs of the planet

4 August 1969
Mariner 7 takes the first high-resolution photographs of the moon Phobos during a Mars fly-by

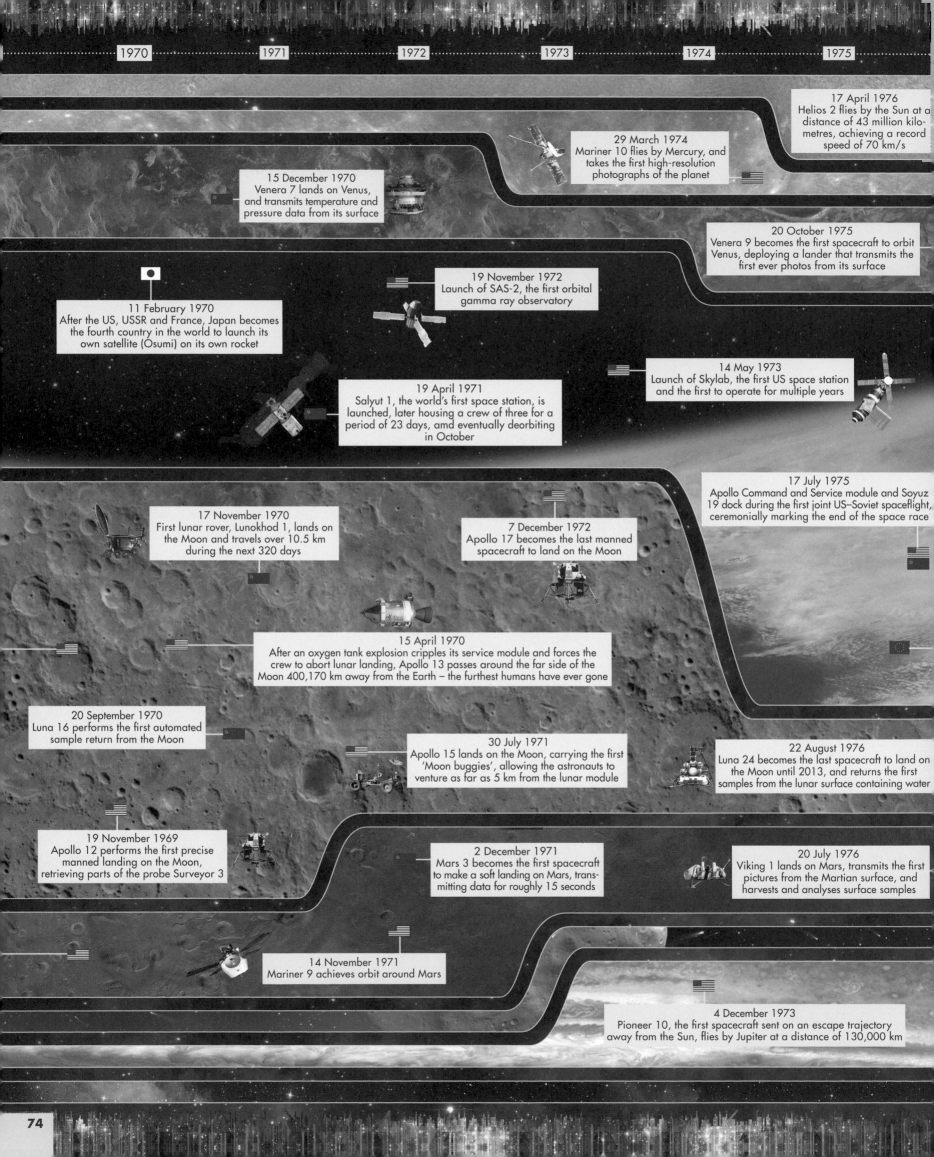

17 April 1976
Helios 2 flies by the Sun at a distance of 43 million kilometres, achieving a record speed of 70 km/s

29 March 1974
Mariner 10 flies by Mercury, and takes the first high-resolution photographs of the planet

15 December 1970
Venera 7 lands on Venus, and transmits temperature and pressure data from its surface

20 October 1975
Venera 9 becomes the first spacecraft to orbit Venus, deploying a lander that transmits the first ever photos from its surface

19 November 1972
Launch of SAS-2, the first orbital gamma ray observatory

11 February 1970
After the US, USSR and France, Japan becomes the fourth country in the world to launch its own satellite (Osumi) on its own rocket

14 May 1973
Launch of Skylab, the first US space station and the first to operate for multiple years

19 April 1971
Salyut 1, the world's first space station, is launched, later housing a crew of three for a period of 23 days, amd eventually deorbiting in October

17 July 1975
Apollo Command and Service module and Soyuz 19 dock during the first joint US–Soviet spaceflight, ceremonially marking the end of the space race

17 November 1970
First lunar rover, Lunokhod 1, lands on the Moon and travels over 10.5 km during the next 320 days

7 December 1972
Apollo 17 becomes the last manned spacecraft to land on the Moon

15 April 1970
After an oxygen tank explosion cripples its service module and forces the crew to abort lunar landing, Apollo 13 passes around the far side of the Moon 400,170 km away from the Earth – the furthest humans have ever gone

20 September 1970
Luna 16 performs the first automated sample return from the Moon

30 July 1971
Apollo 15 lands on the Moon, carrying the first 'Moon buggies', allowing the astronauts to venture as far as 5 km from the lunar module

22 August 1976
Luna 24 becomes the last spacecraft to land on the Moon until 2013, and returns the first samples from the lunar surface containing water

19 November 1969
Apollo 12 performs the first precise manned landing on the Moon, retrieving parts of the probe Surveyor 3

2 December 1971
Mars 3 becomes the first spacecraft to make a soft landing on Mars, transmitting data for roughly 15 seconds

20 July 1976
Viking 1 lands on Mars, transmits the first pictures from the Martian surface, and harvests and analyses surface samples

14 November 1971
Mariner 9 achieves orbit around Mars

4 December 1973
Pioneer 10, the first spacecraft sent on an escape trajectory away from the Sun, flies by Jupiter at a distance of 130,000 km

1 March 1982
Venera 13 takes the first soil samples from the Venusian surface and records the first sounds from another planet

4 December 1978
Pioneer Venus Orbiter begins the first multi-year orbital exploration of Venus, transmitting data until 1992. It releases four probes into the Venusian atmosphere, one of which survives the landing

26 January 1978
Launch of the International Ultraviolet Explorer, the first real-time operated orbital observatory, and the most powerful space telescope until Hubble

25 January 1983
Launch of the Infrared Astronomical Satellite (IRAS), the first-ever space telescope to perform a survey of the entire night sky at infrared wavelengths

11 July 1979
Skylab space station reenters Earth's atmosphere and disintegrates. Large amount of debris falls close to populated areas of Western Australia

29 September 1977
Launch of Salyut 6, the first space station featuring two docking ports. Unlike any space station before, this allowed Salyut 6 to be frequently resupplied and refuelled by unmanned cargo spacecraft

12 April 1981
First flight of the Space Shuttle Columbia, the first of the Space Shuttle family and the world's first reusable spacecraft, capable of carrying a crew of seven and payloads as heavy as 27 tons into low Earth orbit

22 December 1979
Europe's first launch vehicle, the French rocket Ariane 1, launches successfully from the Kourou spaceport

30 May 1975
The European Space Agency (ESA) is established in Paris, with 10 founding member countries

7 February 1984
Bruce McCandless II performs the first untethered spacewalk in history during the tenth mission of the Space Shuttle programme

22 February 1978
Launch of Navstar 1, the first GPS satellite

9 September 1982
Launch of Conestoga 1 – the first privately funded rocket to reach space (during a suborbital flight)

1 September 1979
Pioneer 11 flies by Saturn at a distance of 21,000 km

12 November 1980
Voyager 1 flies by Saturn, and has a close encounter with its largest moon, Titan, the only non-planetary object in the solar system to have an atmosphere

5 March 1979
Voyager 1 flies by Jupiter and takes the first high-resolution pictures of its four largest moons, discovering volcanism on Io

13 June 1983
Pioneer 10 becomes the first spacecraft to pass beyond the orbit of Neptune

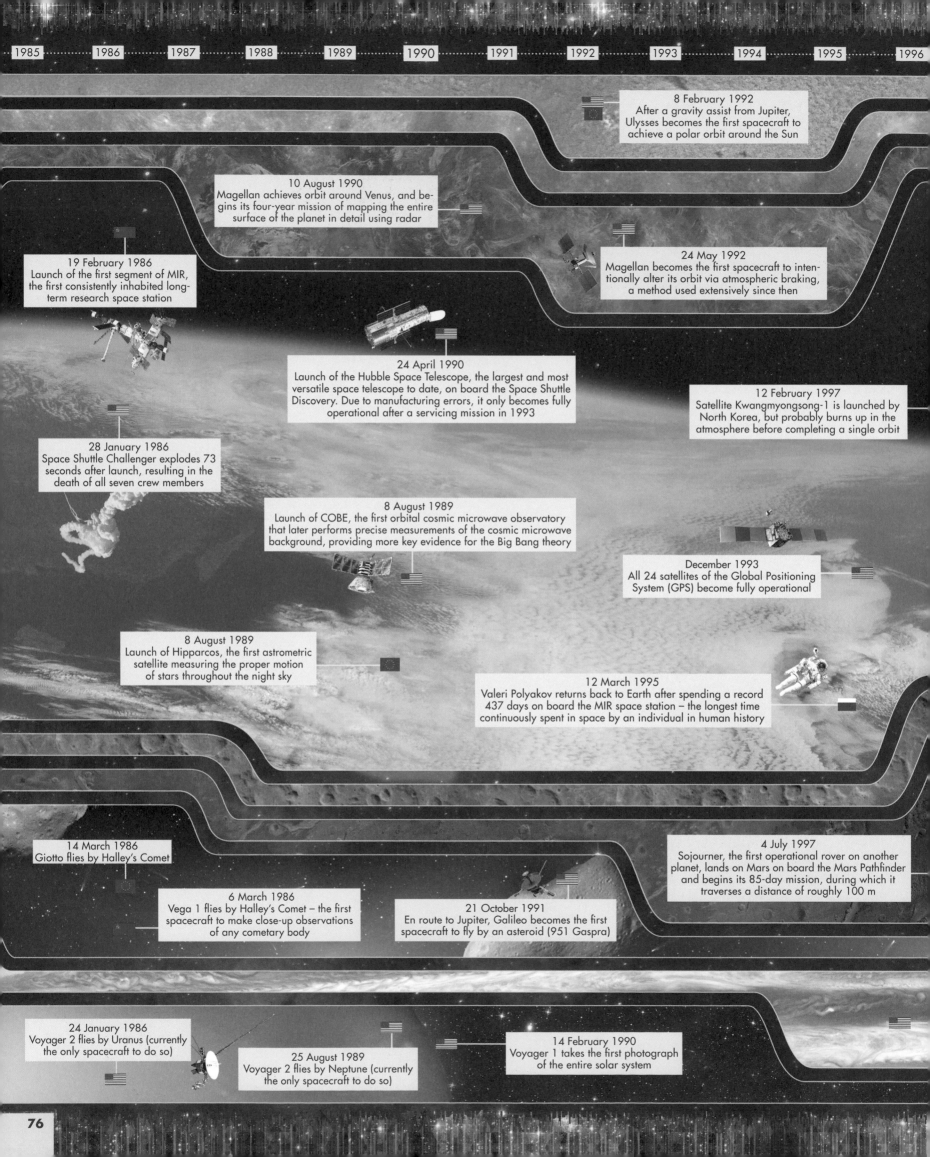

8 February 1992
After a gravity assist from Jupiter, Ulysses becomes the first spacecraft to achieve a polar orbit around the Sun

10 August 1990
Magellan achieves orbit around Venus, and begins its four-year mission of mapping the entire surface of the planet in detail using radar

24 May 1992
Magellan becomes the first spacecraft to intentionally alter its orbit via atmospheric braking, a method used extensively since then

19 February 1986
Launch of the first segment of MIR, the first consistently inhabited long-term research space station

24 April 1990
Launch of the Hubble Space Telescope, the largest and most versatile space telescope to date, on board the Space Shuttle Discovery. Due to manufacturing errors, it only becomes fully operational after a servicing mission in 1993

12 February 1997
Satellite Kwangmyongsong-1 is launched by North Korea, but probably burns up in the atmosphere before completing a single orbit

28 January 1986
Space Shuttle Challenger explodes 73 seconds after launch, resulting in the death of all seven crew members

8 August 1989
Launch of COBE, the first orbital cosmic microwave observatory that later performs precise measurements of the cosmic microwave background, providing more key evidence for the Big Bang theory

December 1993
All 24 satellites of the Global Positioning System (GPS) become fully operational

8 August 1989
Launch of Hipparcos, the first astrometric satellite measuring the proper motion of stars throughout the night sky

12 March 1995
Valeri Polyakov returns back to Earth after spending a record 437 days on board the MIR space station – the longest time continuously spent in space by an individual in human history

14 March 1986
Giotto flies by Halley's Comet

4 July 1997
Sojourner, the first operational rover on another planet, lands on Mars on board the Mars Pathfinder and begins its 85-day mission, during which it traverses a distance of roughly 100 m

6 March 1986
Vega 1 flies by Halley's Comet – the first spacecraft to make close-up observations of any cometary body

21 October 1991
En route to Jupiter, Galileo becomes the first spacecraft to fly by an asteroid (951 Gaspra)

24 January 1986
Voyager 2 flies by Uranus (currently the only spacecraft to do so)

25 August 1989
Voyager 2 flies by Neptune (currently the only spacecraft to do so)

14 February 1990
Voyager 1 takes the first photograph of the entire solar system

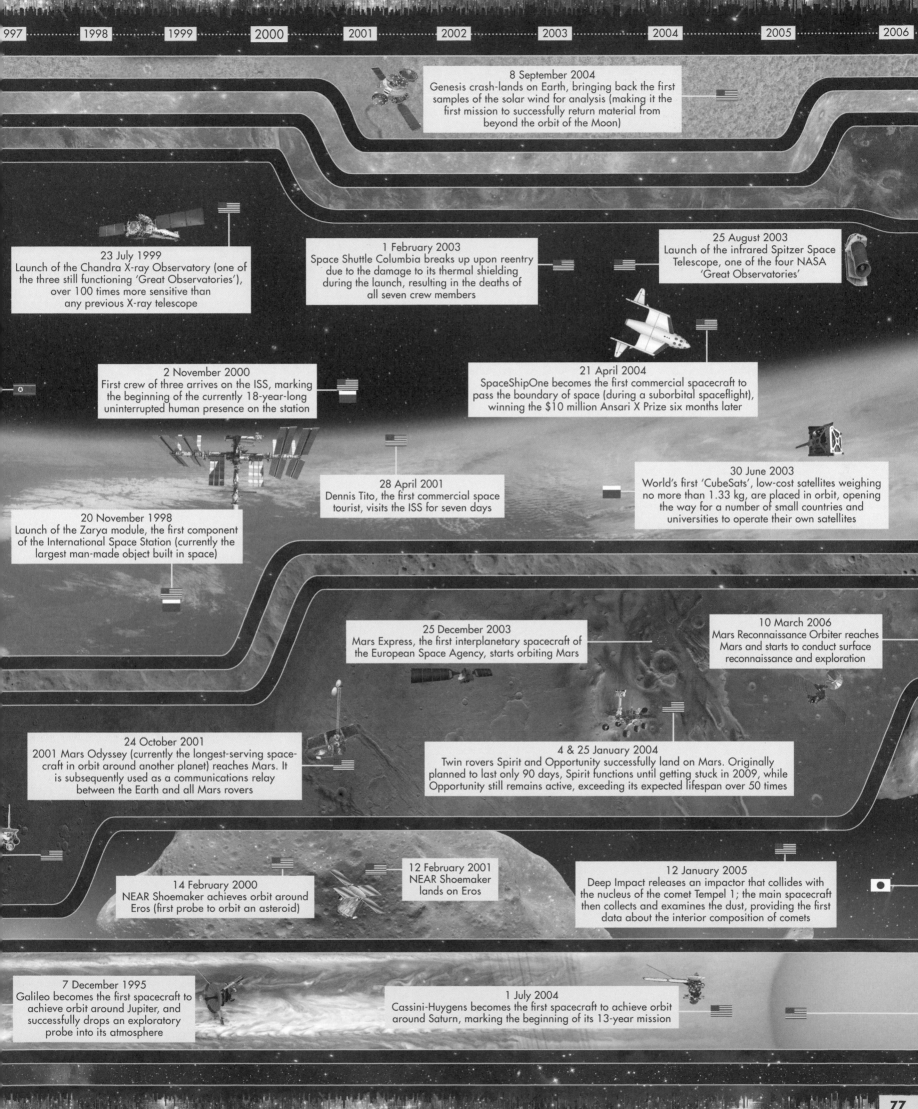

8 September 2004
Genesis crash-lands on Earth, bringing back the first samples of the solar wind for analysis (making it the first mission to successfully return material from beyond the orbit of the Moon)

23 July 1999
Launch of the Chandra X-ray Observatory (one of the three still functioning 'Great Observatories'), over 100 times more sensitive than any previous X-ray telescope

1 February 2003
Space Shuttle Columbia breaks up upon reentry due to the damage to its thermal shielding during the launch, resulting in the deaths of all seven crew members

25 August 2003
Launch of the infrared Spitzer Space Telescope, one of the four NASA 'Great Observatories'

2 November 2000
First crew of three arrives on the ISS, marking the beginning of the currently 18-year-long uninterrupted human presence on the station

21 April 2004
SpaceShipOne becomes the first commercial spacecraft to pass the boundary of space (during a suborbital spaceflight), winning the $10 million Ansari X Prize six months later

28 April 2001
Dennis Tito, the first commercial space tourist, visits the ISS for seven days

30 June 2003
World's first 'CubeSats', low-cost satellites weighing no more than 1.33 kg, are placed in orbit, opening the way for a number of small countries and universities to operate their own satellites

20 November 1998
Launch of the Zarya module, the first component of the International Space Station (currently the largest man-made object built in space)

25 December 2003
Mars Express, the first interplanetary spacecraft of the European Space Agency, starts orbiting Mars

10 March 2006
Mars Reconnaissance Orbiter reaches Mars and starts to conduct surface reconnaissance and exploration

24 October 2001
2001 Mars Odyssey (currently the longest-serving spacecraft in orbit around another planet) reaches Mars. It is subsequently used as a communications relay between the Earth and all Mars rovers

4 & 25 January 2004
Twin rovers Spirit and Opportunity successfully land on Mars. Originally planned to last only 90 days, Spirit functions until getting stuck in 2009, while Opportunity still remains active, exceeding its expected lifespan over 50 times

14 February 2000
NEAR Shoemaker achieves orbit around Eros (first probe to orbit an asteroid)

12 February 2001
NEAR Shoemaker lands on Eros

12 January 2005
Deep Impact releases an impactor that collides with the nucleus of the comet Tempel 1; the main spacecraft then collects and examines the dust, providing the first data about the interior composition of comets

7 December 1995
Galileo becomes the first spacecraft to achieve orbit around Jupiter, and successfully drops an exploratory probe into its atmosphere

1 July 2004
Cassini-Huygens becomes the first spacecraft to achieve orbit around Saturn, marking the beginning of its 13-year mission

18 March 2011
MESSENGER becomes the first spacecraft to orbit Mercury. After using its entire manoeuvring propellant, it impacts the planet four years later

11 April 2006
Venus Express, the first exploratory mission to Venus under the European Space Agency, achieves orbit

8 December 2010
IKAROS, the first spacecraft to utilize solar sails as its main method of propulsion, flies by Venus

25 May 2012
SpaceX Dragon becomes the first private spacecraft to dock with the International Space Station. SpaceX has been contracted by NASA to conduct ISS resupply missions

7 March 2009
Launch of Kepler, a space observatory designed to discover Earth-sized planets orbiting other stars. It discovered more than 1,200 confirmed exoplanets, many of them habitable, before being decommissioned in 2018

21 July 2011
Space Shuttle Atlantis concludes its final flight, marking the end of the Space Shuttle programme after over 30 years and 135 missions. As a result, NASA loses the capability to carry humans into orbit

14 May 2009
Herschel Space Observatory becomes operational. With a mirror 3.5 m in diameter, it is the largest infrared telescope ever launched

1 December 2013
Chinese spacecraft Chang'e 3 lands on the Moon carrying a lunar rover, Yutu, making China the third country in the world to successfully land an object on the Moon

25 May 2008
Stationary spacecraft Phoenix lands on the surface of Mars

26 November 2011
Curiosity, the largest and most advanced space rover to date, lands on Mars. The entire (extremely complex) landing procedure is dubbed 'Seven Minutes of Terror' by NASA

5 November 2013
Mangalyaan probe, India's first interplanetary spacecraft, successfully achieves Mars orbit. With a record low budget of $71 million, it is the least expensive interplanetary mission to date

13 December 2012
Chang'e 2 flies closely by the asteroid 4179 Toutatis, taking detailed pictures

15 January 2006
Stardust probe collects dust samples from the tail of the comet Wild 2, and returns them back to Earth for further analysis

16 July 2011
Dawn probe reaches the asteroid Vesta, the second largest object in the asteroid belt

19 November 2005
Japanese spacecraft Hayabusa (Falcon), propelled by ion engines, lands on near-Earth asteroid Itokawa and collects surface samples, bringing them back to Earth in 2010

6 August 2014
Rosetta becomes the first spacecraft to achieve orbit around a comet (67P/Churyumov–Gerasimenko). A few months later, its lander probe, Philae, lands on the surface of the comet

14 January 2005
Huygens probe lands on Titan (the most distant landing by any man-made object), takes pictures from the surface and confirms the existence of hydrocarbon lakes and rivers on its surface

25 August 2012
Voyager 1 passes the heliopause, becoming the first probe to reach interstellar space

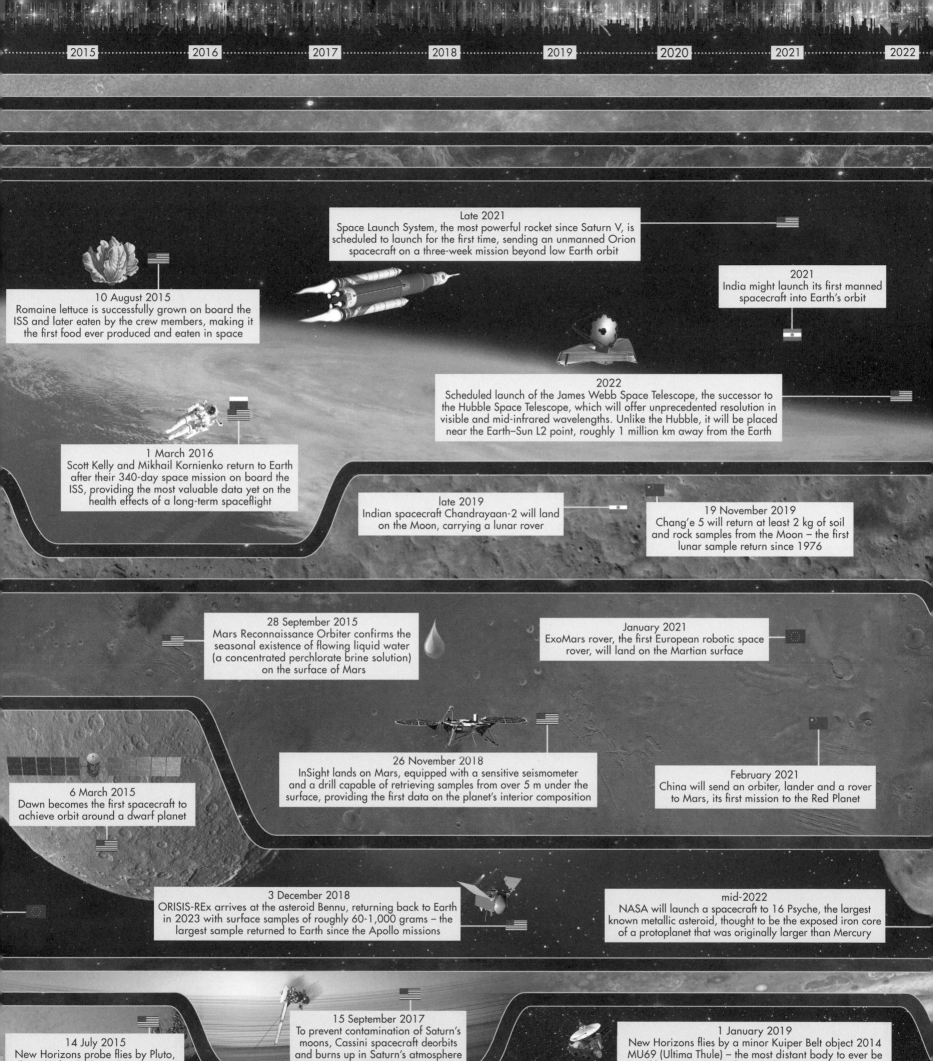

Late 2021
Space Launch System, the most powerful rocket since Saturn V, is scheduled to launch for the first time, sending an unmanned Orion spacecraft on a three-week mission beyond low Earth orbit

10 August 2015
Romaine lettuce is successfully grown on board the ISS and later eaten by the crew members, making it the first food ever produced and eaten in space

2021
India might launch its first manned spacecraft into Earth's orbit

2022
Scheduled launch of the James Webb Space Telescope, the successor to the Hubble Space Telescope, which will offer unprecedented resolution in visible and mid-infrared wavelengths. Unlike the Hubble, it will be placed near the Earth–Sun L2 point, roughly 1 million km away from the Earth

1 March 2016
Scott Kelly and Mikhail Kornienko return to Earth after their 340-day space mission on board the ISS, providing the most valuable data yet on the health effects of a long-term spaceflight

late 2019
Indian spacecraft Chandrayaan-2 will land on the Moon, carrying a lunar rover

19 November 2019
Chang'e 5 will return at least 2 kg of soil and rock samples from the Moon – the first lunar sample return since 1976

28 September 2015
Mars Reconnaissance Orbiter confirms the seasonal existence of flowing liquid water (a concentrated perchlorate brine solution) on the surface of Mars

January 2021
ExoMars rover, the first European robotic space rover, will land on the Martian surface

26 November 2018
InSight lands on Mars, equipped with a sensitive seismometer and a drill capable of retrieving samples from over 5 m under the surface, providing the first data on the planet's interior composition

February 2021
China will send an orbiter, lander and a rover to Mars, its first mission to the Red Planet

6 March 2015
Dawn becomes the first spacecraft to achieve orbit around a dwarf planet

3 December 2018
ORISIS-REx arrives at the asteroid Bennu, returning back to Earth in 2023 with surface samples of roughly 60-1,000 grams – the largest sample returned to Earth since the Apollo missions

mid-2022
NASA will launch a spacecraft to 16 Psyche, the largest known metallic asteroid, thought to be the exposed iron core of a protoplanet that was originally larger than Mercury

15 September 2017
To prevent contamination of Saturn's moons, Cassini spacecraft deorbits and burns up in Saturn's atmosphere

14 July 2015
New Horizons probe flies by Pluto, taking the first high-resolution pictures of the dwarf planet and its moons Charon, Nix and Hydra

1 January 2019
New Horizons flies by a minor Kuiper Belt object 2014 MU69 (Ultima Thule) – the most distant body to ever be visited by any space probe – and continues on an escape trajectory from the solar system

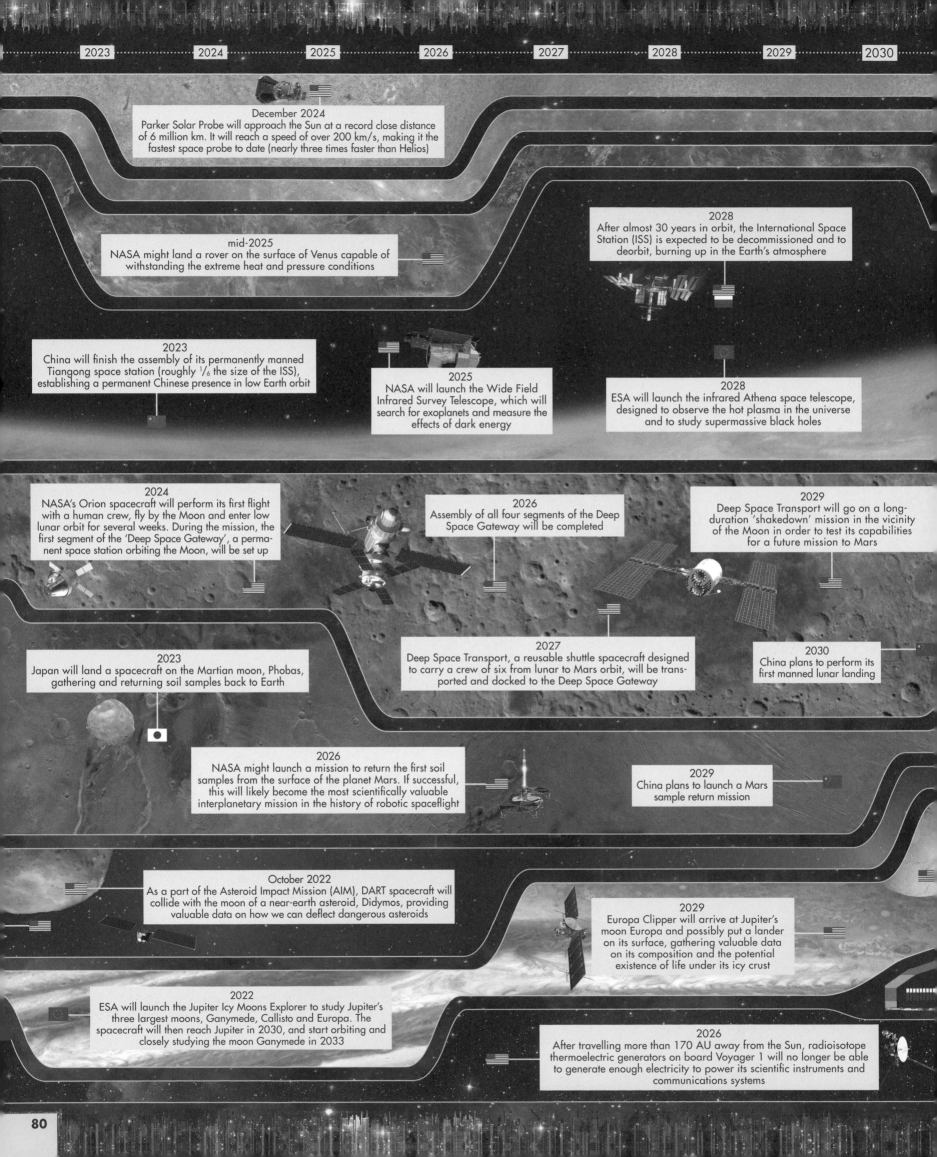

December 2024
Parker Solar Probe will approach the Sun at a record close distance of 6 million km. It will reach a speed of over 200 km/s, making it the fastest space probe to date (nearly three times faster than Helios)

2028
After almost 30 years in orbit, the International Space Station (ISS) is expected to be decommissioned and to deorbit, burning up in the Earth's atmosphere

mid-2025
NASA might land a rover on the surface of Venus capable of withstanding the extreme heat and pressure conditions

2023
China will finish the assembly of its permanently manned Tiangong space station (roughly 1/6 the size of the ISS), establishing a permanent Chinese presence in low Earth orbit

2025
NASA will launch the Wide Field Infrared Survey Telescope, which will search for exoplanets and measure the effects of dark energy

2028
ESA will launch the infrared Athena space telescope, designed to observe the hot plasma in the universe and to study supermassive black holes

2024
NASA's Orion spacecraft will perform its first flight with a human crew, fly by the Moon and enter low lunar orbit for several weeks. During the mission, the first segment of the 'Deep Space Gateway', a permanent space station orbiting the Moon, will be set up

2026
Assembly of all four segments of the Deep Space Gateway will be completed

2029
Deep Space Transport will go on a long-duration 'shakedown' mission in the vicinity of the Moon in order to test its capabilities for a future mission to Mars

2023
Japan will land a spacecraft on the Martian moon, Phobas, gathering and returning soil samples back to Earth

2027
Deep Space Transport, a reusable shuttle spacecraft designed to carry a crew of six from lunar to Mars orbit, will be transported and docked to the Deep Space Gateway

2030
China plans to perform its first manned lunar landing

2026
NASA might launch a mission to return the first soil samples from the surface of the planet Mars. If successful, this will likely become the most scientifically valuable interplanetary mission in the history of robotic spaceflight

2029
China plans to launch a Mars sample return mission

October 2022
As a part of the Asteroid Impact Mission (AIM), DART spacecraft will collide with the moon of a near-earth asteroid, Didymos, providing valuable data on how we can deflect dangerous asteroids

2029
Europa Clipper will arrive at Jupiter's moon Europa and possibly put a lander on its surface, gathering valuable data on its composition and the potential existence of life under its icy crust

2022
ESA will launch the Jupiter Icy Moons Explorer to study Jupiter's three largest moons, Ganymede, Callisto and Europa. The spacecraft will then reach Jupiter in 2030, and start orbiting and closely studying the moon Ganymede in 2033

2026
After travelling more than 170 AU away from the Sun, radioisotope thermoelectric generators on board Voyager 1 will no longer be able to generate enough electricity to power its scientific instruments and communications systems

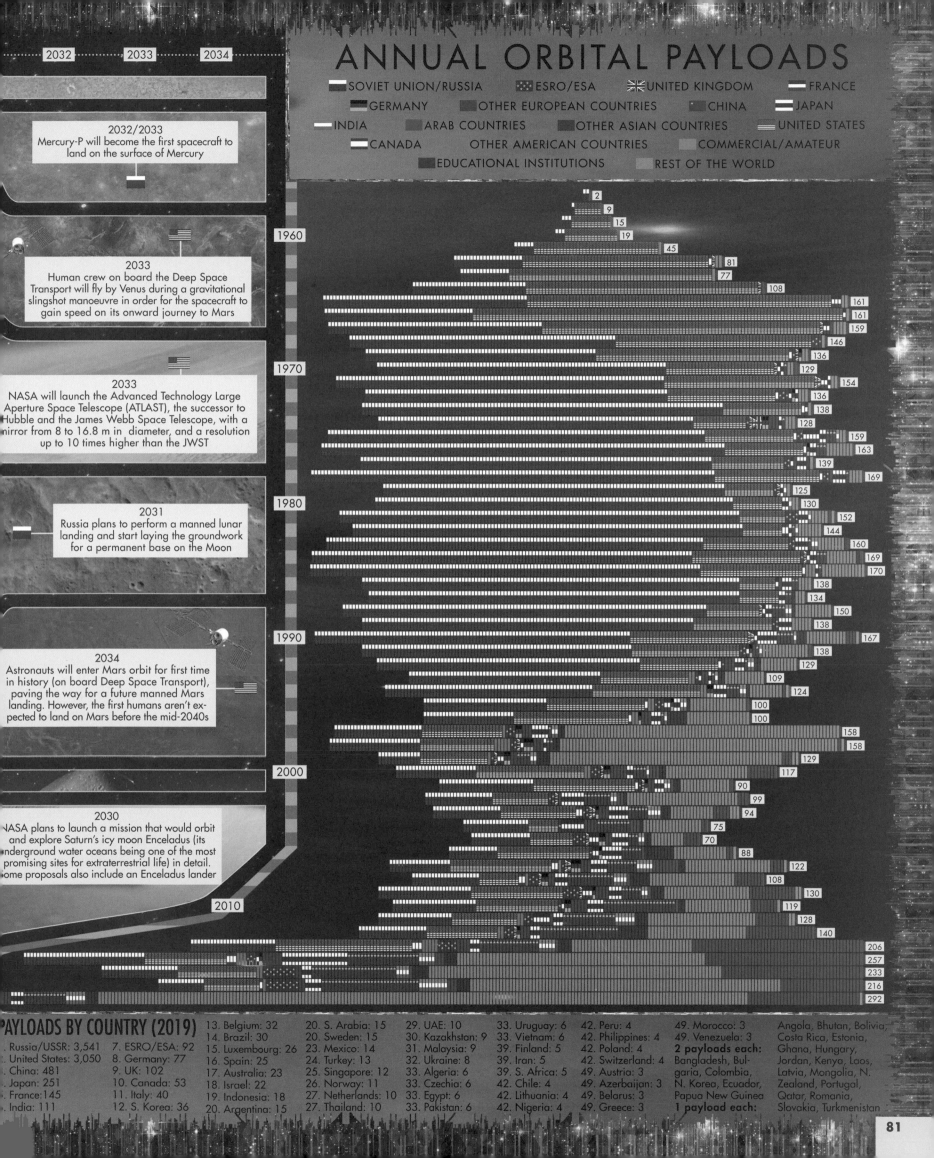

ANNUAL ORBITAL PAYLOADS

Legend:
- SOVIET UNION/RUSSIA
- ESRO/ESA
- UNITED KINGDOM
- FRANCE
- GERMANY
- OTHER EUROPEAN COUNTRIES
- CHINA
- JAPAN
- INDIA
- ARAB COUNTRIES
- OTHER ASIAN COUNTRIES
- UNITED STATES
- CANADA
- OTHER AMERICAN COUNTRIES
- COMMERCIAL/AMATEUR
- EDUCATIONAL INSTITUTIONS
- REST OF THE WORLD

2032 — 2033 — 2034

2032/2033
Mercury-P will become the first spacecraft to land on the surface of Mercury

2033
Human crew on board the Deep Space Transport will fly by Venus during a gravitational slingshot manoeuvre in order for the spacecraft to gain speed on its onward journey to Mars

2033
NASA will launch the Advanced Technology Large Aperture Space Telescope (ATLAST), the successor to Hubble and the James Webb Space Telescope, with a mirror from 8 to 16.8 m in diameter, and a resolution up to 10 times higher than the JWST

2031
Russia plans to perform a manned lunar landing and start laying the groundwork for a permanent base on the Moon

2034
Astronauts will enter Mars orbit for first time in history (on board Deep Space Transport), paving the way for a future manned Mars landing. However, the first humans aren't expected to land on Mars before the mid-2040s

2030
NASA plans to launch a mission that would orbit and explore Saturn's icy moon Enceladus (its underground water oceans being one of the most promising sites for extraterrestrial life) in detail. Some proposals also include an Enceladus lander

Timeline years: 1960, 1970, 1980, 1990, 2000, 2010

Annual totals (right edge, top to bottom): 2, 9, 15, 19, 45, 81, 77, 108, 161, 161, 159, 146, 136, 129, 154, 136, 138, 128, 159, 163, 139, 169, 125, 130, 152, 144, 160, 169, 170, 138, 134, 150, 138, 167, 138, 129, 109, 124, 100, 100, 158, 158, 129, 117, 90, 99, 94, 75, 70, 88, 122, 108, 130, 119, 128, 140, 206, 257, 233, 216, 292

PAYLOADS BY COUNTRY (2019)

Rank	Country	Payloads
1	Russia/USSR	3,541
2	United States	3,050
3	China	481
4	Japan	251
5	France	145
6	India	111
7	ESRO/ESA	92
8	Germany	77
9	UK	102
10	Canada	53
11	Italy	40
12	S. Korea	36
13	Belgium	32
14	Brazil	30
15	Luxembourg	26
16	Spain	25
17	Australia	23
18	Israel	22
19	Indonesia	18
20	Argentina	15
20	S. Arabia	15
20	Sweden	15
23	Mexico	14
24	Turkey	13
25	Singapore	12
26	Norway	11
27	Netherlands	10
27	Thailand	10
29	UAE	10
30	Kazakhstan	9
31	Malaysia	9
32	Ukraine	8
33	Algeria	6
33	Czechia	6
33	Egypt	6
33	Pakistan	6
33	Uruguay	6
33	Vietnam	6
39	Finland	5
39	Iran	5
39	S. Africa	5
42	Chile	4
42	Lithuania	4
42	Nigeria	4
42	Peru	4
42	Philippines	4
42	Poland	4
42	Switzerland	4
49	Morocco	3
49	Venezuela	3
49	Austria	3
49	Azerbaijan	3
49	Belarus	3
49	Greece	3

2 payloads each: Bangladesh, Bulgaria, Colombia, N. Korea, Ecuador, Papua New Guinea

1 payload each: Angola, Bhutan, Bolivia, Costa Rica, Estonia, Ghana, Hungary, Jordan, Kenya, Laos, N. Zealand, Portugal, Qatar, Romania, Slovakia, Turkmenistan

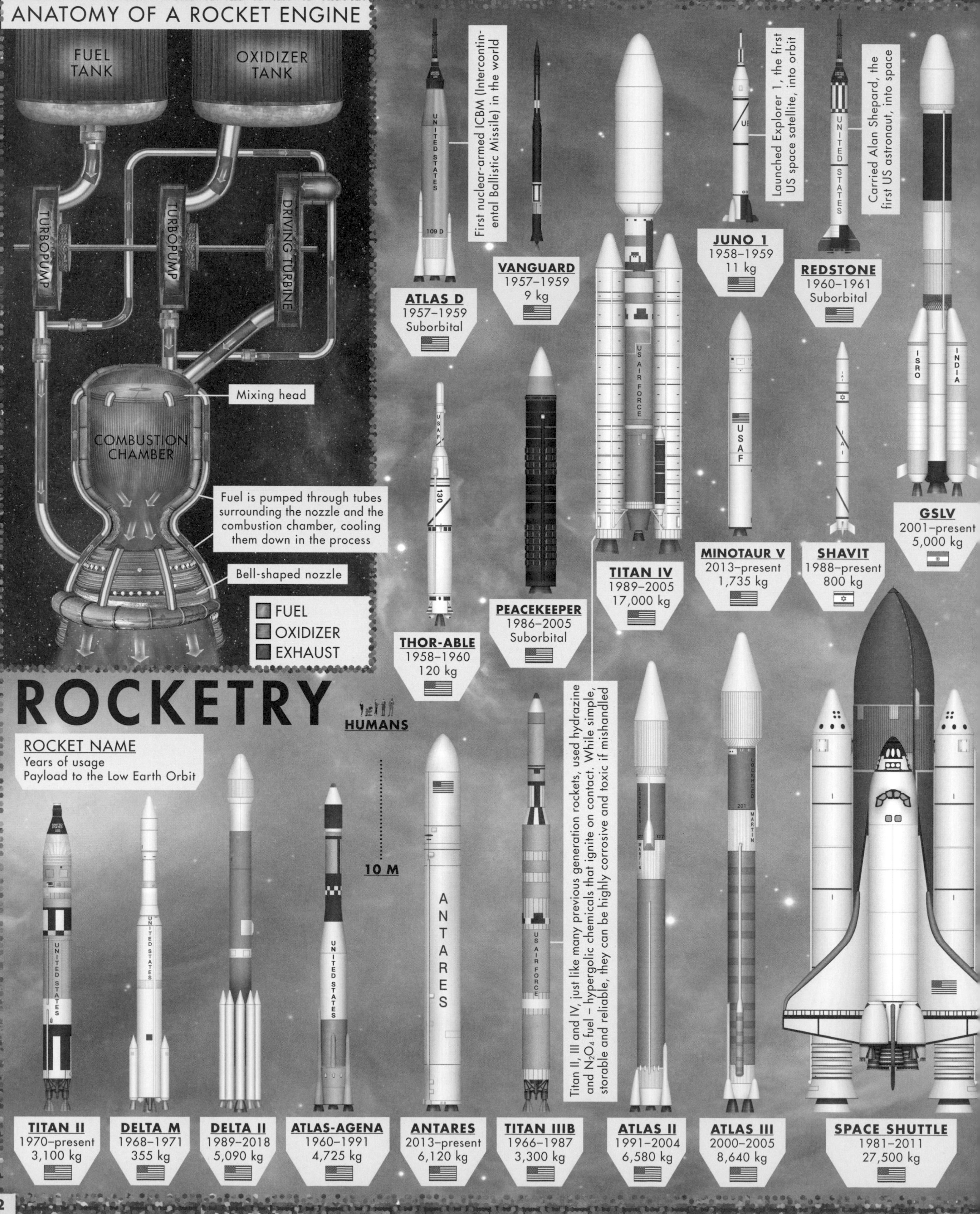

ANATOMY OF A ROCKET ENGINE

FUEL TANK

OXIDIZER TANK

TURBOPUMP

TURBOPUMP

DRIVING TURBINE

Mixing head

COMBUSTION CHAMBER

Fuel is pumped through tubes surrounding the nozzle and the combustion chamber, cooling them down in the process

Bell-shaped nozzle

- FUEL
- OXIDIZER
- EXHAUST

ROCKETRY

HUMANS

10 M

ATLAS D
1957–1959
Suborbital

First nuclear-armed ICBM (Intercontinental Ballistic Missile) in the world

VANGUARD
1957–1959
9 kg

THOR-ABLE
1958–1960
120 kg

PEACEKEEPER
1986–2005
Suborbital

TITAN IV
1989–2005
17,000 kg

US AIR FORCE

JUNO 1
1958–1959
11 kg

Launched Explorer 1, the first US space satellite, into orbit

MINOTAUR V
2013–present
1,735 kg

USAF

REDSTONE
1960–1961
Suborbital

UNITED STATES

Carried Alan Shepard, the first US astronaut, into space

SHAVIT
1988–present
800 kg

GSLV
2001–present
5,000 kg

ISRO INDIA

TITAN II
1970–present
3,100 kg

DELTA M
1968–1971
355 kg

DELTA II
1989–2018
5,090 kg

ATLAS-AGENA
1960–1991
4,725 kg

ANTARES
2013–present
6,120 kg

ANTARES

TITAN IIIB
1966–1987
3,300 kg

US AIR FORCE

Titan II, III and IV, just like many previous generation rockets, used hydrazine and N$_2$O$_4$ fuel – hypergolic chemicals that ignite on contact. While simple, storable and reliable, they can be highly corrosive and toxic if mishandled

ATLAS II
1991–2004
6,580 kg

ATLAS III
2000–2005
8,640 kg

SPACE SHUTTLE
1981–2011
27,500 kg

PSLV
1993–2008
3,800 kg

ATLAS-CENTAUR
1962–2004
3,630 kg

TITAN IIIC
1965–1982
13,100 kg

PEGASUS
1990–2016
443 kg

Unlike any other orbital rocket, Pegasus is launched directly from a flying carrier aircraft. This makes it one of the lightest and most efficient rockets in use

LITTLE JOE II
1963–1966
Suborbital

DELTA E
1965–1971
540 kg

MINOTAUR 1
2000–present
580 kg

Most manned spacecraft feature a so-called 'launch escape system' – a solid-fuelled rocket, designed to ignite in case of an emergency and carry the capsule to a safe distance away from the launch vehicle

MINUTEMAN
1962–present
Suborbital

Space Launch System (SLS) is a superheavy rocket being developed by NASA. It's designed for crewed missions into deep space, with a goal of eventually paving the way for a manned mission to Mars

During the launch process, expended rocket stages are subsequently detached using explosive separators

JUNO II
1958–1961
41 kg

TAURUS
1994–present
1,350 kg

Until the first launch of Falcon Heavy in 2018, Delta IV Heavy was the most powerful rocket in active operation

At launch, the total energy output of the Saturn V rocket surpassed 150 Gigawatts – the equivalent of over 10 Itaipu Dams!

ATLAS V
1991–2004
17,440 kg

VULCAN
Mid 2020s
35,000 kg

DELTA IV
2002–present
11,470 kg

SATURN I
1961–1965
9,070 kg

DELTA IV HEAVY
2004–present
28,790 kg

SATURN V
1967–1973
140,000 kg

SLS BLOCK 1B
2020s
105,000 kg

83

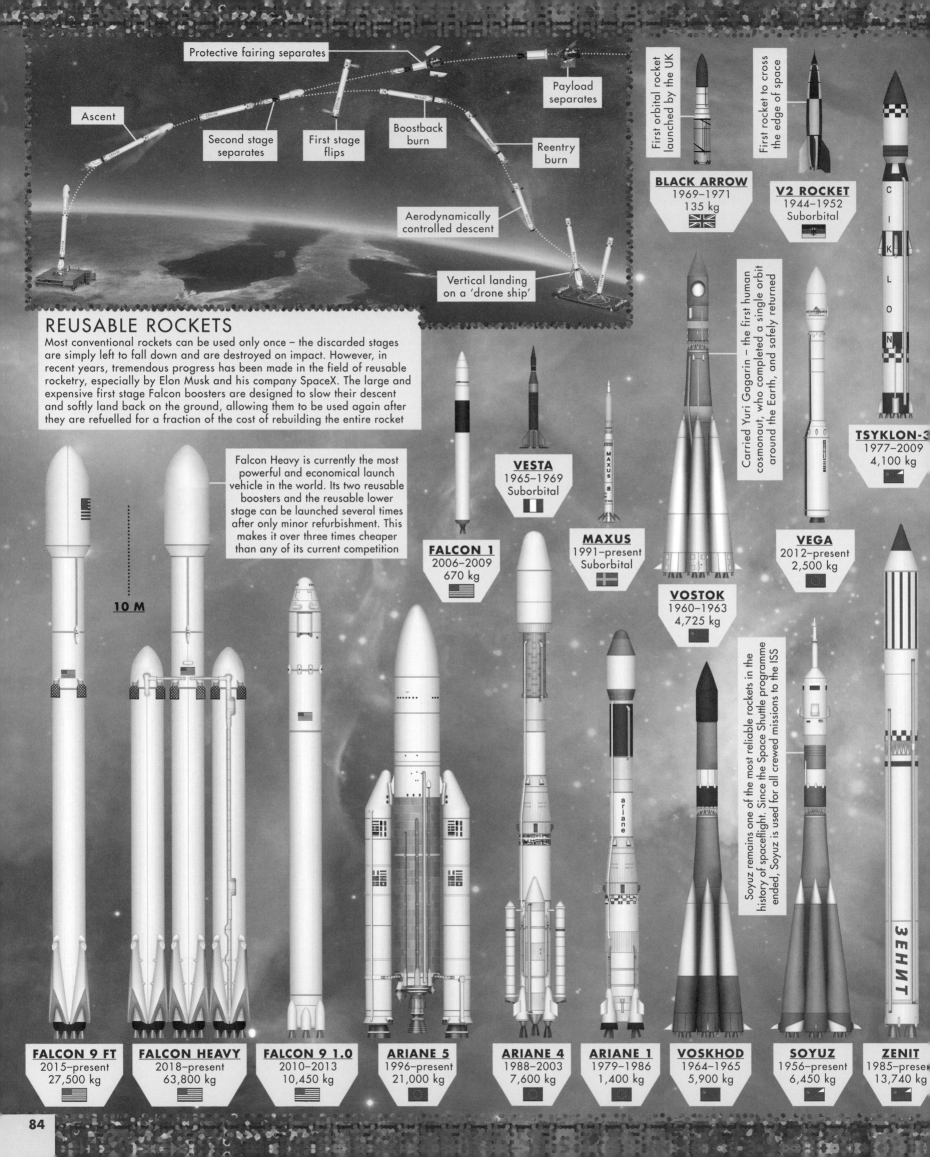

Protective fairing separates

Ascent

Second stage separates

First stage flips

Boostback burn

Payload separates

Reentry burn

Aerodynamically controlled descent

Vertical landing on a 'drone ship'

First orbital rocket launched by the UK

First rocket to cross the edge of space

Carried Yuri Gagarin – the first human cosmonaut, who completed a single orbit around the Earth, and safely returned

Soyuz remains one of the most reliable rockets in the history of spaceflight. Since the Space Shuttle programme ended, Soyuz is used for all crewed missions to the ISS

REUSABLE ROCKETS

Most conventional rockets can be used only once – the discarded stages are simply left to fall down and are destroyed on impact. However, in recent years, tremendous progress has been made in the field of reusable rocketry, especially by Elon Musk and his company SpaceX. The large and expensive first stage Falcon boosters are designed to slow their descent and softly land back on the ground, allowing them to be used again after they are refuelled for a fraction of the cost of rebuilding the entire rocket

Falcon Heavy is currently the most powerful and economical launch vehicle in the world. Its two reusable boosters and the reusable lower stage can be launched several times after only minor refurbishment. This makes it over three times cheaper than any of its current competition

10 M

BLACK ARROW
1969–1971
135 kg

V2 ROCKET
1944–1952
Suborbital

TSYKLON-3
1977–2009
4,100 kg

VESTA
1965–1969
Suborbital

MAXUS
1991–present
Suborbital

FALCON 1
2006–2009
670 kg

VOSTOK
1960–1963
4,725 kg

VEGA
2012–present
2,500 kg

FALCON 9 FT
2015–present
27,500 kg

FALCON HEAVY
2018–present
63,800 kg

FALCON 9 1.0
2010–2013
10,450 kg

ARIANE 5
1996–present
21,000 kg

ARIANE 4
1988–2003
7,600 kg

ARIANE 1
1979–1986
1,400 kg

VOSKHOD
1964–1965
5,900 kg

SOYUZ
1956–present
6,450 kg

ZENIT
1985–presen
13,740 kg

RT-2
1968–1976
Suborbital

Carried the first artificial satellite – Sputnik 1, into orbit

SPUTNIK
1957–1958
1,327 kg

TOPOL-M
1997–present
Suborbital

ВЕРТИКАЛЬ

R-5-V
1959–1983
Suborbital

ВЕРТИКАЛЬ

VERTIKAL
1973–1983
Suborbital

КОСМОС

KOSMOS-21
1961–1977
450 kg

AFRICAN ELEPHANT

BLUE ORIGIN

NEW SHEPARD
2015–present
Suborbital

Big Falcon Rocket (BFR) is envisioned to fulfil the primary goal of SpaceX – to help bring cargo and eventually a human crew to the planet Mars, setting the stage for future colonization. Both stages are designed to be fully reusable, reducing expenses massively

DIAMANT
1965–1975
160 kg

The first orbital launch rocket not built by either US or the USSR

R-2-A
1957–1962
Suborbital

US submarine-launched nuclear ballistic missile

BLACK KNIGHT
1958–1965
Suborbital

TRIDENT II
1983–present
Suborbital

КОСМОС

Blue Origin is a private spaceflight company founded by Jeff Bezos – the richest person in the world. Its suborbital reusable rocket New Shepard is expected to start taking tourists on commercial flights by 2020

ЭНЕРГИЯ

Energia remains the most powerful rocket ever successfully launched by the USSR. It was designed to carry the Soviet version of the US Space Shuttle – Buran

ПРОТОН

АНГАРА·5

KOSMOS-3M
1967–2010
1,500 kg

N1 was a Soviet rocket designed to compete with the Saturn V, to launch the first manned mission to the Moon. The project was rushed, and all four launch attempts ended in a catastrophic failure

КОСМОС

BLUE ORIGIN

ENERGIA-BURAN
1987–1988
100,000 kg

PROTON
1965–present
22,800 kg

ANGARA A5
2014–present
24,500 kg

N1/L3
1969–1972
95,000 kg

NEW GLENN
Mid-2020s
45,000 kg

BIG FALCON ROCKET
Mid-2020s
150,000 kg

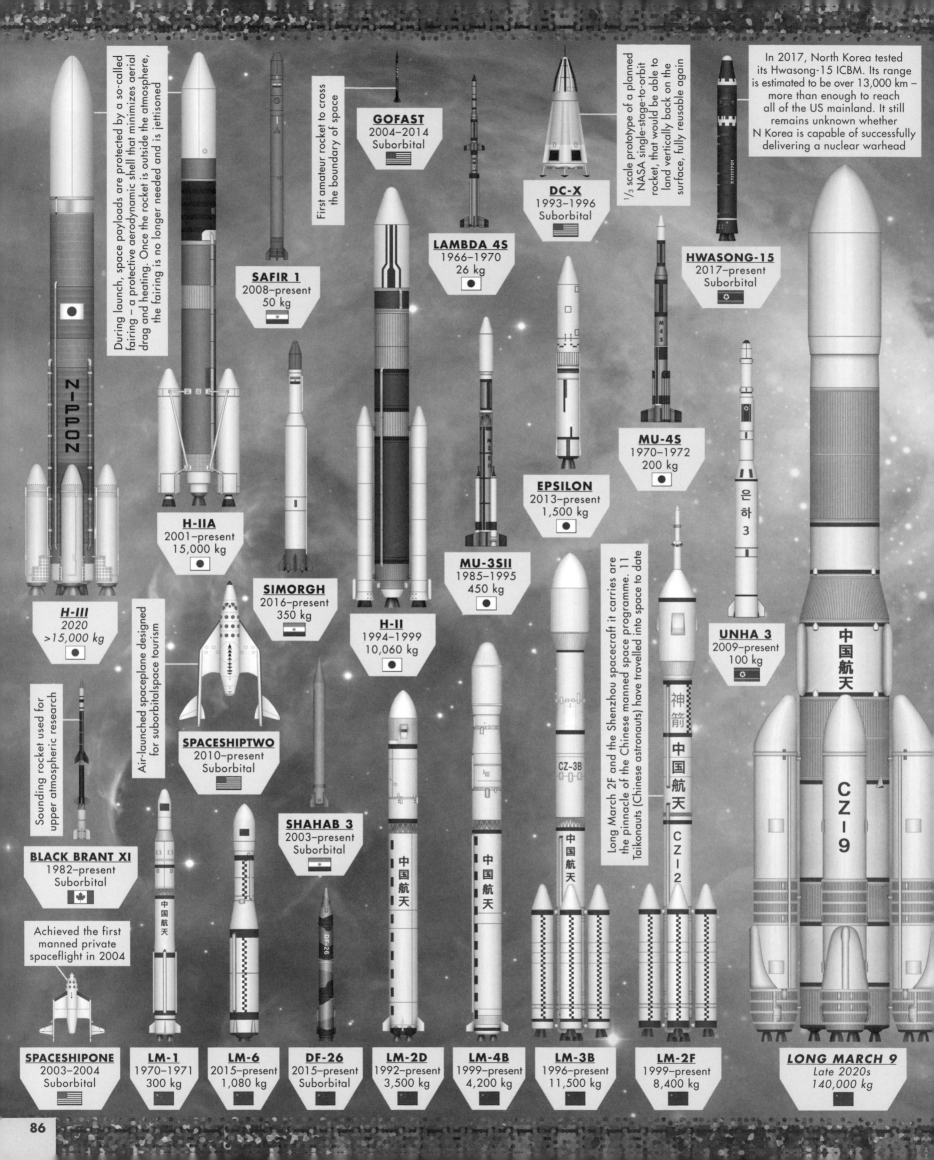

During launch, space payloads are protected by a so-called fairing – a protective aerodynamic shell that minimizes aerial drag and heating. Once the rocket is outside the atmosphere, the fairing is no longer needed and is jettisoned

GOFAST
2004–2014
Suborbital

First amateur rocket to cross the boundary of space

DC-X
1993–1996
Suborbital

¹/₃ scale prototype of a planned NASA single-stage-to-orbit rocket, that would be able to land vertically back on the surface, fully reusable again

In 2017, North Korea tested its Hwasong-15 ICBM. Its range is estimated to be over 13,000 km – more than enough to reach all of the US mainland. It still remains unknown whether N Korea is capable of successfully delivering a nuclear warhead

SAFIR 1
2008–present
50 kg

LAMBDA 4S
1966–1970
26 kg

HWASONG-15
2017–present
Suborbital

MU-4S
1970–1972
200 kg

EPSILON
2013–present
1,500 kg

H-IIA
2001–present
15,000 kg

SIMORGH
2016–present
350 kg

MU-3SII
1985–1995
450 kg

H-II
1994–1999
10,060 kg

H-III
2020
>15,000 kg

Air-launched spaceplane designed for suborbital space tourism

SPACESHIPTWO
2010–present
Suborbital

UNHA 3
2009–present
100 kg

Sounding rocket used for upper atmospheric research

Long March 2F and the Shenzhou spacecraft it carries are the pinnacle of the Chinese manned space programme. 11 Taikonauts (Chinese astronauts) have travelled into space to date

BLACK BRANT XI
1982–present
Suborbital

SHAHAB 3
2003–present
Suborbital

Achieved the first manned private spaceflight in 2004

SPACESHIPONE
2003–2004
Suborbital

LM-1
1970–1971
300 kg

LM-6
2015–present
1,080 kg

DF-26
2015–present
Suborbital

LM-2D
1992–present
3,500 kg

LM-4B
1999–present
4,200 kg

LM-3B
1996–present
11,500 kg

LM-2F
1999–present
8,400 kg

LONG MARCH 9
Late 2020s
140,000 kg

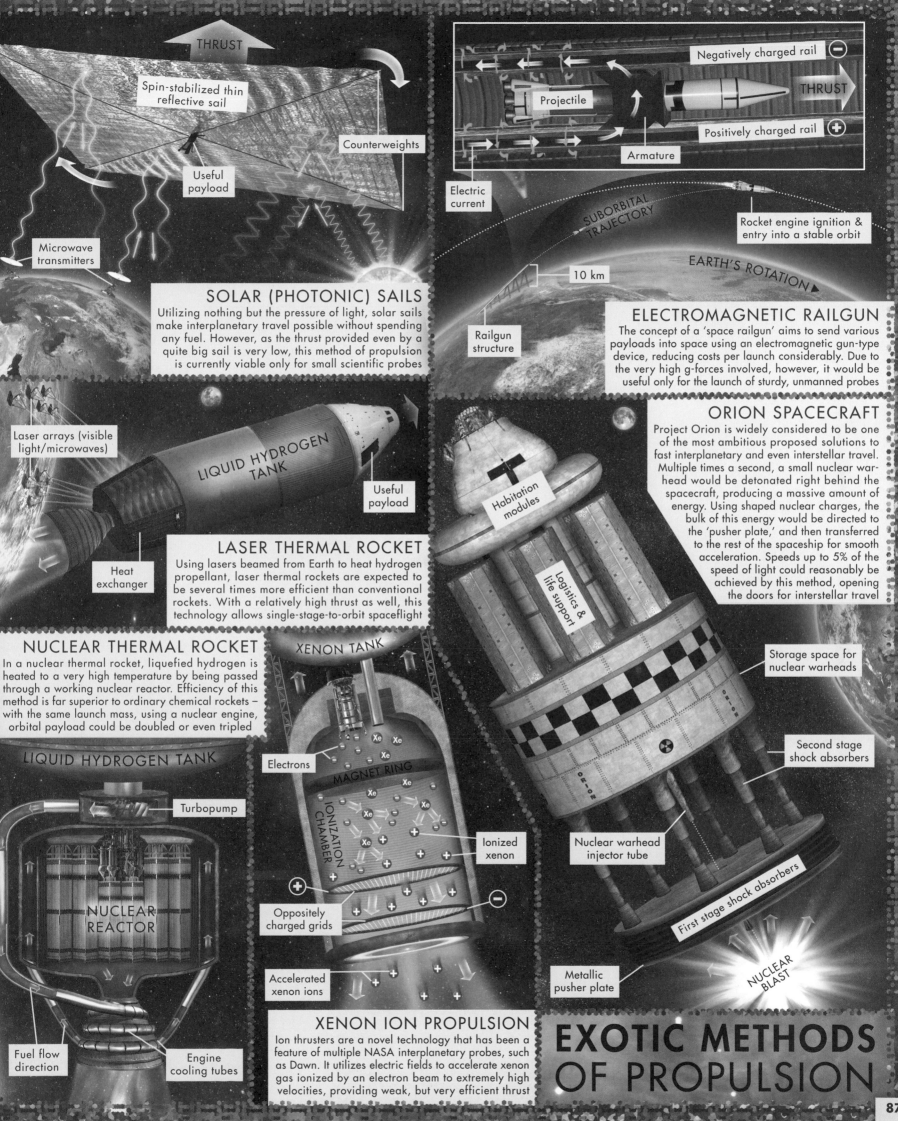

SOLAR (PHOTONIC) SAILS

Utilizing nothing but the pressure of light, solar sails make interplanetary travel possible without spending any fuel. However, as the thrust provided even by a quite big sail is very low, this method of propulsion is currently viable only for small scientific probes

THRUST

Spin-stabilized thin reflective sail

Counterweights

Useful payload

Microwave transmitters

ELECTROMAGNETIC RAILGUN

The concept of a 'space railgun' aims to send various payloads into space using an electromagnetic gun-type device, reducing costs per launch considerably. Due to the very high g-forces involved, however, it would be useful only for the launch of sturdy, unmanned probes

Negatively charged rail ⊖

THRUST

Projectile

Positively charged rail ⊕

Armature

Electric current

SUBORBITAL TRAJECTORY

Rocket engine ignition & entry into a stable orbit

EARTH'S ROTATION ▶

10 km

Railgun structure

LASER THERMAL ROCKET

Using lasers beamed from Earth to heat hydrogen propellant, laser thermal rockets are expected to be several times more efficient than conventional rockets. With a relatively high thrust as well, this technology allows single-stage-to-orbit spaceflight

Laser arrays (visible light/microwaves)

LIQUID HYDROGEN TANK

Useful payload

Heat exchanger

ORION SPACECRAFT

Project Orion is widely considered to be one of the most ambitious proposed solutions to fast interplanetary and even interstellar travel. Multiple times a second, a small nuclear warhead would be detonated right behind the spacecraft, producing a massive amount of energy. Using shaped nuclear charges, the bulk of this energy would be directed to the 'pusher plate,' and then transferred to the rest of the spaceship for smooth acceleration. Speeds up to 5% of the speed of light could reasonably be achieved by this method, opening the doors for interstellar travel

Habitation modules

Logistics & life support

Storage space for nuclear warheads

Second stage shock absorbers

Nuclear warhead injector tube

First stage shock absorbers

Metallic pusher plate

NUCLEAR BLAST

NUCLEAR THERMAL ROCKET

In a nuclear thermal rocket, liquefied hydrogen is heated to a very high temperature by being passed through a working nuclear reactor. Efficiency of this method is far superior to ordinary chemical rockets – with the same launch mass, using a nuclear engine, orbital payload could be doubled or even tripled

LIQUID HYDROGEN TANK

Turbopump

NUCLEAR REACTOR

Fuel flow direction

Engine cooling tubes

XENON TANK

Electrons

MAGNET RING

IONIZATION CHAMBER

Xe

Ionized xenon

Oppositely charged grids

Accelerated xenon ions

XENON ION PROPULSION

Ion thrusters are a novel technology that has been a feature of multiple NASA interplanetary probes, such as Dawn. It utilizes electric fields to accelerate xenon gas ionized by an electron beam to extremely high velocities, providing weak, but very efficient thrust

EXOTIC METHODS OF PROPULSION

JAMES WEBB SPACE TELESCOPE
(Mid-infrared and visible light range telescope, planned to be launched in 2022)

HUBBLE SPACE TELESCOPE
(Visible light range space telescope, operational since 1993)

500-METRE APERTURE
(Short wavelength radiotelescope located
With a diameter of 500 metres

MAGELLAN TELESCOPES
(Twin visible light range telescopes in Atacama, Chile, fully operational since 2000 and 2002)

GIANT MAGELLAN TELESCOPE
(Visible light and near-infrared range telescope under construction near La Serena, Chile, to become operational in 2024)

LARGE BINOCULAR TELESCOPE
(Visible light range telescope located on Mount Graham, Arizona, operational since 2005)

SOUTHERN AFRICAN LARGE TELESCOPE
(Visible light and near-infrared range telescope located in Sutherland, SA, operational since 2005)

LARGE SYNOPTIC SURVEY TELESCOPE
(Visible light range telescope under construction in Cerro Pachón, Chile to become operational in 2021)

Earth's atmospheric turbulence continuously distorts the light coming from distant stars, limiting the resolving power of ground telescopes. As a result, space telescopes of the same size tend to be far more powerful. To mitigate the atmospheric distortions, many modern ground telescopes employ the principle of adaptive optics. A laser beam is used to continuously measure the atmospheric distortions, while a deformable corrective mirror slightly changes its shape in unison to cancel out the perturbations. This way, the resolution of ground telescopes can be improved more than tenfold

SPITZER SPACE TELESCOPE
(Infrared space telescope operational since 2003)

GAIA
(Visible light range space telescope measuring the precise positions of stars, operational since 2013)

GREAT PARIS EXHIBITION TELESCOPE
(Largest refracting telescope ever, however it was never operational)

HERSCHEL SPACE OBSERVATORY
(Far-infrared range space telescope operational from 2009 to 2013)

YERKES TELESCOPE
(Refracting visible light telescope located in Williams Bay, WI, operational since 1893)

MMT OBSERVATORY
(Visible light range telescope located on Mt Hopkins, Arizona operational since 1979)

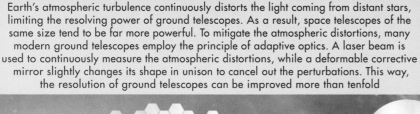

VERY LARGE TELESCOPE
(Four visible light range telescopes located in Atacama Desert, Chile, operational since 1998)

HOBBY EBERLY TELESCOPE
(Visible and near-infrared light range telescope located in the Davis Mountains, Texas, operational since 1996)

30-METRE TELESCOPE
(Near-UV, visible light and near-infrared range telescope under construction on Mauna Kea, Hawaii, to become operational in 2027)

KECK TELESCOPE
(Twin visible light range telescopes located on Mauna Kea, Hawaii, fully operational since 1996)

KEPLER
(Visible light range space telescope designed to detect exoplanets, operational since 2013)

HALE TELESCOPE
(Visible light range telescope located on Mt Palomar, CA, operational since 1949)

SUBARU TELESCOPE
(Visible light range telescope on Mauna Kea, Hawaii, operational since 1998)

VISTA
(Near-infrared range telescope located in Paranal Observatory, Chile operational since 2009)

OBSERVATORIES

SPHERICAL TELESCOPE

n Guizhou, China, operational since 2016.
it is the largest telescope ever built)

For spatial reasons, any radiotelescopes other than the colossal FAST are not included here. In terms of size, the Arecibo Telescope located in Puerto Rico with a dish 300 m in diameter is a close runner-up. Other notable radiotelescopes include the Green Bank Telescope in West Virginia, with the largest fully steerable dish in the world (100 m in diameter)

PAN-STARRS
(Visible light range telescope array located on Maui, Hawaii, partially operational since 2008)

SHANE TELESCOPE
(Visible light range telescope located near San Jose, California, operational since 1959)

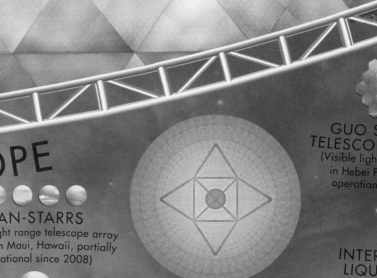

JAMES CLERK MAXWELL TELESCOPE
(Far-infrared and microwave range telescope located on Mauna Kea, Hawaii, operational since 1987)

GUO SHOUJING TELESCOPE (LAMOST)
(Visible light range telescope in Hebei Province, China, operational since 2008)

INTERNATIONAL LIQUID MIRROR TELESCOPE
(Visible light range telescope located in Devasthal, India, to become operational in 2020)

ATLAST
(Near-ultraviolet, visible light and near-infrared range space telescope, under conceptual development)

SOAR
(UV to infrared range telescope located in Chile, operational since 2007)

When a body of a liquid rotates, it naturally assumes a paraboloidal shape. This principle is commonly used in rotating furnaces for casting monolithic telescope mirrors. Using mercury and a rotating container, it is even possible to make a telescope with the reflective surface of liquid metal serving as a perfect mirror; however, due to their nature, such telescopes have to permanently point straight up. Large Zenith Telescope was the largest liquid mirror telescope in the world until its decommissioning in 2017

LARGE ZENITH TELESCOPE
(Visible light range telescope located in BC, Canada, operational from 2003 to 2017)

CHINESE GIANT SOLAR TELESCOPE
(Infrared and visible light range telescope planned to be constructed in Yunnan, China)

GEMINI NORTH
(Visible light range telescope located on Mauna Kea, Hawaii, operational since 2000)

GRAN TELESCOPIO CANARIAS
(Visible light range telescope located on La Palma, Spain, operational since 2009)

AFRICAN ELEPHANT

VOLKSWAGEN BEETLE

HUMANS

GEMINI SOUTH
(Visible light range telescope located on Cerro Pachón, Chile, operational since 2000)

....
10 METRES

The 100-inch Hooker Telescope is one of the most important in the history of astronomy. In the 1920s it was used by Edwin Hubble to conclusively prove that the universe extends billions of light years beyond the Milky Way, and that distant galaxies are rapidly drifting away from each other, providing important evidence in favour of the Big Bang model

SAN PEDRO MARTIR TELESCOPE
(Visible light range telescope in Baja California, Mexico, to become operational in 2023)

HOOKER TELESCOPE
(Visible light range telescope located on Mt Wilson, California, operational since 1917)

4,100 K	7,200 K
VISIBLE LIGHT	
700 nm	390 nm

EUROPEAN EXTREMELY LARGE TELESCOPE
(Visible light and near-infrared range telescope under construction in Cerro Amazonas, Chile, to become operational in 2025)

ELECTROMAGNETIC SPECTRUM

TEMPERATURE	0.01 K	0.1 K	1 K	10 K	100 K	1,000 K	10,000 K	100,000 K	1 MK	10 MK	100 MK	1 GK

RADIO WAVES		MICROWAVES		INFRARED			ULTRAVIOLET		X-RAYS		GAMMA RAYS

WAVELENGTH	1 m	10 cm	1 cm	1 mm	100 μm	10 μm	1 μm	100 nm	10 nm	1 nm	100 pm	10 pm	1 pm

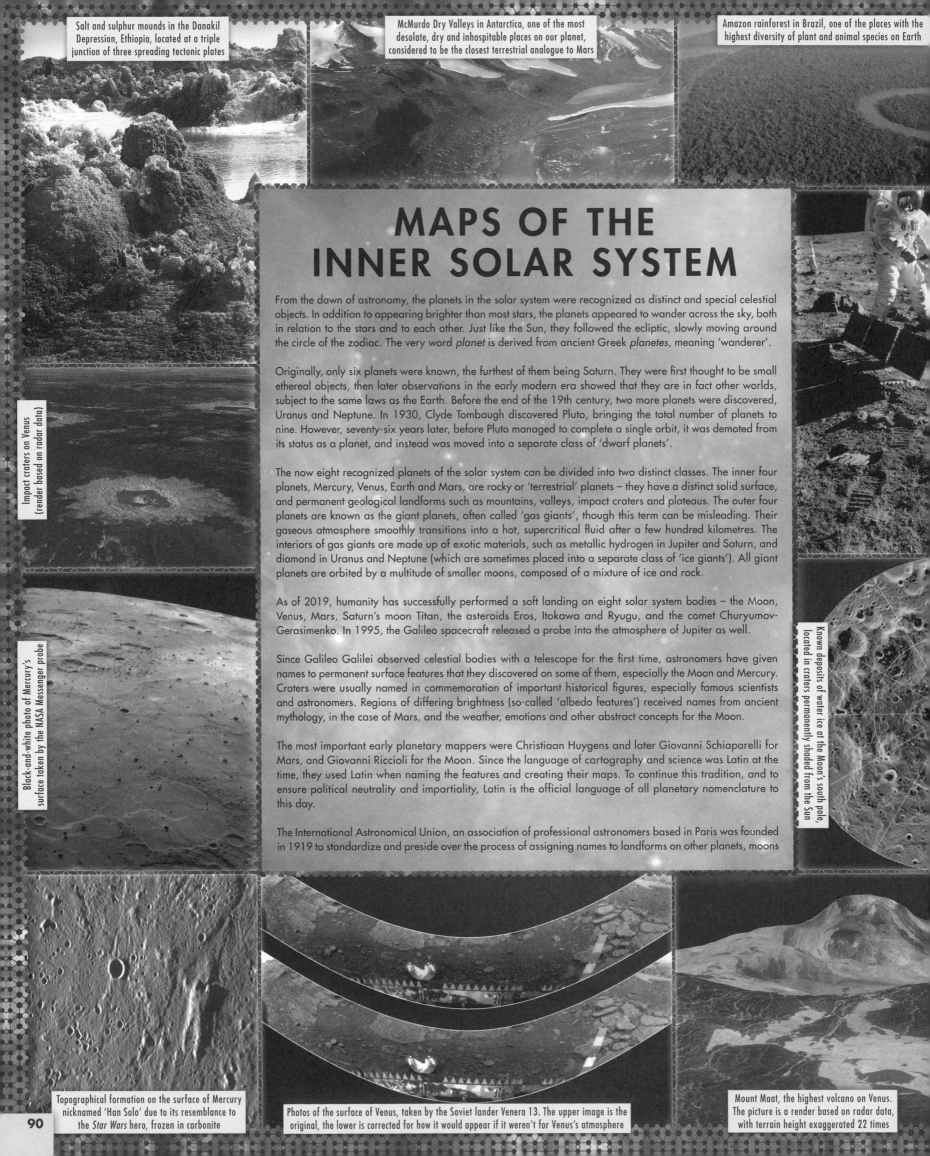

Salt and sulphur mounds in the Danakil Depression, Ethiopia, located at a triple junction of three spreading tectonic plates

McMurdo Dry Valleys in Antarctica, one of the most desolate, dry and inhospitable places on our planet, considered to be the closest terrestrial analogue to Mars

Amazon rainforest in Brazil, one of the places with the highest diversity of plant and animal species on Earth

Impact craters on Venus (render based on radar data)

Black-and-white photo of Mercury's surface taken by the NASA Messenger probe

MAPS OF THE INNER SOLAR SYSTEM

From the dawn of astronomy, the planets in the solar system were recognized as distinct and special celestial objects. In addition to appearing brighter than most stars, the planets appeared to wander across the sky, both in relation to the stars and to each other. Just like the Sun, they followed the ecliptic, slowly moving around the circle of the zodiac. The very word *planet* is derived from ancient Greek *planetes*, meaning 'wanderer'.

Originally, only six planets were known, the furthest of them being Saturn. They were first thought to be small ethereal objects, then later observations in the early modern era showed that they are in fact other worlds, subject to the same laws as the Earth. Before the end of the 19th century, two more planets were discovered, Uranus and Neptune. In 1930, Clyde Tombaugh discovered Pluto, bringing the total number of planets to nine. However, seventy-six years later, before Pluto managed to complete a single orbit, it was demoted from its status as a planet, and instead was moved into a separate class of 'dwarf planets'.

The now eight recognized planets of the solar system can be divided into two distinct classes. The inner four planets, Mercury, Venus, Earth and Mars, are rocky or 'terrestrial' planets – they have a distinct solid surface, and permanent geological landforms such as mountains, valleys, impact craters and plateaus. The outer four planets are known as the giant planets, often called 'gas giants', though this term can be misleading. Their gaseous atmosphere smoothly transitions into a hot, supercritical fluid after a few hundred kilometres. The interiors of gas giants are made up of exotic materials, such as metallic hydrogen in Jupiter and Saturn, and diamond in Uranus and Neptune (which are sometimes placed into a separate class of 'ice giants'). All giant planets are orbited by a multitude of smaller moons, composed of a mixture of ice and rock.

As of 2019, humanity has successfully performed a soft landing on eight solar system bodies – the Moon, Venus, Mars, Saturn's moon Titan, the asteroids Eros, Itokawa and Ryugu, and the comet Churyumov-Gerasimenko. In 1995, the Galileo spacecraft released a probe into the atmosphere of Jupiter as well.

Since Galileo Galilei observed celestial bodies with a telescope for the first time, astronomers have given names to permanent surface features that they discovered on some of them, especially the Moon and Mercury. Craters were usually named in commemoration of important historical figures, especially famous scientists and astronomers. Regions of differing brightness (so-called 'albedo features') received names from ancient mythology, in the case of Mars, and the weather, emotions and other abstract concepts for the Moon.

The most important early planetary mappers were Christiaan Huygens and later Giovanni Schiaparelli for Mars, and Giovanni Riccioli for the Moon. Since the language of cartography and science was Latin at the time, they used Latin when naming the features and creating their maps. To continue this tradition, and to ensure political neutrality and impartiality, Latin is the official language of all planetary nomenclature to this day.

The International Astronomical Union, an association of professional astronomers based in Paris was founded in 1919 to standardize and preside over the process of assigning names to landforms on other planets, moons

Known deposits of water ice at the Moon's south pole, located in craters permanently shaded from the Sun

Topographical formation on the surface of Mercury nicknamed 'Han Solo' due to its resemblance to the *Star Wars* hero, frozen in carbonite

Photos of the surface of Venus, taken by the Soviet lander Venera 13. The upper image is the original, the lower is corrected for how it would appear if it weren't for Venus's atmosphere

Mount Maat, the highest volcano on Venus. The picture is a render based on radar data, with terrain height exaggerated 22 times

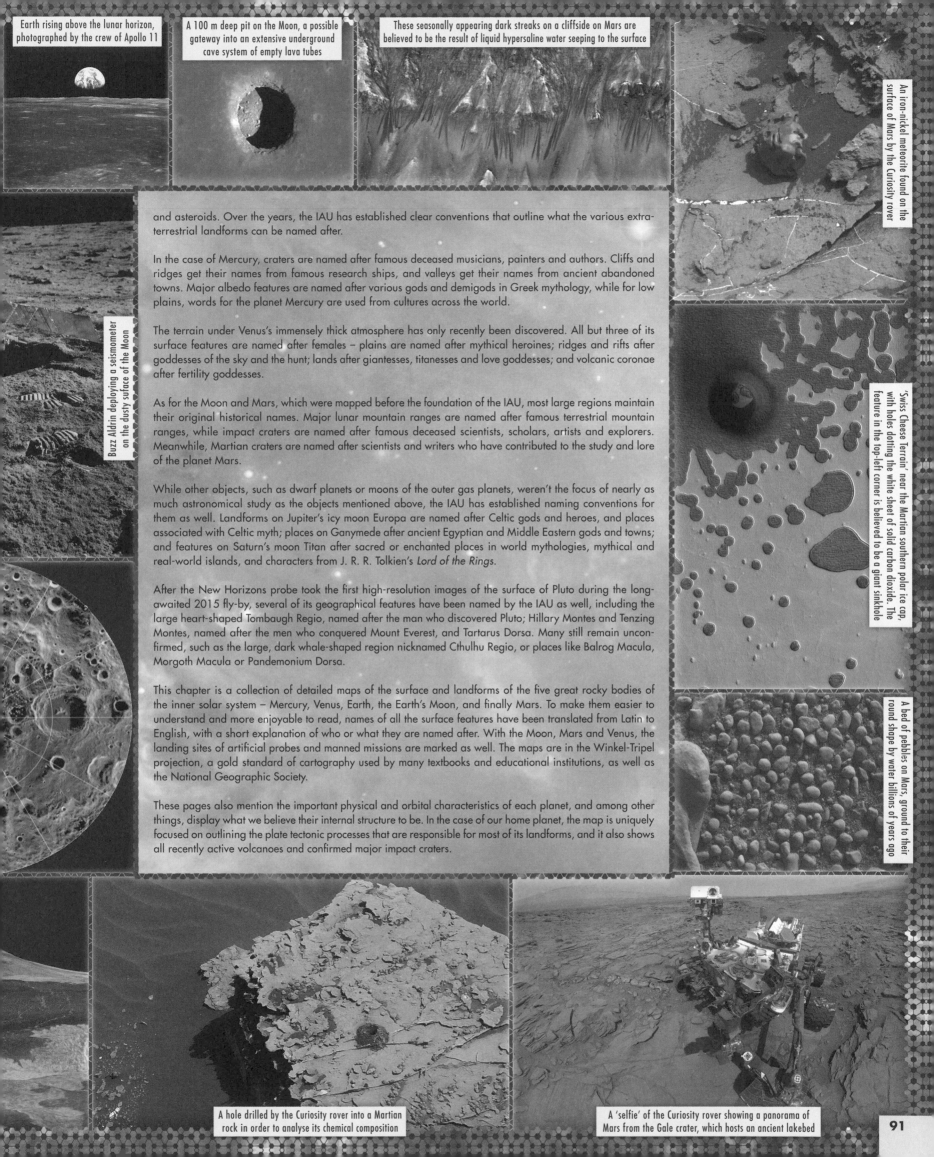

Earth rising above the lunar horizon, photographed by the crew of Apollo 11

A 100 m deep pit on the Moon, a possible gateway into an extensive underground cave system of empty lava tubes

These seasonally appearing dark streaks on a cliffside on Mars are believed to be the result of liquid hypersaline water seeping to the surface

An iron-nickel meteorite found on the surface of Mars by the Curiosity rover

Buzz Aldrin deploying a seismometer on the dusty surface of the Moon

'Swiss Cheese Terrain' near the Martian southern polar ice cap, with holes dotting the white sheet of solid carbon dioxide. The feature in the top-left corner is believed to be a giant sinkhole

A bed of pebbles on Mars, ground to their round shape by water billions of years ago

and asteroids. Over the years, the IAU has established clear conventions that outline what the various extra-terrestrial landforms can be named after.

In the case of Mercury, craters are named after famous deceased musicians, painters and authors. Cliffs and ridges get their names from famous research ships, and valleys get their names from ancient abandoned towns. Major albedo features are named after various gods and demigods in Greek mythology, while for low plains, words for the planet Mercury are used from cultures across the world.

The terrain under Venus's immensely thick atmosphere has only recently been discovered. All but three of its surface features are named after females – plains are named after mythical heroines; ridges and rifts after goddesses of the sky and the hunt; lands after giantesses, titanesses and love goddesses; and volcanic coronae after fertility goddesses.

As for the Moon and Mars, which were mapped before the foundation of the IAU, most large regions maintain their original historical names. Major lunar mountain ranges are named after famous terrestrial mountain ranges, while impact craters are named after famous deceased scientists, scholars, artists and explorers. Meanwhile, Martian craters are named after scientists and writers who have contributed to the study and lore of the planet Mars.

While other objects, such as dwarf planets or moons of the outer gas planets, weren't the focus of nearly as much astronomical study as the objects mentioned above, the IAU has established naming conventions for them as well. Landforms on Jupiter's icy moon Europa are named after Celtic gods and heroes, and places associated with Celtic myth; places on Ganymede after ancient Egyptian and Middle Eastern gods and towns; and features on Saturn's moon Titan after sacred or enchanted places in world mythologies, mythical and real-world islands, and characters from J. R. R. Tolkien's Lord of the Rings.

After the New Horizons probe took the first high-resolution images of the surface of Pluto during the long-awaited 2015 fly-by, several of its geographical features have been named by the IAU as well, including the large heart-shaped Tombaugh Regio, named after the man who discovered Pluto; Hillary Montes and Tenzing Montes, named after the men who conquered Mount Everest, and Tartarus Dorsa. Many still remain uncon-firmed, such as the large, dark whale-shaped region nicknamed Cthulhu Regio, or places like Balrog Macula, Morgoth Macula or Pandemonium Dorsa.

This chapter is a collection of detailed maps of the surface and landforms of the five great rocky bodies of the inner solar system – Mercury, Venus, Earth, the Earth's Moon, and finally Mars. To make them easier to understand and more enjoyable to read, names of all the surface features have been translated from Latin to English, with a short explanation of who or what they are named after. With the Moon, Mars and Venus, the landing sites of artificial probes and manned missions are marked as well. The maps are in the Winkel-Tripel projection, a gold standard of cartography used by many textbooks and educational institutions, as well as the National Geographic Society.

These pages also mention the important physical and orbital characteristics of each planet, and among other things, display what we believe their internal structure to be. In the case of our home planet, the map is uniquely focused on outlining the plate tectonic processes that are responsible for most of its landforms, and it also shows all recently active volcanoes and confirmed major impact craters.

A hole drilled by the Curiosity rover into a Martian rock in order to analyse its chemical composition

A 'selfie' of the Curiosity rover showing a panorama of Mars from the Gale crater, which hosts an ancient lakebed

Molten iron and nickel middle core producing a weak magnetic field

Tectonically inactive silicate crust, 100–200 km thick

Thin silicate mantle, 300–400 km thick

Solid iron and nickel inner core

Outer core layer composed of solid iron sulphide

PLAIN OF SUISEI
("MERCURY" IN JAPANESE)

LAND OF THE CADUCEUS
(STAFF OF THE GREEK GOD HERMES)

GOETHE
(GERMAN WRITER)

STRINDBERG
(SWEDISH WRITER)

RUBENS
(FRENCH ARTIST)

BOZNANSKA
(POLISH PAINTER)

ABEDIN
(BENGALI PAINTER)

HOKUSAI
(JAPANESE ARTIST)

DESERT OF ADMETUS
(LEGENDARY TROJAN WARRIOR)

AKUTAGAWA
(JAPANESE AUTHOR)

STRAVINSKY
(RUSSIAN COMPOSER)

VYASA
(INDIAN POET)

LAND OF

PLAIN OF SOBKOU
(EGYPTIAN CROCODILE GOD)

SCARLATTI
(ITALIAN COMPOSER)

LARROCHA
(SPANISH PIANIST)

AURORA

VELASQUEZ
(SPANISH PAINTER)

Cliffs of Victoria (British research vessel)

SOUSA
(US COMPOSER)

PLAIN OF BUDH
(HINDU GOD OF MERCURY)

LAND OF GALLIA

LAND OF THE PLEIADES
(SEVEN NYMPHS, DAUGHTERS OF THE TITAN ATLAS)

DESERT OF THE HOURS

AKSAKOV
(RUSSIAN AUTHOR)

JOBIM
(BRAZILIAN COMPOSER)

PRAXITELES
(GREEK SCULPTOR)

Ridge of Antoniadi
(GREEK ASTRONOMER)

WREN
(ENGLISH ARCHITECT)

Cliffs of Endeavour
(British survey ship)

CATULLUS
(ROMAN POET)

LERMONTOV
(RUSSIAN WRITER)

Cliffs of Santa Maria
(Ship of Columbus)

BALZAC
(FRENCH WRITER)

VIVALDI
(ITALIAN COMPOSER)

PETIPA
(RUSSIAN DANCER)

DESERT OF LYCAON
(LEGENDARY GREEK KING WHO WAS TRANS-
FORMED INTO A WOLF AS A PUNISHMENT)

TYAGARAJA
(INDIAN COMPOSER)

MENA
(SPANISH POET)

POLYGNOTUS
(GREEK PAINTER)

LAND OF TRICRENA
(LEGENDARY FOUNTAINS IN WHICH
MOUNTAIN NYMPHS WASHED THE
NEWBORN GOD HERMES)

DESERT OF JUPITER
(ROMAN GOD OF THE SKY AND THUNDER)

DESERT OF ARGIPHONTES
(GREEK GOD HERMES – 'SLAYER OF ARGUS')

ZEAMI
(JAPANESE PLAYWRIGHT)

MACHAUT
(FRENCH POET)

PLAIN OF LUGUS
(GALLIC GOD OF MERCURY)

DESERT OF MAIA
(MOTHER OF THE GREEK MESSENGER GOD HERMES)

CEZANNE
(FRENCH PAINTER)

WATERS
(US MUSICIAN)

SNORRI
(ICELANDIC POET)

MURASAKI
(JAPANESE WRITER)

ELLINGTON
(US MUSICIAN)

Cliffs of Duyfken
(Dutch ship)

BEETHOVEN
(GERMAN COMPOSER)

DESERT OF MARS

RAPHAEL
(ITALIAN PAINTER)

RENOIR
(FRENCH ARTIST)

IBSEN
(NORWEGIAN PLAYWRIGHT)

PLAIN OF TURMS
(ETRUSCAN MESSENGER GOD)

HAYDN
(AUSTRIAN COMPOSER)

PETRARCH
(ITALIAN POET)

DEBUSSY
(FRENCH COMPOSER)

BARTOK
(HUNGARIAN COMPOSER)

(ROMAN GOD OF WAR)

SIMONIDES
(GREEK POET)

Cliffs of Vostok
(Russian explor-
ation ship)

CHEKHOV
(RUSSIAN PLAYWRIGHT)

LAND OF HES

DESERT OF PROMETHEUS
(TITAN IN GREEK MYTHOLOGY)

CARDUCCI
(ITALIAN POET)

COPLEY
(US PAINTER)

DESERT OF HERMES TRISMEGISTOS
(LEGENDARY AUTHOR OF THE HERMETIC TEXTS, OFTEN
ASSOCIATED WITH THE GODS HERMES AND THOTH)

Cliffs of Meteor
(British exploration ship)

MICHELANGELO
(ITALIAN ARTIST)

SUR DAS
(HINDU POET)

Cliffs of Fram
(Norwegian polar
expeditionary ship)

SHEVCHENKO
(RUSSIAN PAINTER)

HESIOD
(GREEK POET)

HAWTHORNE
(US NOVELIST)

RABELAIS
(FRENCH WRITER)

MA-CHIH-YUAN
(CHINESE PLAYWRIGHT)

Cliffs of Adventure
(James Cook's ship)

CALLICRATES
(GREEK ARCHITECT)

PUSHKIN
(RUSSIAN POET)

CHOPIN
(POLISH COMPOSER)

WAGNER
(GERMAN COMPOSER)

BACH
(GERMAN COMPOSER)

SPITTELER
(SWISS POET)

LAND

AUSTR

SADI
(PERSIAN POET)

BOCCACCIO
(ITALIAN RENAIS-
SANCE AUTHOR)

ROERICH
(RUSSIAN PAINTER)

RADIUS: 2,439 km (0.383 Earths)
MASS: 3.301×10²³ kg (0.055 Earths)
SURFACE GRAVITY: 38% that of Earth
ORBITAL PERIOD: 87 days, 23 hours, 15 mins
AVERAGE ORBITAL SPEED: 47.362 km/s
TEMPERATURE RANGE: 173 °C to 427 °C at the equator

MERCURY

Mercury is tidally locked with the Sun in a special 3:2 spin-
orbit resonance – it rotates 3 times around its axis for every
2 revolutions it makes around the Sun. This makes a solar
day on Mercury last twice as long as a Mercurian year

PROKOFIEV
(RUSSIAN COMPOSER)

YOSHIKAWA
(JAPANESE NOVELIST)

LISMER
(CANADIAN PAINTER)

GAUDI
(CATALAN ARCHITECT)

STIEGLITZ
(US PHOTOGRAPHER)

MENDELSSOHN
(GERMAN COMPOSER)

Valley of Cahokia
(Ancient city in Illinois)

VERDI
(ITALIAN COMPOSER)

PLAIN OF SUISEI
(MERCURY IN JAPANESE)

Valley of Timgad
(Roman city in Algeria)

OSKISON
(NATIVE AMERICAN AUTHOR)

PLAIN OF
STILBON
(GREEK GOD OF THE
PLANET MERCURY)

PLAIN

LAND OF
APOLLO
(GREEK GOD OF THE ARTS)

DESERT OF
APHRODITE
(GREEK LOVE GODDESS)

COPLAND
(AMERICAN COMPOSER)

PLAIN OF MEARCAIR
(MERCURY IN GAELIC)

POE
(US POET)

BASIN

Bright spot of Nathair
('Snake' in Irish Gaelic)

FONTEYN
(ENGLISH BALLERINA)

Valleys of Pantheon
(Temple in Rome)

Mountains
of Heat

RACHMANINOFF
(RUSSIAN COMPOSER)

Cliffs of Unity
(British ship)

RADITLADI
(BOTSWANIAN AUTHOR)

OF HEAT

Cliffs of Calypso
(British research vessel)

HAFIZ
(PERSIAN POET)

LAND
OF THE FIVE

DESERT OF
THE PHOENIX

PLAIN
OF ODIN
(NORSE GOD OF WISDOM)

SEUSS
(US CHILDREN'S AUTHOR)

LAND OF
APARANGI
(MERCURY IN MAORI)

FAULKNER
(US AUTHOR)

EMINESCU
(ROMANIAN POET)

XIAO ZHAO
(CHINESE ARTIST)

MOZART
(AUSTRIAN COMPOSER)

LAND OF
PHAETHON
(GREEK DEMIGOD,
SON OF HELIOS)

PICASSO
(SPANISH ARTIST)

FIRDOUSI
(PERSIAN POET)

DESERT OF CRIOPHORUS
('RAM-BEARER' GREEK GOD HERMES)

QI BAISHI
(CHINESE PAINTER)

BAGRYANA
(BULGARIAN POET)

Cliffs of Blossom
(British survey ship)

Cliffs of the Beagle
(British research vessel)

DESERT OF WINGS

STEICHEN
(US PHOTOGRAPHER)

PLATEAU OF
CATUILLA
(PERUVIAN GOD
OF MERCURY)

PLAIN
OF TIR
(MERCURY IN PERSIAN)

DESERT
OF MAIA
(MOTHER OF THE GREEK
MESSENGER GOD HERMES)

NABOKOV
(RUSSIAN WRITER)

HOLST
(ENGLISH COMPOSER)

DAVID
(FRENCH PAINTER)

LAND OF
PIERIA
(REGION IN GREECE)

DESERT OF ATLAS
(TITAN IN GREEK MYTHOLOGY)

TOLSTOY
(RUSSIAN WRITER)

Cliffs of Nautilus
(US survey ship)

OF THE
PERIDES

MPHS OF THE EVENING
(US survey ship)

Cliffs of Enterprise
(US survey ship)

REMBRANDT
(DUTCH PAINTER)

AMARAL
(BRAZILIAN PAINTER)

DESERT OF HELIOS
(GREEK SUN GOD)

BASHO
(JAPANESE POET)

CASTIGLIONE
(ITALIAN PAINTER)

DESERT
OF
PERSEPHONE
(GREEK GODDESS
OF THE UNDER-
WORLD, MARRIED TO HADES)

DESERT OF
PROMETHEUS
(TITAN IN GREEK MYTHOLOGY)

Cliffs of Belgica
(Norwegian
research ship)

DONELAITIS
(LITHUANIAN POET)

GRAINGER
(AUSTRALIAN
COMPOSER)

DOSTOEVSKY
(RUSSIAN WRITER)

CARLETON
(IRISH WRITER)

LAND OF
CYLLENE
(BIRTHPLACE OF HERMES)

NERUDA
(CHILEAN POET)

Cliffs of Hero
(US exploration ship)

PLAIN OF
UTARIDI
('MERCURY' IN SWAHILI)

ALVER
(ESTONIAN POET)

OF
ALIA

DISNEY
(US FILM PRODUCER)

Cliffs of Terror
(British exploration ship)

MAGRITTE
(BELGIAN ARTIST)

LEOPARDI
(ITALIAN POET)

Cliffs of Elfanin
(US research ship)

(GREEK ARCHITECT)

ICTINUS

PETOFI
(HUNGARIAN POET)

L'ENGLE
(US WRITER)

TERRAIN HEIGHT

APPARENT SIZE
OF THE SUN AS
SEEN FROM:

VENUS

EARTH

CERES

MARS

MERCURY

JUPITER SATURN URANUS PLUTO

The 'basin of heat' (Caloris), is one of the largest
confirmed impact craters in the solar system
(1,550 km in diameter). It was created by an
impactor roughly 100 km in diameter.
However, as Mercury is located deep within
the Sun's gravity well, the impact velocity
and the energy released was many times
higher than if this had occurred on Earth

1,000 KM

4,000 m
3,000 m
2,000 m
1,000 m
0 m
−1,000 m
−2,000 m
−4,000 m

Liquid outer core, size unknown

PARTIALLY LIQUID MANTLE

The surface of Venus is only a few hundred million years old and mostly shaped by volcanism – the planet even shows signs of ongoing volcanic activity

Solid silicate crust, 50 to 70 km thick

Solid inner core, size unknown

As Venus doesn't have a reliable way to continuously remove heat generated in its interior, it is theorized that every few hundred million years, the temperature of the mantle reaches a critical point, rapidly melting and recycling most of Venus's crust in a cataclysmic event

PLAIN OF VINMARA
(VANUATUAN SWAN MAIDEN)

Rift of Ahsonnutli
(Navajo spirit of light)

PLAIN OF THE SNOW
(CHARACTER IN RUSSIAN FAIR

CORONA OF POMONA
(ROMAN GODDESS OF FRUITS)

LAN
(MES

LAND OF METIS
(GREEK TITANESS)

Mountains of Freyja
(Norse love goddess)

CORONA OF FERONIA
(ANCIENT ITALIAN GODDESS OF SPRING AND FLOWERS)

Mts of Akna
(Inuit fertility goddess)

PLAIN OF LAKSHMI
(HINDU GODDESS OF WEALTH AND BEAUTY)

CORONA OF COATLICUE
(AZTEC EARTH-MOTHER GODDESS)

Mountains of Danu
(Irish mother goddess)

CORONA OF DEMETER
(GREEK FERTILITY GODDESS)

CORONA OF ASHNA
(MESOPOTAMIAN HARVEST GODDESS)

YABLOCHKINA
(RUSSIAN ACTRESS)

Ridge of Pandrosos (Greek dew maiden)

Ridge of Iris (Greek rainbow goddess)

Mount Mokosha
(Slavic goddess of women)

TESSERA OF NEMESIS
(GREEK GODDESS OF RETRIBUTION)

Mount Sekhmet
(Egyptian war goddess)

PLAIN OF KAWELU
(HAWAIIAN MYTHICAL HEROINE)

PLAIN OF
(INUIT SEA GOD

PLAIN OF GUINEVERE
(WIFE OF KING ARTHUR)

Mount Rheia
(Greek titaness)

BETA LAND

Venera 9 (1975)

(S

Rift of Bellona
(Roman war goddess)

TESSERA OF ASTERIA
(GREEK TITANESS)

Rift of Hecate
(Greek sorcery goddess)

Mount Theia
(Greek titaness)

Venera 10 (1975)

Mount Sif
(Norse earth goddess)

PLAIN OF GANIKI
(SIBERIAN RIVER SPIRIT)

LAND OF ULFRUN
(NORSE GIANTESS)

Rift of Canis
(Saami forest maiden)

Rift of Tkashi-mapa
(Georgian forest goddess)

Rift of Zewana
(Polish goddess of the hunt)

Rift of Zverine
(Baltic hunting goddess)

CORONA OF TARANGA
(POLYNESIAN FERTILITY GODDESS)

SEYMOUR
(ENGLISH QUEEN)

Mount Gu
(Babyloni goddess healin

PLAIN OF RUSALKA
(SLAVIC WATER NYMPH)

Mount Tuulikki
(Finnish spirit)

PLAIN OF UNDINE
(LITHUANIAN MERMAID)

Mt Sapas
(Canaanite sun goddess)

PLAIN OF HINEMOA
(MAORI LEGENDARY HEROINE)

Rift of Devana
(Slavic hunting goddess)

Venera 13 (1981)

Venera 7 (1970)

CORON OF CHANG
(CHINESE MO GODDE

LAND OF ATLA
(NORSE GIANTESS)

Venera 12 (1978)

TESSERA OF DOLYA
(EAST SLAVIC GODDESS OF GOOD FATE)

PLAIN OF NAVKA
(EAST SLAVIC MERMAID)

Mount Maat
(Egyptian justice goddess)

Lava field of Ningyo
(Japanese fish goddess)

CORONA OF MARAM
(ETHIOPIAN FERTILITY GODDESS)

LAND OF PHOEBE
(GREEK TITANESS)

Venera 8 (1972)

BENDER
(US ARTIST)

PLAIN OF KANYKEY
(KIRGHIZ EPIC HEROIN

Venera 11 (1978)

CORONA OF ZEMINA
(BALTIC FERTILITY GODDESS)

CORONA OF ATETE
(ETHIOPIAN FERTILITY GODDESS)

PLAIN OF GUNDA
(ABKHAZIAN EPIC HEROINE)

PLAIN OF DZERASSA
(OSSETIAN MYTHICAL HEROINE)

CORONA OF IWERIDD
(CELTIC EARTH GODDESS)

CORONA OF QETESH
(EGYPTIAN FERTILITY GODDESS)

Mount Ushas
(Hindu dawn goddess)

CORONA OF E
(BIBLICAL FIRST WOMA

PLAIN OF WAWALAG
(MYTHICAL AUSTRALIAN SISTERS)

Ridge of Aditi
(Indian sky goddess)

Rift of Parga
(Nenets forest witch)

LAND OF DIONE
(GREEK TITANESS)

CORON
(CHEROKEE CO

ISABELLA
(SPANISH QUEEN)

PLAIN OF HELEN
(WIFE OF KING MENELAUS FROM HOMER'S ILIAD)

LAND OF THEMIS
(GREEK TITANESS)

Mount Tefnut
(Egyptian rain goddess)

Mount Hathor
(Egyptian fertility goddess)

PLAIN OF LAV
(WIFE OF THE TROJAN HER

CORONA OF SHIWANOKIA
(NATIVE AMERICAN FERTILITY GODDESS)

PLAIN OF NSOMEKA
(BANTU MYTHICAL HEROINE)

LAND OF IMDR
(NORSE GIANTESS)

Mount Idunn
(Norse youth goddess)

Ridge of Tinianavyt
(Wife of a Siberian sky god)

LAND OF ISHKUS
(NATIVE AMERICAN FOREST GIANTESS)

MEITNER
(AUSTRIAN PHYSICIST)

CORONA
(IROQUOIS EAR

Ridge of Saule
(Baltic sea goddess)

Rift of Morrigan
(Celtic war goddess)

CORONA OF KAMUI-HUKI
(AINU EARTH GODDESS)

LAND OF NERINGA
(LITHUANIAN SEA GIANTESS)

CORONA OF QUETZALPETLA
(AZTEC FERTILITY GODDESS)

PLAIN OF NUPTADI
(NATIVE AMERICAN FOLK HEROINE)

PLAIN
AIBAN
(UZBEK EP

RADIUS: 6,051 km (0.95 Earths)
MASS: 4.86x10²⁴ kg (0.815 Earths)
SURFACE GRAVITY: 90.4% that of Earth
SOLAR DAY LENGTH: 116 days, 18 hours
AVERAGE ORBITAL SPEED: 35.02 km/s
TEMPERATURE: 460 °C over the entire Venusian surface

VENUS

ORBITAL PERIOD: 224.7 days
SURFACE PRESSURE: 9.3 MPa

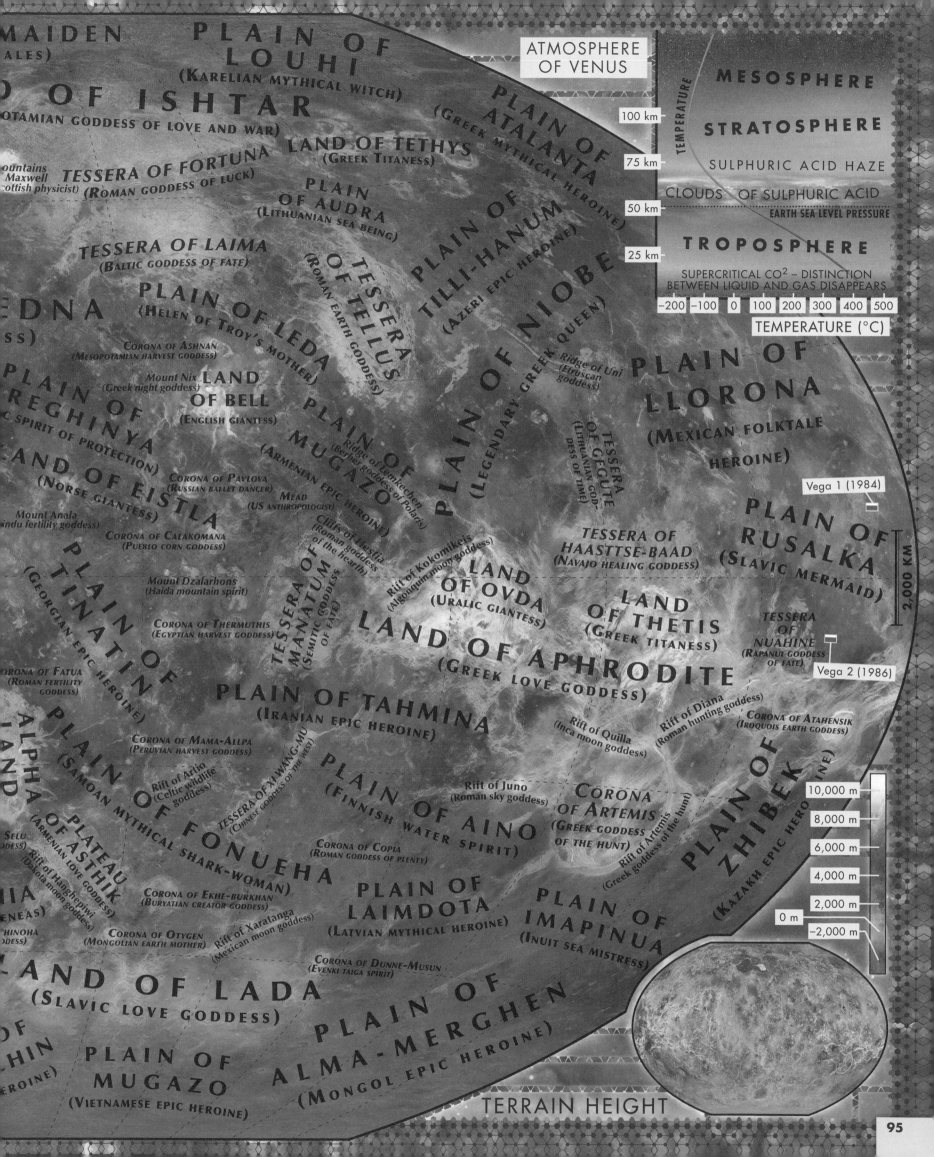

MAIDEN
(ALES)

PLAIN OF LOUHI
(KARELIAN MYTHICAL WITCH)

OF ISHTAR
OTAMIAN GODDESS OF LOVE AND WAR)

ountains
Maxwell
cottish physicist)

TESSERA OF FORTUNA
(ROMAN GODDESS OF LUCK)

LAND OF TETHYS
(GREEK TITANESS)

PLAIN OF ATALANTA
(GREEK MYTHICAL HEROINE)

ATMOSPHERE OF VENUS

TEMPERATURE

MESOSPHERE

100 km

STRATOSPHERE

75 km

SULPHURIC ACID HAZE

CLOUDS OF SULPHURIC ACID

50 km

EARTH SEA LEVEL PRESSURE

PLAIN OF AUDRA
(LITHUANIAN SEA BEING)

TESSERA OF LAIMA
(BALTIC GODDESS OF FATE)

PLAIN OF LEDA
(HELEN OF TROY'S MOTHER)

EDNA

S)

TESSERA
OF
TELLUS
(ROMAN EARTH GODDESS)

PLAIN OF TILLI-HANUM
(AZERI EPIC HEROINE)

25 km

TROPOSPHERE

SUPERCRITICAL CO$_2$ – DISTINCTION
BETWEEN LIQUID AND GAS DISAPPEARS

-200 -100 0 100 200 300 400 500

TEMPERATURE (°C)

CORONA OF ASHNAN
(MESOPOTAMIAN HARVEST GODDESS)

Mount Nix
(Greek night goddess)
LAND
OF BELL
(ENGLISH GIANTESS)

PLAIN OF MUGAZO
Ridge of Lemkechen
(Berber goddess of Polaris)
(ARMENIAN EPIC HEROINE)

PLAIN OF NIOBE
(LEGENDARY GREEK QUEEN)

Ridge of Uni
(Etruscan goddess)

PLAIN OF LLORONA
(MEXICAN FOLKTALE HEROINE)

PLAIN OF REGHINYA
C SPIRIT OF PROTECTION)

TESSERA
OF
GEGUTE
(LITHUANIAN GOD-
DESS OF TIME)

Vega 1 (1984)

AND OF EISTLA
(NORSE GIANTESS)

CORONA OF PAVLOVA
(RUSSIAN BALLET DANCER)

MEAD
(US anthropologist)

Cliffs of Hestia
(Roman goddess
of the hearth)

Rift of Kokomikeis
(Algonquin moon goddess)

LAND OF OVDA
(URALIC GIANTESS)

TESSERA OF HAASTTSE-BAAD
(NAVAJO HEALING GODDESS)

PLAIN OF RUSALKA
(SLAVIC MERMAID)

2,000 KM

Mount Anala
indu fertility goddess)

CORONA OF CALAKOMANA
(PUEBLO CORN GODDESS)

Mount Dzalarhons
(Haida mountain spirit)

TESSERA OF MANATUM
(Semitic goddess of fate)

LAND OF THETIS
(GREEK TITANESS)

TESSERA
OF
NUAHINE
(RAPANUI GODDESS
OF FATE)

PLAIN OF TINATIN
(GEORGIAN EPIC HEROINE)

CORONA OF THERMUTHIS
(EGYPTIAN HARVEST GODDESS)

LAND OF APHRODITE
(GREEK LOVE GODDESS)

Vega 2 (1986)

ORONA OF FATUA
(ROMAN FERTILITY
GODDESS)

PLAIN OF TAHMINA
(IRANIAN EPIC HEROINE)

Rift of Quilla
(Inca moon goddess)

Rift of Diana
(Roman hunting goddess)

CORONA OF ATAHENSIK
(IROQUOIS EARTH GODDESS)

CORONA OF MAMA-ALLPA
(PERUVIAN HARVEST GODDESS)

ALPHA
IAD

TESSERA OF XI-WANG-MU
(CHINESE GODDESS OF THE WEST)

Rift of Arlio
(Celtic wildlife goddess)

Rift of Juno
(Roman sky goddess)

CORONA OF ARTEMIS
(GREEK GODDESS OF THE HUNT)

10,000 m

8,000 m

PLAIN OF ASTHIK
(ARMENIAN LOVE GODDESS)

Selu
DESS)

Rift of Hanghepiwi
(Dakota moon goddess)

PLAIN OF FONUEHA
(SAMOAN MYTHICAL SHARK-WOMAN)

PLAIN OF AINO
(FINNISH WATER SPIRIT)

CORONA OF COPIA
(ROMAN GODDESS OF PLENTY)

Rift of Artemis
(Greek goddess of the hunt)

PLAIN OF ZHIBEK
(KAZAKH EPIC HEROINE)

6,000 m

4,000 m

HIA
ENEAS)

HINOHA
DESS)

CORONA OF EKHE-BURKHAN
(BURYATIAN CREATOR GODDESS)

CORONA OF OTYGEN
(MONGOLIAN EARTH MOTHER)

Rift of Xaratanga
(Mexican moon goddess)

PLAIN OF LAIMDOTA
(LATVIAN MYTHICAL HEROINE)

PLAIN OF IMAPINUA
(INUIT SEA MISTRESS)

2,000 m

0 m

-2,000 m

CORONA OF DUNNE-MUSUN
(EVENKI TAIGA SPIRIT)

LAND OF LADA
(SLAVIC LOVE GODDESS)

HIN
ROINE)

PLAIN OF MUGAZO
(VIETNAMESE EPIC HEROINE)

PLAIN OF ALMA-MERGHEN
(MONGOL EPIC HEROINE)

TERRAIN HEIGHT

95

Viscous, flowing astenosphere, 50 to 200 km thick

Solid lithosphere, 10 to 200 km thick, and broken into dozens of drifting tectonic plates

Stiff mesospheric silicate mantle, 2,650 km thick

Solid iron and nickel inner core, 2,400 km in diameter

Molten iron and nickel outer core, 2,150 km thick

IMPACT CRATERS
Diameter (km):
- <15
- 15–50
- 50–100
- >100

GEOLOGICAL ACTIVITY

Divergent plate boundary   Transform/mixed boundary

Convergent plate boundary   Active subduction zone

Tectonic plate motion   Active volcano

RADIUS: 6,378.1 km
MASS: 5.9724x10^{24} kg
LENGTH OF DAY: 24 hours
SEA LEVEL PRESSURE: 101.325 kPa
AVERAGE ORBITAL SPEED: 29.78 km/s
ORBITAL PERIOD: 365 days, 6 hours, 9 mins
TEMPERATURE RANGE: 15 °C to 40 °C at the equator

EARTH

Ellesmere Island   Hiawatha   GREENLAND ICE SHEET

Tunnunik   Victoria Island   Haughton   Baffin Island

Iceland

Brooks Range   LAURENTIAN PLATEAU   Reykjanes Ridge

Alaska Range

Aleutian Islands

Aleutian Trench

Rocky Mountains

Juan de Fuca Ridge

Cascade Range

Saint Martin   Manicouagan   Charlevoix   Newfoundland

ROCHECHOUA

Beaverhead   GREAT LAKES   Sudbury

San Andreas Fault

Barringer   NORTH AMERICAN PLATE

Mid-Atlantic Ridge

MEXICAN PLATEAU

Appalachians

Azore Islands

Atlas

Hawaiian Islands

Cuba   CHICXULUB

Canary Islands

Cape Verde Islands

CARIBBEAN PLATE

Middle America Trench

COCOS PLATE

Guiana Highlands

Guinea Highlands

PACIFIC PLATE

AMAZON BASIN

Line Islands

NAZCA PLATE

Andes

Peru-Chile Trench

Brazilian Highlands

Araguainha

Mid-Atlantic Ridge

East Pacific Rise

Chile Ridge

Andes

PAMPAS

SOUTH AMERICAN PLATE

Tonga Trench

SCOTIA PLATE

ANTARCTIC PLATE

Larsen Ice Shelf

PALMER LAND

WEST ANTARCTICA

Ronne Ice Shelf

Ross Ice Shelf

Transantarctic Mountains

ANTARCTIC

NORTH AMERICAN PLATE

Svalbard

Severnaya Zemlya

New Siberian Islands

Novaya Zemlya

Byrranga Mountains

·MJOLNIR

·POPIGAI

·KARA

EL'GYGYTGYN·

Verkhoyansk Range

EURASIAN PLATE

OKHOTSK PLATE

AMUR PLATE

Aleutian Trench

Scandinavian Mts

·SILJAN

Ural Mountains

·PUCHEZH-KATUNKI

·KALUGA

Altaï Mountains

Sikhote-Alin Mountains

Kuril Trench

RIES·

·KAMENSK

Tian Shan

Honshu

Alps

Carpathians

·KARAKUL

Qinling

PACIFIC PLATE

BLACK SEA

ANATOLIAN PLATE

CASPIAN SEA

Kunlun Mountains

TIBETAN PLATEAU

Hengduan Mountains

YANGTZE PLATE

PHILIPPINE SEA PLATE

Hindu Kush

Zagros Mountains

Himalayas

Arakan Mts

Hainan

SAQQAR

DECCAN PLATEAU

Luzon

Toggar Mountains

·OASIS

RED SEA

ARABIAN PLATE

INDIAN PLATE

Minda-nao

Matiana Trench

AFRICAN PLATE

Ethiopian Highlands

Sri Lanka

SUNDA PLATE

CONGO BASIN

Borneo

New Guinea

Luizi

Great Rift Valley

SOMALIAN PLATE

Madagascar

Central Indian Ridge

Ninety East Ridge

Sumatra

Sunda Trench

Java

New Caledonia

Great Barrier Reef

WOODLEIGH

MOROKWENG·

·VREDEFORT

Southwest Indian Ridge

TOOKOONOOKA

KERGUELEN

Pacific-Antarctic Ridge

AUSTRALIAN PLATE

·ACRAMAN

Great Dividing Range

ZEALANDIA

Tasmania

South Island

ANTARCTIC PLATE

EAST ANTARCTICA

ICE SHEET

Amery Ice Shelf

Ross Ice Shelf

AGE OF OCEANIC LITHOSPHERE

2,000 KM

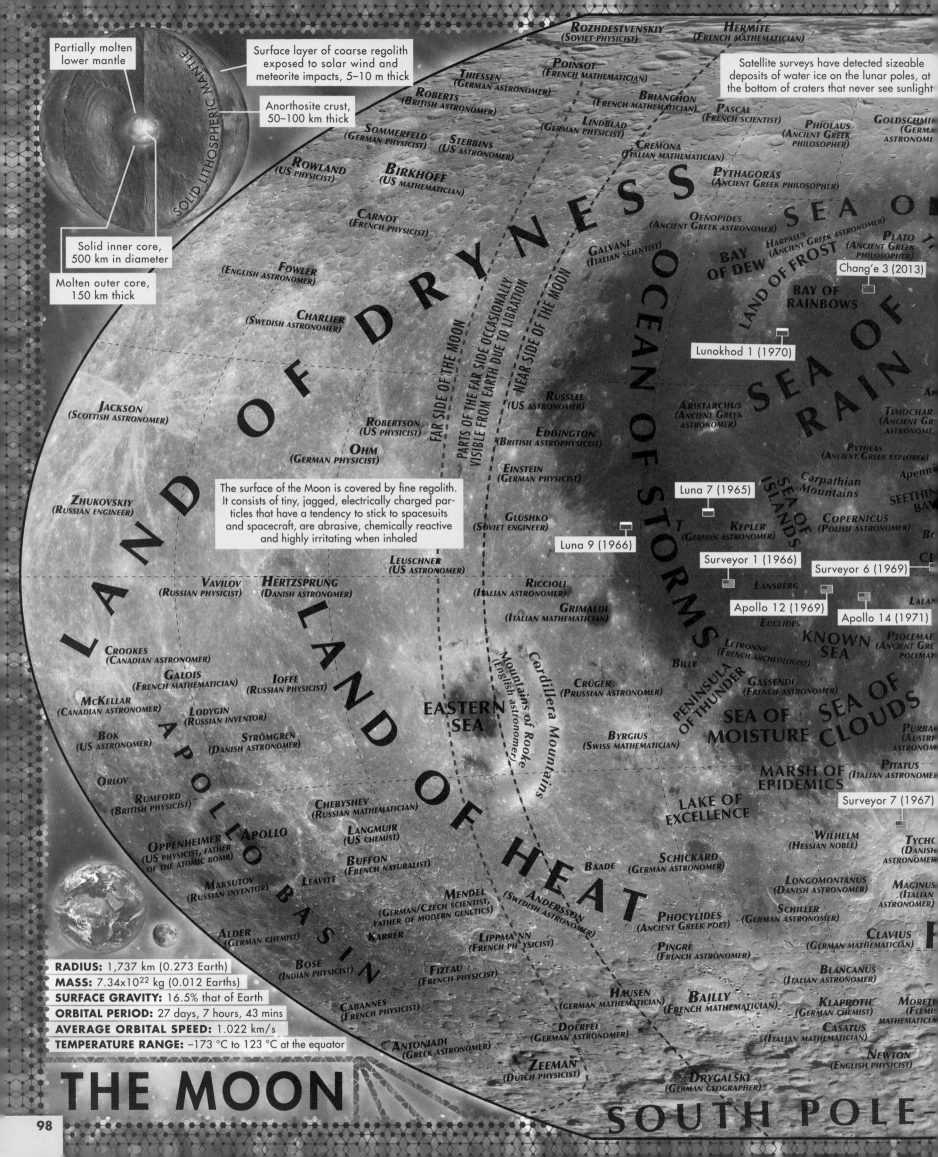

Partially molten lower mantle

Surface layer of coarse regolith exposed to solar wind and meteorite impacts, 5–10 m thick

MANTLE

SOLID LITHOSPHERIC

Anorthosite crust, 50–100 km thick

Solid inner core, 500 km in diameter

Molten outer core, 150 km thick

Satellite surveys have detected sizeable deposits of water ice on the lunar poles, at the bottom of craters that never see sunlight

ROZHDESTVENSKIY
(SOVIET PHYSICIST)

HERMITE
(FRENCH MATHEMATICIAN)

THIESSEN
(GERMAN ASTRONOMER)

POINSOT
(FRENCH MATHEMATICIAN)

ROBERTS
(BRITISH ASTRONOMER)

BRIANCHON
(FRENCH MATHEMATICIAN)

PASCAL
(FRENCH SCIENTIST)

PHIOLAUS
(ANCIENT GREEK PHILOSOPHER)

GOLDSCHMI
(GERMA
ASTRONOME

LINDBLAD
(GERMAN PHYSICIST)

SOMMERFELD
(GERMAN PHYSICIST)

STEBBINS
(US ASTRONOMER)

CREMONA
(ITALIAN MATHEMATICIAN)

ROWLAND
(US PHYSICIST)

BIRKHOFF
(US MATHEMATICIAN)

PYTHAGORAS
(ANCIENT GREEK PHILOSOPHER)

SEA OF

CARNOT
(FRENCH PHYSICIST)

OENOPIDES
(ANCIENT GREEK ASTRONOMER)

GALVANI
(ITALIAN SCIENTIST)

BAY
OF DEW

HARPALUS
(ANCIENT GREEK ASTRONOMER)

PLATO
(ANCIENT GREEK PHILOSOPHER)

LAND OF FROST

FOWLER
(ENGLISH ASTRONOMER)

Chang'e 3 (2013)

BAY OF
RAINBOWS

CHARLIER
(SWEDISH ASTRONOMER)

Lunokhod 1 (1970)

JACKSON
(SCOTTISH ASTRONOMER)

ROBERTSON
(US PHYSICIST)

RUSSELL
(US ASTRONOMER)

ARISTARCHUS
(ANCIENT GREEK ASTRONOMER)

SEA OF
RAIN

TIMOCHAR
(ANCIENT GR
ASTRONOME

OHM
(GERMAN PHYSICIST)

EDDINGTON
(BRITISH ASTROPHYSICIST)

PYTHEAS
(ANCIENT GREEK EXPLORER)

EINSTEIN
(GERMAN PHYSICIST)

Luna 7 (1965)

Carpathian
Mountains

Apenn

The surface of the Moon is covered by fine regolith. It consists of tiny, jagged, electrically charged particles that have a tendency to stick to spacesuits and spacecraft, are abrasive, chemically reactive and highly irritating when inhaled

ZHUKOVSKIY
(RUSSIAN ENGINEER)

GLUSHKO
(SOVIET ENGINEER)

KEPLER
(GERMAN ASTRONOMER)

COPERNICUS
(POLISH ASTRONOMER)

SEETHI
BA

Luna 9 (1966)

Surveyor 1 (1966)

Surveyor 6 (1969)

LEUSCHNER
(US ASTRONOMER)

VAVILOV
(RUSSIAN PHYSICIST)

HERTZSPRUNG
(DANISH ASTRONOMER)

RICCIOLI
(ITALIAN ASTRONOMER)

LANSBERG

Apollo 12 (1969)

Apollo 14 (1971)

LALAN

CROOKES
(CANADIAN ASTRONOMER)

GRIMALDI
(ITALIAN MATHEMATICIAN)

EUCLIDES

KNOWN
SEA

PTOLEMAE
(ANCIENT GRE
POLYMAT

GALOIS
(FRENCH MATHEMATICIAN)

IOFFE
(RUSSIAN PHYSICIST)

LETRONNE
(FRENCH ARCHEOLOGIST)

MCKELLAR
(CANADIAN ASTRONOMER)

LODYGIN
(RUSSIAN INVENTOR)

CRÜGER
(PRUSSIAN ASTRONOMER)

BILLY

GASSENDI
(FRENCH ASTRONOMER)

BOK
(US ASTRONOMER)

STRÖMGREN
(DANISH ASTRONOMER)

EASTERN
SEA

PENINSULA
OF THUNDER

PURBA
(AUSTR
ASTRONOM

ORLOV

Mountains of Rooke
(English astronomer)

Cordillera Mountains

SEA OF
MOISTURE

SEA OF
CLOUDS

RUMFORD
(BRITISH PHYSICIST)

CHEBYSHEV
(RUSSIAN MATHEMATICIAN)

BYRGIUS
(SWISS MATHEMATICIAN)

PITATUS
(ITALIAN ASTRONOMER)

MARSH OF
EPIDEMICS

APOLLO

LANGMUIR
(US CHEMIST)

LAKE OF
EXCELLENCE

Surveyor 7 (1967)

OPPENHEIMER
(US PHYSICIST, FATHER
OF THE ATOMIC BOMB)

BUFFON
(FRENCH NATURALIST)

BAADE

SCHICKARD
(GERMAN ASTRONOMER)

WILHELM
(HESSIAN NOBLE)

TYCHO
(DANISH
ASTRONOMER

MAKSUTOV
(RUSSIAN INVENTOR)

LEAVITT

LONGOMONTANUS
(DANISH ASTRONOMER)

MAGINUS
(ITALIAN
ASTRONOMER)

MENDEL
(GERMAN/CZECH SCIENTIST,
FATHER OF MODERN GENETICS)

ANDERSSON
(SWEDISH ASTRONOMER)

SCHILLER
(GERMAN ASTRONOMER)

ALDER
(GERMAN CHEMIST)

KARRER

PHOCYLIDES
(ANCIENT GREEK POET)

CLAVIUS
(GERMAN MATHEMATICIAN)

LIPPMANN
(FRENCH PHYSICIST)

PINGRE
(FRENCH ASTRONOMER)

BLANCANUS
(ITALIAN ASTRONOMER)

BOSE
(INDIAN PHYSICIST)

FIZEAU
(FRENCH PHYSICIST)

RADIUS: 1,737 km (0.273 Earth)
MASS: 7.34x10²² kg (0.012 Earths)
SURFACE GRAVITY: 16.5% that of Earth
ORBITAL PERIOD: 27 days, 7 hours, 43 mins
AVERAGE ORBITAL SPEED: 1.022 km/s
TEMPERATURE RANGE: −173 °C to 123 °C at the equator

CABANNES
(FRENCH PHYSICIST)

HAUSEN
(GERMAN MATHEMATICIAN)

BAILLY
(FRENCH MATHEMATICIAN)

KLAPROTH
(GERMAN CHEMIST)

MORETU
(FLEMIS
MATHEMATICIA

ANTONIADI
(GREEK ASTRONOMER)

DOERFEL
(GERMAN ASTRONOMER)

CASATUS
(ITALIAN MATHEMATICIAN)

THE MOON

ZEEMAN
(DUTCH PHYSICIST)

DRYGALSKI
(GERMAN GEOGRAPHER)

NEWTON
(ENGLISH PHYSICIST)

SOUTH POLE

PEARY
(US ARCTIC EXPLORER)
BYRD
(US ANTARCTIC EXPLORER)
NANSEN
(NORWEGIAN EXPLORER)
PLASKETT
(CANADIAN ASTRONOMER)
DE SITTER
(DUTCH ASTRONOMER)
SEARES
(US ASTRONOMER)
KARPINSKIY
(RUSSIAN GEOLOGIST)
BAILLAUD
(FRENCH ASTRONOMER)
PETERMANN
(GERMAN CARTOGRAPHER)
SCHWARZSCHILD
(GERMAN PHYSICIST)
BARROW
(ENGLISH MATHEMATICIAN)
GAMOW
(SOVIET-US COSMOLOGIST)
AVOGADRO
(ITALIAN SCIENTIST)
W. BOND
US ASTRONOMER)
ARNOLD
(GERMAN ASTRONOMER)
HAYN
(GERMAN ASTRONOMER)
STORMER
(NORWEGIAN MATHEMATICIAN)
GARTNER
(GERMAN ASTRONOMER)
COMPTON
(US PHYSICIST)
VAN RHIJN
(DUTCH ASTRONOMER)
D'ALEMBERT
(FRENCH MATHEMATICIAN)

ELEMENTAL COMPOSITION

50%   40%   30%   20%   10%   0%

Oxygen
Silicon
Aluminium
Iron
Calcium
Sodium
Potassium
Magnesium
Titanium

☐ Lunar highlands
☐ Lunar lowlands
☐ Earth

LAND OF

COLD
ARISTOTLE
(ANCIENT GREEK
PHILOSOPHER)
Ips
LAKE
OF DEATH
ENDYMION
(GREEK MYTHICAL SHEPHERD)
ATLAS
(MYTHICAL
GREEK TITAN)
LAKE OF HOPE
MILLIKAN
(US PHYSICIST)
BRIDGMAN
(US PHYSICIST)
CAMPBELL
(US ASTRONOMER)
WIENER
(GERMAN PHYSICIST)
VON NEUMANN
(US MATHEMATICIAN)

LIVELINESS

Caucasus Mountains
MACROBIUS
(ROMAN SCHOLAR)
LAKE OF
DREAMS
(HERCULES
(MYTHICAL GREEK HERO)
GEMINUS
(ANCIENT GREEK ASTRONOMER)
GAUSS
(GERMAN MATHEMATICIAN)
BALDET
(FRENCH ASTRONOMER)
SHAJN
(SOVIET ASTRONOMER)

Apollo 15 (1971)
POSIDONIUS
(ANCIENT GREEK ASTRONOMER)
BEROSUS
(BABYLONIAN SCHOLAR)
SEA OF
MOSCOW

des
SEA OF
SERENITY
CLEOMEDES
(GREEK ASTRONOMER)
Lunokhod 2 (1973)
LOMONOSOV
(RUSSIAN POLYMATH)
LAKE OF LUXURY

ntains
Apollo 17 (1972)
MACROBIUS
(ROMAN SCHOLAR)
SEA OF
CRISES
JOLIOT
(FRENCH PHYSICIST)
OLCOTT
(US ASTRONOMER)
HOFFMEISTER
(GERMAN ASTRONOMER)
VIRTANEN
(FINNISH CHEMIST)

SEA OF
VAPOURS
MANILIUS
(ROMAN POET)
PLINIUS
(ROMAN NATURALIST)
PROCLUS
(BYZANTINE PHILOSOPHER)
GODDARD
(US ROCKET SCIENTIST)
FRINGE
SEA
LOBACHEVSKY
(RUSSIAN MATHEMATICIAN)
RECHT
(US ASTRONOMER)
KOHLSCHÜTTER
(GERMAN ASTRONOMER)

SEA OF
TRANQUILLITY
LAND OF
CHEERFULNESS
Luna 24 (1976)
KING
(US PHYSICIST)
MENDELEEV
(RUSSIAN CHEMIST)
MANDELSHTAM
(SOVIET PHYSICIST)
VALIER
(AUSTRIAN ROCKET PIONEER)

RAL
AGRIPPA
TARUNTIUS
SEA OF
WAVES
Luna 16 (1970)
FOAMING
SEA
SEA OF
W.H.SMITH
(BRITISH ASTRONOMER)
WYLD
(US ROCKET SCIENTIST)
GLAZENAP
(SOVIET ASTRONOMER)

500 KM

Apollo 11 (1969)
LAND OF
MANNA
SEA OF
FERTILITY
NECHO
(EGYPTIAN PHARAOH)
VENTRIS
(ENGLISH PHILOLOGIST)

Apollo 16 (1972)
LAND OF
HEALTH
THEOPHILUS
(PATRIARCH OF
ALEXANDRIA)
SEA OF
NECTAR
LAND OF
PYRENEES
GOCLENIUS
(GERMAN SCHOLAR)
LANGRENUS
(DUTCH ASTRONOMER)
PASTEUR
(FRENCH MICROBIOLOGIST)
KEELER
(US ASTRONOMER)

LA PEROUSE
(FRENCH EXPLORER)
FRACASTORIUS
(ITALIAN ASTRONOMER)
GIBBS
(US PHYSICIST)
SKLODOWSKA
(POLISH PHYSICIST)
HILBERT
(GERMAN MATHEMATICIAN)
GAGARIN
(SOVIET ASTRONAUT,
FIRST HUMAN IN SPACE)
AITKEN
(US ASTRONOMER)

PETAVIUS
(FRENCH THEOLOGIST)
TSIOLKOVSKY
(RUSSIAN ROCKET SCIENTIST)

PICCOLOMINI
(ITALIAN ASTRONOMER)
HUMBOLDT
(PRUSSIAN PHILOSOPHER)
LAKE OF
SOLITUDE
SCALIGER
(FRENCH CLASSICAL SCHOLAR)

STEVINUS
(FLEMISH ENGINEER)
ABEL
(NORWEGIAN MATHEMATICIAN)

STÖFLER
(GERMAN ASTRONOMER)
RHEITA
(CZECH ASTRONOMER)
JULES VERNE
(FRENCH WRITER)
THOMSON
(ENGLISH PHYSICIST)

MAUROLYCUS
(SICILIAN ASTRONOMER)
LAND OF
FERTILITY
SOUTHERN
SEA
LAMB
(ENGLISH
MATHEMATICIAN)
KOCH
(GERMAN MICROBIOLOGIST)
SEA OF
CLEVERNESS
LEIBNIZ
(GERMAN POLYMATH)

ZACH
(HUNGARIAN ASTRONOMER)
OKEN
(GERMAN NATURALIST)
LYOT
(FRENCH ASTRONOMER)
LEBEDEV
(RUSSIAN PHYSICIST)
PAULI
(SWISS/US PHYSICIST)
HOPMANN
(GERMAN ASTRONOMER)

First quarter
Waxing
gibbous
Waxing
crescent

CURTIUS
(GERMAN JESUIT
ASTRONOMER)
MANZINUS
(ITALIAN ASTRONOMER)
BIELA
(GERMAN ASTRONOMER)
JEANS
(ENGLISH PHYSICIST)
KÜGLER
(GERMAN CHEMIST)

PONTECOULANT
(FRENCH ASTRONOMER)
MOULTON
(US ASTRONOMER)
PLANCK
(AUSTRIAN PHYSICIST)
POINCARE
(FRENCH POLYMATH)
Full moon

MACROBIUS
(ROMAN SCHOLAR)
HELMHOLTZ
(GERMAN PHYSICIST)
SIKORSKY
(RUSSIAN AVIATION PIONEER)
Waning
gibbous

BOUSSINGAULT
(FRENCH CHEMIST)
DEMONAX
(GREEK PHILOSOPHER)
HALL
(US ASTRONOMER)
SCHRÖDINGER
(AUSTRIAN PHYSICIST)
LYMAN
(US PHYSICIST)
Third quarter
New
moon
Waning
crescent

PHASES
OF THE MOON

AITKEN BASIN

Molten outer core, thickness unknown

In the summer months, streaks of liquid hypersaline water can emerge from the Martian soil

Tectonically inactive crust, 60 km thick

SOLID MANTLE

Olympia Sand

Phoenix (2008)

Abalos Sand Sea (Island in the North Sea)

Polar Canyon

Hyperborean (Legendary Northern land)

Solid inner core, 1,800 km in diameter

Martian interior is considerably richer in sulphur than Earth's

N O R T H E R N

Scandinavian Hills

LOMONOSOV (Russian polymath)

MILANKOVIC (Serbian astronomer)

ARCADIAN PLAIN (Region in Greece)

White Valleys

Valleys of Tantalus (Legendary Greek king)

KUNOWSKY (German astronomer)

Valleys of Mareotis (Region in Lower Egypt)

PEREPELKIN (Soviet astronomer)

Valleys of Tempe (Valley in Thessaly)

TIMOSHENKO (Soviet astronomer)

SYTINSKAYA (Soviet astronomer)

PLAIN OF ACIDALIA (Legendary fountain of the Greek goddess Aphrodite)

Valleys of Acheron (Greek legendary river of the underworld)

▲White Mountain

LAND OF TEMPE (Valley in Thessaly)

Mesas of Cydonia (Ancient city in Crete)

Ridges of Lycus (Legendary son of Poseidon)

Mount Uranius (Greek god of time)

SHARONOV (Soviet astronomer)

CURIE (French-Polish physicist)

Olympus Mons is one of the highest mountains in the solar system, rising over 27 km above the surrounding landscape

AMAZONIAN PLAIN (Legendary tribe of warrior women)

Valley of Marte ('Mars' in Spanish)

Thundering Mountain

Valleys of Kasei ('Mars' in Japanese)

FESENKOV (Russian physicist)

Viking 1 (1976)

GOLDEN PLAIN

OYAMA (US biochemist)

ARA

Mount Olympus (Olympus Mons)

Cliffs of Olympus (Olympus – highest mountain in Greece, legendary seat of the gods)

Mount Tharsis (Legendary land, mentioned in the Bible)

Sojourner (1997)

TROUVELOT (French astronomer)

PETTIT (US astronomer)

Mountains of Tharsis

Mount Ascra (Birthplace of Hesiod)

PLATEAU OF TIUNA

PLATEAU OF XANTHUS (Multiple figures in Greek mythology)

Valley of Simud ('Mars' in Sumerian)

SAGAN (US astronomer)

Opportunity (2004)

Gordian Ridge (Ancient Anatolian city)

Mount Odysseus (Greek legendary hero)

Valley of Shalbatana ('Mars' in Akkadian)

Chaos of Aram (Biblical land)

LAND MER

NICHOLSON (US astronomer)

Mount Biblis (Legendary Greek heroine)

Mount Peacock

Canyon of Hebe (Greek goddess of youth)

MUTCH (US planetary scientist)

Valley of Youth

Chaos of Iani (Old name of the river Brahmaputra)

LUCANIAN PLATEAU (Region in Southern Italy)

Valleys of Mangala ('Mars' in Sanskrit)

Valleys of Medusa (Greek legendary monster)

Mount Arsia (Forest near Ancient Rome)

Canyon of Ophir (Biblical land)

Canyon of Melas (Son of Poseidon)

Canyon of Candor

Canyon of Ganges (River in India)

Canyon of Echo (Greek legendary nymph)

Labyrinth of the Night

Mariner Valleys (Mariner 9 US space probe)

Canyon of Eos (Greek goddess of dawn)

VINOGRADOV (Soviet geochemist)

JONES (English astronomer)

BURTON (British astronomer)

PLATEAU OF DAEDALUS (Legendary Greek aviator)

PLATEAU OF SYRIA

PLATEAU OF SINAI (Peninsula in Egypt)

PLATEAU OF THE SUN

PLATEAU OF THAUMASIA (Legendary land of wonders)

Valley of Nirgal ('Mars' in Babylonian)

HOLDEN (US astronomer)

Mars 6 (1974)

WILLIAMS (US astronomer)

Valleys of Memnon (Legendary king of Ethiopia)

BERNARD (French scientist)

Bright Cliffs

Bright Valleys

Valleys of Nectar

BOND (US astronomer)

LAN

DEJNEV (Russian geographer)

COLUMBUS (Italian explorer)

KOVAL'SKY (Polish-Russian astronomer)

Raven Valleys

PLATEAU OF BOSPHORUS (Strait forming a boundary between Europe and Asia)

HALE (US astronomer)

VOGEL (German physicist)

MARINER (US Mars probe)

NICHOLSON (Polish-Russian astronomer)

Valleys of the Sirens

LAND OF THE SIRENS

PLATEAU OF ICARUS (Son of Daedalus)

Mountains of the Nereids (Greek legendary sea nymphs)

NO

LOHSE (German astronomer)

NEWTON (British physicist)

Mars 3 (1971)

SLIPHER (US astronomer)

PLAIN OF ARGYRE (Legendary island of silver)

GALLE (German astronomer)

(BIBLICA

COPERNICUS (Polish astronomer)

(Greek mythological sea monsters who used their singing to lure sailors to their deaths)

HUSSEY (US astronomer)

LOWELL (US astronomer)

Mountains of the Graces (Three daughters of Zeus)

GRE (British astronome

KUIPER (Dutch astronomer)

DOKUCHAEV (Russian geologist)

LAND OF THE MUSES

DARWIN (English naturalist)

KEELER (US astronomer)

CHAMBERLIN (US geologist)

(Greek goddesses of poetry, music, science and the arts)

PHILLIPS (English geologist)

MARALDI (Italian astronomer)

LYELL (Scottish geologist)

CHARLIER (Swedish astronomer)

AGASSIZ (US glaciologist)

SCHMIDT (German astronomer)

RADIUS: 3,389 km (0.533 Earths)
MASS: 6.41×10²³ kg (0.107 Earths)
SURFACE GRAVITY: 16.5% that of Earth
ORBITAL PERIOD: 686 days, 23 hours
AVERAGE ORBITAL SPEED: 24.01 km/s
TEMPERATURE RANGE: −100 °C to 26 °C at the equator
NATURAL SATELLITES: 2
DAY LENGTH: 24 hours, 37 mins
SURFACE PRESSURE: 0.4 to 1 kPa

REYNOLDS (Irish physicist)

SILVER PLATEAU

Pits of Sysiphus (Legendary Greek king)

MARS

Narrow Pits

Ultimate Cliffs

SOUTHERN

Mars features two permanent polar ice caps, consisting mostly of water ice and solid carbon dioxide. In fact, there is enough water locked inside the Martian caps to evenly cover the planet with an ocean over 20 m deep

ATMOSPHERIC COMPOSITION

CARBON DIOXIDE: 94–96%, varies seasonally

CARBON MONOXIDE: 0.07%

OXYGEN: 0.14%

ARGON: 1.6%

NITROGEN: 2.8%

WATER VAPOUR: 0.03%

...and Sea
(Greek mythology)

LOWLANDS

The Mesas of Cydonia are home to the famous 'Face on Mars' – a hill resembling a giant human face, later shown to be just an optical illusion

UTOPIAN

PLAIN

LYOT
(FRENCH ASTRONOMER)

Viking 2 (1976)

MIE
(GERMAN PHYSICIST)

Mesas of the Deuteronile

Mesas of the Protonile

MOREUX
(FRENCH ASTRONOMER)

Ismenian Valleys
(Ismene, character in several ancient Greek plays)

Mesas of the Sidro-nile

Phlegraean Mountains
(Volcanic area near Naples)

Mount Hecate
(Greek goddess of dark magic)

Mount Elysium

Elysian Valleys

Mountain of Dawn

...BIAN ...AND

CASSINI
(ITALIAN ASTRONOMER)

FLAMMARION
(FRENCH ASTRONOMER)

BALDET
(FRENCH ASTRONOMER)

ANTONIADI
(GREEK ASTRONOMER)

Valleys of the Nile

PERIDIER
(FRENCH ASTRONOMER)

ELYSIAN PLAIN
(ANCIENT GREEK AFTERLIFE REALM FOR THE BLESSED)

Mountains of Tartarus
(Greek mythological abyss for the wicked souls)

Patera of Orcus
(Roman underworld god)

SABAEAN LAND
(BIBLICAL LAND OF SHEBA, LOCATED IN SOUTH ARABIA)

HENRY
(FRENCH ASTRONOMER)

PLAIN OF ISIS
(EGYPTIAN GODDESS OF MAGIC AND MOTHERHOOD)

PLATEAU OF SIDRA
(MEDITERRANEAN GULF OFF THE COAST OF LIBYA)

SCHROETER
(GERMAN ASTRONOMER)

SCHIAPARELLI
(ITALIAN ASTRONOMER)

Lybyan Mountains

FOURNIER
(FRENCH ASTRONOMER)

Oenotrian Escarpment
(Ancient Italic tribe)

JARRY-DESLOGES
(FRENCH ASTRONOMER)

PLATEAU OF AMENTI
(EGYPTIAN REALM OF THE DEAD)

PLATEAU OF NUMBNESS

InSight (2018)

Mount Aeolus
(Mount Sharp)

KNOBEL
(ENGLISH ASTRONOMER)

Mesas of Aeolus
(Greek god of wind)

Mount Apollinaris
(Spring near Rome)

...OF THE ...DIAN

DAWES
(ENGLISH ASTRONOMER)

HUYGENS
(DUTCH ASTRONOMER)

Curiosity (2012)

WIEN
(GERMAN PHYSICIST)

Spirit (2004)

FLAUGERGUES
(FRENCH ASTRONOMER)

DENNING
(BRITISH ASTRONOMER)

TYRRHENIAN LAND
(SEA OFF THE WESTERN COAST OF ITALY)

PLATEAU OF HESPERIA
('WESTERN LAND,' ANCIENT ROMAN TERM FOR IBERIA)

HERSCHEL
(BRITISH-GERMAN ASTRONOMER)

DE VAUCOULEURS
(FRENCH ASTRONOMER)

WISLICENUS
(GERMAN CHEMIST)

...WCOMB
...ASTRONOMER)

BAKHUYSEN
(DUTCH ASTRONOMER)

SCHAEBERLE
(GERMAN ASTRONOMER)

GRAFF
(GERMAN ASTRONOMER)

Valley of Cairo

Douro Valley
(River in Spain)

Valley of Ma'adim
(Mars in Hebrew)

...D OF

Hellespontic Mountains
(Ancient name for the strait of Dardanelles)

TERBY
(BELGIAN ASTRONOMER)

SAVICH
(RUSSIAN ASTRONOMER)

MÜLLER
(GERMAN ASTRONOMER)

MOLESWORTH
(BRITISH ASTRONOMER)

LE VERRIER
(FRENCH MATHEMATICIAN)

The Plain of Hellas, a giant crater, is the lowest-lying region on Mars. It features the highest air pressure and summer air temperatures

PLAIN OF HELLAS
(ANCIENT NAME FOR GREECE)

Mars 2 (1971)

Valley of Harmakhis
(Egyptian god Horus)

ARRHENIUS
(SWEDISH CHEMIST)

Valley of Reull
('Planet' in Gaellic)

MARTZ
(US PHYSICIST)

20,000 m

15,000 m

10,000 m

5,000 m

0 m

−3,000 m

−6,000 m

...AH

KAISER
(DUTCH ASTRONOMER)

...PATRIARCH)

PROCTOR
(BRITISH ASTRONOMER)

Chaos of Hellas

LAND OF PROMETHEUS
(TITAN IN GREEK MYTHOLOGY)

KEPLER
(GERMAN ASTRONOMER)

RUSSELL
(US ASTRONOMER)

HENRY MOORE
(BRITISH SCULPTOR)

GLEDHILL
(BRITISH ASTRONOMER)

WALLACE
(BRITISH NATURALIST)

CIMMERIAN LAND
(ANCIENT TERM FOR CRIMEA)

...ountains of Sysiphus
...legendary Greek king)

Patera of pine trees

BARNARD
(US ASTRONOMER)

SECCHI
(ITALIAN ASTRONOMER)

Escarpment of Eridanus
(Greek mythological river)

WELLS
(ENGLISH WRITER)

PLATEAU OF MALEA
(PENINSULA IN GREECE)

MITCHEL
(US ASTRONOMER)

GILBERT
(US GEOLOGIST)

HUXLEY
(BRITISH NATURALIST)

BYRD
(US EXPLORER)

SOUTH
(BRITISH ASTRONOMER)

MAIN
(ENGLISH ASTRONOMER)

LIAIS
(FRENCH ASTRONOMER)

BURROUGHS
(US FICTION WRITER)

RAYLEIGH
(BRITISH PHYSICIST)

Deep Space 2 (1999)

PLATEAU OF PROMETHEUS
(TITAN IN GREEK MYTHOLOGY)

POLAR PLAIN

Ultimate Cliffs

TERRAIN HEIGHT

1,000 KM

101

Z=28.62   Z=10.01   Z= 5.03

TIMELINE OF THE UNIVERSE

While our ancestors strived to understand the universe, they were perhaps even more fascinated by the question of its origin and ultimate fate. Almost all cultures in history have some sort of a 'creation myth' that explained the origin of humanity and the cosmos as a whole.

Some, such as the ancient Greeks and Egyptians, believed that the universe was once a formless chaos, from which the first titans and gods emerged, creating the world from their own bodies. Others, such as the Maya or Hindus, thought the cosmos was cyclical, bound in an eternal cycle of creation and destruction, whilst others believed that the universe was young, having a single creator.

After the heliocentric system and Newtonian mechanics became widely accepted, the general opinion in secular scientific circles shifted towards a mechanistic universe, reducible to the predetermined motion and collision of matter. Many scientists embraced the view that the universe was eternal as well – our existence being a cosmic fluke caused by a random reorganization of particles in an infinitely old and infinitely large universe. In this cosmos, everything that could have ever occurred would have already occurred an infinite number of times in the past, and will recur again in the infinite future.

In the 1920s, however, the Belgian Catholic priest Georges Lemaître examined astronomical observations that showed that the light arriving to us from distant galaxies was red-shifted. This would imply that they are moving away from the Earth – and the further away they are, the faster they are receding. He realized this could be best explained by a universe that is expanding. Tracing the cosmic expansion backwards, all matter in the universe would get concentrated into a single inifinitely small point – implying the universe originated at a finite time in the past. Lemaître published his findings in 1927, calling his proposal 'The Hypothesis of the Primeval Atom'. At first, his ideas were generally met with scepticism from the general scientific community, that still favoured the steady-state model. However, in 1929, Edwin Hubble arrived independently at the same conclusions, and calculated the approximate rate at which the universe is expanding – the Hubble constant.

In 1949, Sir Fred Hoyle dubbed this theory the Big Bang – a term that has stuck to this day. Since the 1990s, advances in telescope technology have provided more valuable evidence of the Big Bang, such as the discovery of the Cosmic Microwave Background – and data suggests that the expansion of the universe might even be accelerating.

Just as the Scale of the Universe chapter explored the space dimension of the universe, aiming to put cosmic distances into perspective, the following pages will take you on a journey through cosmic time. Starting right at the Big Bang and the exotic, inconceivably hot infant universe, we will slowly follow the formation and evolution of the first stars and galaxies in the universe to see how they gradually diversified into their present shapes and sizes.

While still listing the most important and interesting events from the cosmic past as they happened, the middle part of this chapter focuses on the natural history of our planet as well. The reason for including it here is two-fold: throughout its 4.5 billion years of existence, the Earth was affected by the broader trends present in our solar system and the stellar neighbourhood, such as supernovae or the stabilizing presence of the Moon. However,

Illustration showing the formation of galaxies from colliding protogalaxies in the early universe

Probable glow of some of the earliest stars in the universe

Artistic depiction of ULAS J1120+0641, a very distant quasar powered by a black hole of over 2 billion solar masses (Credit: ESO/M. Kornmesser)

Artist's impression of a gamma-ray burst, one of the most powerful events in the universe (ESO/A. Roquette)

Protoplanet Theia impacting the young Earth. This event likely resulted in the formation of the Moon

Artistic depiction of A2744 YD4, one of the youngest and most distant known galaxies

Artistic impression of the young Sun, still surrounded by a protoplanetary disc of gas and dust

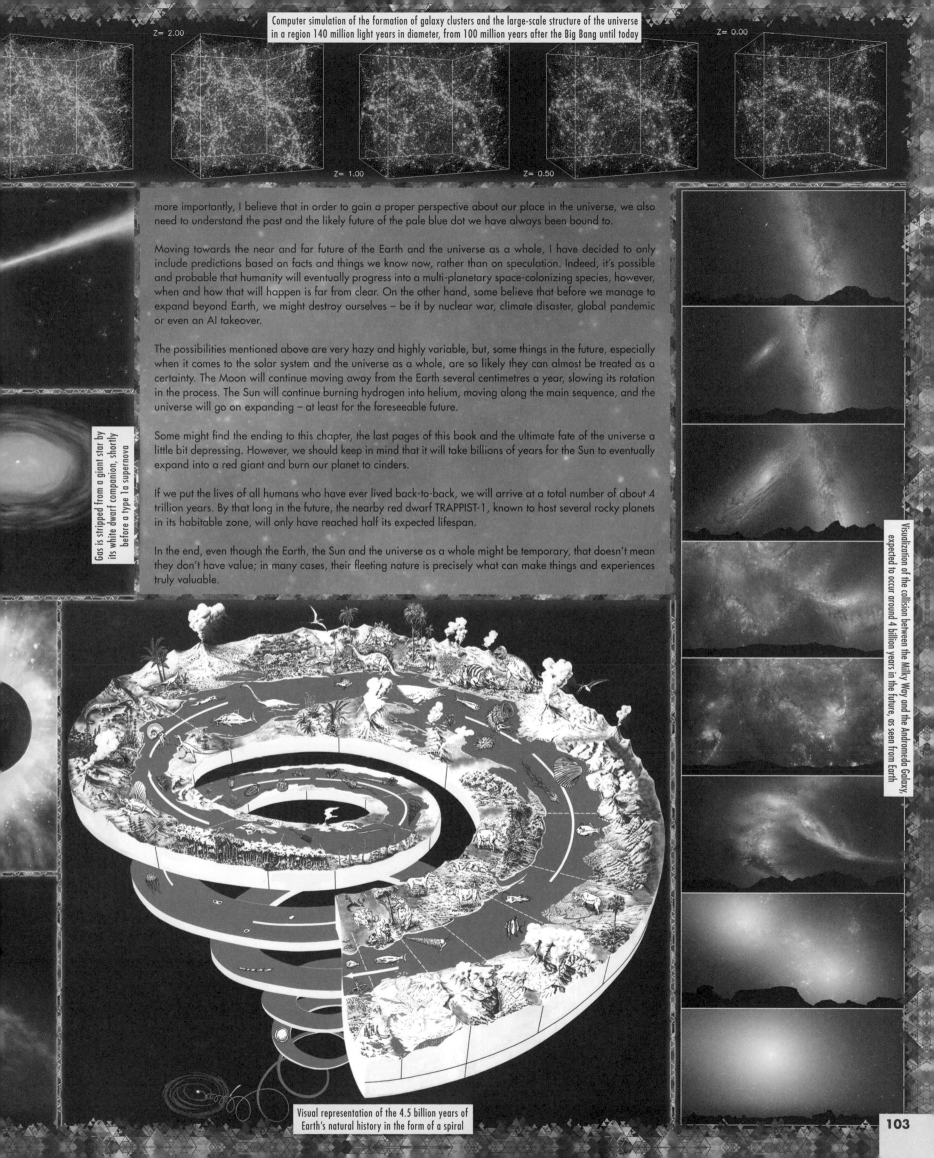

Z= 2.00

Z= 1.00

Z= 0.50

Z= 0.00

more importantly, I believe that in order to gain a proper perspective about our place in the universe, we also need to understand the past and the likely future of the pale blue dot we have always been bound to.

Moving towards the near and far future of the Earth and the universe as a whole, I have decided to only include predictions based on facts and things we know now, rather than on speculation. Indeed, it's possible and probable that humanity will eventually progress into a multi-planetary space-colonizing species, however, when and how that will happen is far from clear. On the other hand, some believe that before we manage to expand beyond Earth, we might destroy ourselves – be it by nuclear war, climate disaster, global pandemic or even an AI takeover.

The possibilities mentioned above are very hazy and highly variable, but, some things in the future, especially when it comes to the solar system and the universe as a whole, are so likely they can almost be treated as a certainty. The Moon will continue moving away from the Earth several centimetres a year, slowing its rotation in the process. The Sun will continue burning hydrogen into helium, moving along the main sequence, and the universe will go on expanding – at least for the foreseeable future.

Some might find the ending to this chapter, the last pages of this book and the ultimate fate of the universe a little bit depressing. However, we should keep in mind that it will take billions of years for the Sun to eventually expand into a red giant and burn our planet to cinders.

If we put the lives of all humans who have ever lived back-to-back, we will arrive at a total number of about 4 trillion years. By that long in the future, the nearby red dwarf TRAPPIST-1, known to host several rocky planets in its habitable zone, will only have reached half its expected lifespan.

In the end, even though the Earth, the Sun and the universe as a whole might be temporary, that doesn't mean they don't have value; in many cases, their fleeting nature is precisely what can make things and experiences truly valuable.

Gas is stripped from a giant star by its white dwarf companion, shortly before a type 1a supernova

Visualization of the collision between the Milky Way and the Andromeda Galaxy, expected to occur around 4 billion years in the future, as seen from Earth

Visual representation of the 4.5 billion years of Earth's natural history in the form of a spiral

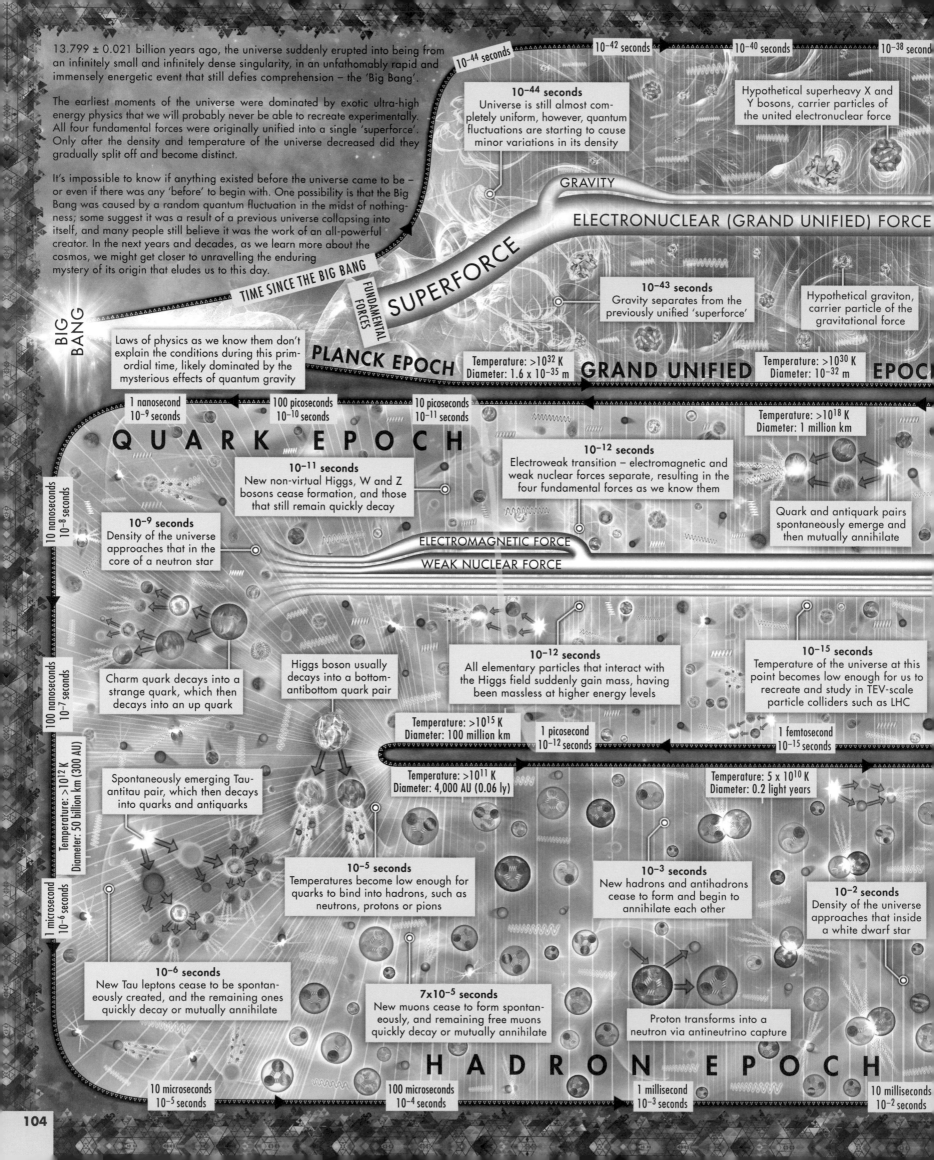

13.799 ± 0.021 billion years ago, the universe suddenly erupted into being from an infinitely small and infinitely dense singularity, in an unfathomably rapid and immensely energetic event that still defies comprehension – the 'Big Bang'.

The earliest moments of the universe were dominated by exotic ultra-high energy physics that we will probably never be able to recreate experimentally. All four fundamental forces were originally unified into a single 'superforce'. Only after the density and temperature of the universe decreased did they gradually split off and become distinct.

It's impossible to know if anything existed before the universe came to be – or even if there was any 'before' to begin with. One possibility is that the Big Bang was caused by a random quantum fluctuation in the midst of nothingness; some suggest it was a result of a previous universe collapsing into itself, and many people still believe it was the work of an all-powerful creator. In the next years and decades, as we learn more about the cosmos, we might get closer to unravelling the enduring mystery of its origin that eludes us to this day.

10^{-44} seconds

10^{-42} seconds

10^{-40} seconds

10^{-38} seconds

10^{-44} seconds
Universe is still almost completely uniform, however, quantum fluctuations are starting to cause minor variations in its density

Hypothetical superheavy X and Y bosons, carrier particles of the united electronuclear force

GRAVITY

ELECTRONUCLEAR (GRAND UNIFIED) FORCE

SUPERFORCE

TIME SINCE THE BIG BANG

FUNDAMENTAL FORCES

BIG BANG

10^{-43} seconds
Gravity separates from the previously unified 'superforce'

Hypothetical graviton, carrier particle of the gravitational force

Laws of physics as we know them don't explain the conditions during this primordial time, likely dominated by the mysterious effects of quantum gravity

PLANCK EPOCH

Temperature: $>10^{32}$ K
Diameter: 1.6×10^{-35} m

GRAND UNIFIED

Temperature: $>10^{30}$ K
Diameter: 10^{-32} m

EPOCH

1 nanosecond
10^{-9} seconds

100 picoseconds
10^{-10} seconds

10 picoseconds
10^{-11} seconds

Temperature: $>10^{18}$ K
Diameter: 1 million km

QUARK EPOCH

10^{-11} seconds
New non-virtual Higgs, W and Z bosons cease formation, and those that still remain quickly decay

10^{-12} seconds
Electroweak transition – electromagnetic and weak nuclear forces separate, resulting in the four fundamental forces as we know them

Quark and antiquark pairs spontaneously emerge and then mutually annihilate

10^{-9} seconds
Density of the universe approaches that in the core of a neutron star

10 nanoseconds
10^{-8} seconds

ELECTROMAGNETIC FORCE

WEAK NUCLEAR FORCE

Charm quark decays into a strange quark, which then decays into an up quark

100 nanoseconds
10^{-7} seconds

Higgs boson usually decays into a bottom-antibottom quark pair

10^{-12} seconds
All elementary particles that interact with the Higgs field suddenly gain mass, having been massless at higher energy levels

10^{-15} seconds
Temperature of the universe at this point becomes low enough for us to recreate and study in TEV-scale particle colliders such as LHC

Temperature: $>10^{15}$ K
Diameter: 100 million km

1 picosecond
10^{-12} seconds

1 femtosecond
10^{-15} seconds

Spontaneously emerging Tau-antitau pair, which then decays into quarks and antiquarks

Temperature: $>10^{12}$ K
Diameter: 50 billion km (300 AU)

Temperature: $>10^{11}$ K
Diameter: 4,000 AU (0.06 ly)

Temperature: 5×10^{10} K
Diameter: 0.2 light years

10^{-5} seconds
Temperatures become low enough for quarks to bind into hadrons, such as neutrons, protons or pions

10^{-3} seconds
New hadrons and antihadrons cease to form and begin to annihilate each other

10^{-2} seconds
Density of the universe approaches that inside a white dwarf star

1 microsecond
10^{-6} seconds

10^{-6} seconds
New Tau leptons cease to be spontaneously created, and the remaining ones quickly decay or mutually annihilate

7×10^{-5} seconds
New muons cease to form spontaneously, and remaining free muons quickly decay or mutually annihilate

Proton transforms into a neutron via antineutrino capture

HADRON EPOCH

10 microseconds
10^{-5} seconds

100 microseconds
10^{-4} seconds

1 millisecond
10^{-3} seconds

10 milliseconds
10^{-2} seconds

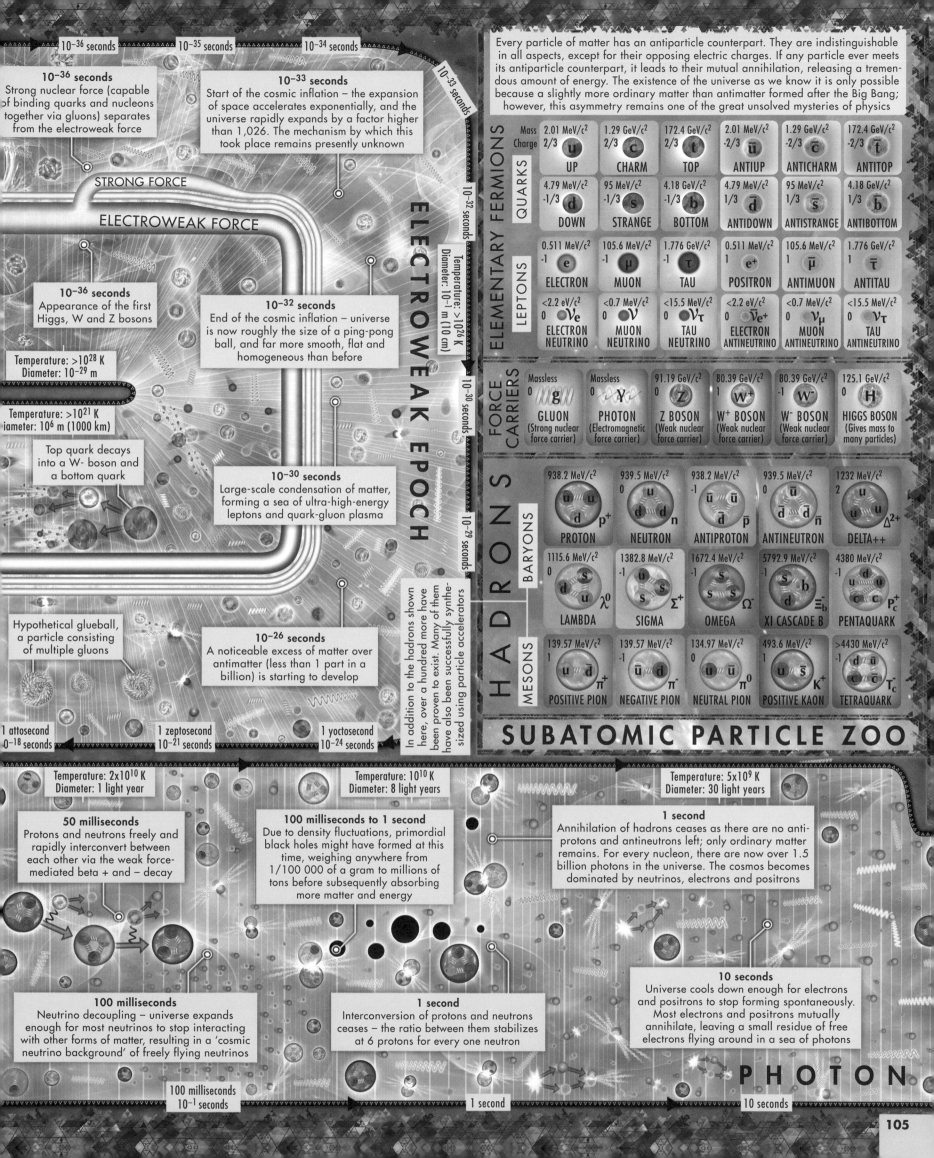

10⁻³⁶ seconds — **10⁻³⁵ seconds** — **10⁻³⁴ seconds** — **10⁻³³ seconds**

10⁻³⁶ seconds
Strong nuclear force (capable of binding quarks and nucleons together via gluons) separates from the electroweak force

10⁻³³ seconds
Start of the cosmic inflation – the expansion of space accelerates exponentially, and the universe rapidly expands by a factor higher than 1,026. The mechanism by which this took place remains presently unknown

STRONG FORCE

ELECTROWEAK FORCE

10⁻³⁶ seconds
Appearance of the first Higgs, W and Z bosons

Temperature: >10²⁸ K
Diameter: 10⁻²⁹ m

10⁻³² seconds
End of the cosmic inflation – universe is now roughly the size of a ping-pong ball, and far more smooth, flat and homogeneous than before

Temperature: >10²¹ K
Diameter: 10⁶ m (1000 km)

Top quark decays into a W- boson and a bottom quark

10⁻³⁰ seconds
Large-scale condensation of matter, forming a sea of ultra-high-energy leptons and quark-gluon plasma

Hypothetical glueball, a particle consisting of multiple gluons

10⁻²⁶ seconds
A noticeable excess of matter over antimatter (less than 1 part in a billion) is starting to develop

In addition to the hadrons shown here, over a hundred more have been proven to exist. Many of them have also been successfully synthesized using particle accelerators

1 attosecond
10⁻¹⁸ seconds — 1 zeptosecond 10⁻²¹ seconds — 1 yoctosecond 10⁻²⁴ seconds

ELECTROWEAK EPOCH

10⁻³³ seconds

10⁻³² seconds

10⁻³⁰ seconds

10⁻²⁹ seconds

Temperature: >10²⁶ K
Diameter: 10⁻¹ m (10 cm)

Every particle of matter has an antiparticle counterpart. They are indistinguishable in all aspects, except for their opposing electric charges. If any particle ever meets its antiparticle counterpart, it leads to their mutual annihilation, releasing a tremendous amount of energy. The existence of the universe as we know it is only possible because a slightly more ordinary matter than antimatter formed after the Big Bang; however, this asymmetry remains one of the great unsolved mysteries of physics

ELEMENTARY FERMIONS

QUARKS

| Mass Charge | 2.01 MeV/c² 2/3 **u** UP | 1.29 GeV/c² 2/3 **c** CHARM | 172.4 GeV/c² 2/3 **t** TOP | 2.01 MeV/c² -2/3 **ū** ANTIUP | 1.29 GeV/c² -2/3 **c̄** ANTICHARM | 172.4 GeV/c² -2/3 **t̄** ANTITOP |
| 4.79 MeV/c² -1/3 **d** DOWN | 95 MeV/c² -1/3 **s** STRANGE | 4.18 GeV/c² -1/3 **b** BOTTOM | 4.79 MeV/c² 1/3 **d̄** ANTIDOWN | 95 MeV/c² 1/3 **s̄** ANTISTRANGE | 4.18 GeV/c² 1/3 **b̄** ANTIBOTTOM |

LEPTONS

| 0.511 MeV/c² -1 **e** ELECTRON | 105.6 MeV/c² -1 **μ** MUON | 1.776 GeV/c² -1 **τ** TAU | 0.511 MeV/c² 1 **e⁺** POSITRON | 105.6 MeV/c² 1 **μ̄** ANTIMUON | 1.776 GeV/c² 1 **τ̄** ANTITAU |
| <2.2 eV/c² 0 **νₑ** ELECTRON NEUTRINO | <0.7 MeV/c² 0 **ν** MUON NEUTRINO | <15.5 MeV/c² 0 **ν_τ** TAU NEUTRINO | <2.2 eV/c² 0 **ν̄ₑ⁺** ELECTRON ANTINEUTRINO | <0.7 MeV/c² 0 **ν_μ** MUON ANTINEUTRINO | <15.5 MeV/c² 0 **ν_τ** TAU ANTINEUTRINO |

FORCE CARRIERS

| Massless 0 **g** GLUON (Strong nuclear force carrier) | Massless 0 **γ** PHOTON (Electromagnetic force carrier) | 91.19 GeV/c² 0 **Z** Z BOSON (Weak nuclear force carrier) | 80.39 GeV/c² 1 **W⁺** W⁺ BOSON (Weak nuclear force carrier) | 80.39 GeV/c² 1 **W⁻** W⁻ BOSON (Weak nuclear force carrier) | 125.1 GeV/c² 0 **H** HIGGS BOSON (Gives mass to many particles) |

HADRONS

BARYONS

| 938.2 MeV/c² 1 **p⁺** PROTON | 939.5 MeV/c² 0 **n** NEUTRON | 938.2 MeV/c² -1 **p̄** ANTIPROTON | 939.5 MeV/c² 0 **n̄** ANTINEUTRON | 1232 MeV/c² 2 **Δ²⁺** DELTA++ |
| 1115.6 MeV/c² 0 **λ⁰** LAMBDA | 1382.8 MeV/c² -1 **Σ⁺** SIGMA | 1672.4 MeV/c² -1 **Ω⁻** OMEGA | 5792.9 MeV/c² -1 **Ξ_b** XI CASCADE B | 4380 MeV/c² **P_c** PENTAQUARK |

MESONS

| 139.57 MeV/c² 1 **π⁺** POSITIVE PION | 139.57 MeV/c² -1 **π⁻** NEGATIVE PION | 134.97 MeV/c² 0 **π⁰** NEUTRAL PION | 493.6 MeV/c² 1 **K⁺** POSITIVE KAON | >4430 MeV/c² -1 **T_c** TETRAQUARK |

SUBATOMIC PARTICLE ZOO

Temperature: 2x10¹⁰ K
Diameter: 1 light year

Temperature: 10¹⁰ K
Diameter: 8 light years

Temperature: 5x10⁹ K
Diameter: 30 light years

50 milliseconds
Protons and neutrons freely and rapidly interconvert between each other via the weak force-mediated beta + and − decay

100 milliseconds to 1 second
Due to density fluctuations, primordial black holes might have formed at this time, weighing anywhere from 1/100 000 of a gram to millions of tons before subsequently absorbing more matter and energy

1 second
Annihilation of hadrons ceases as there are no anti-protons and antineutrons left; only ordinary matter remains. For every nucleon, there are now over 1.5 billion photons in the universe. The cosmos becomes dominated by neutrinos, electrons and positrons

100 milliseconds
Neutrino decoupling – universe expands enough for most neutrinos to stop interacting with other forms of matter, resulting in a 'cosmic neutrino background' of freely flying neutrinos

1 second
Interconversion of protons and neutrons ceases – the ratio between them stabilizes at 6 protons for every one neutron

10 seconds
Universe cools down enough for electrons and positrons to stop forming spontaneously. Most electrons and positrons mutually annihilate, leaving a small residue of free electrons flying around in a sea of photons

100 milliseconds
10⁻¹ seconds

1 second

10 seconds

PHOTON

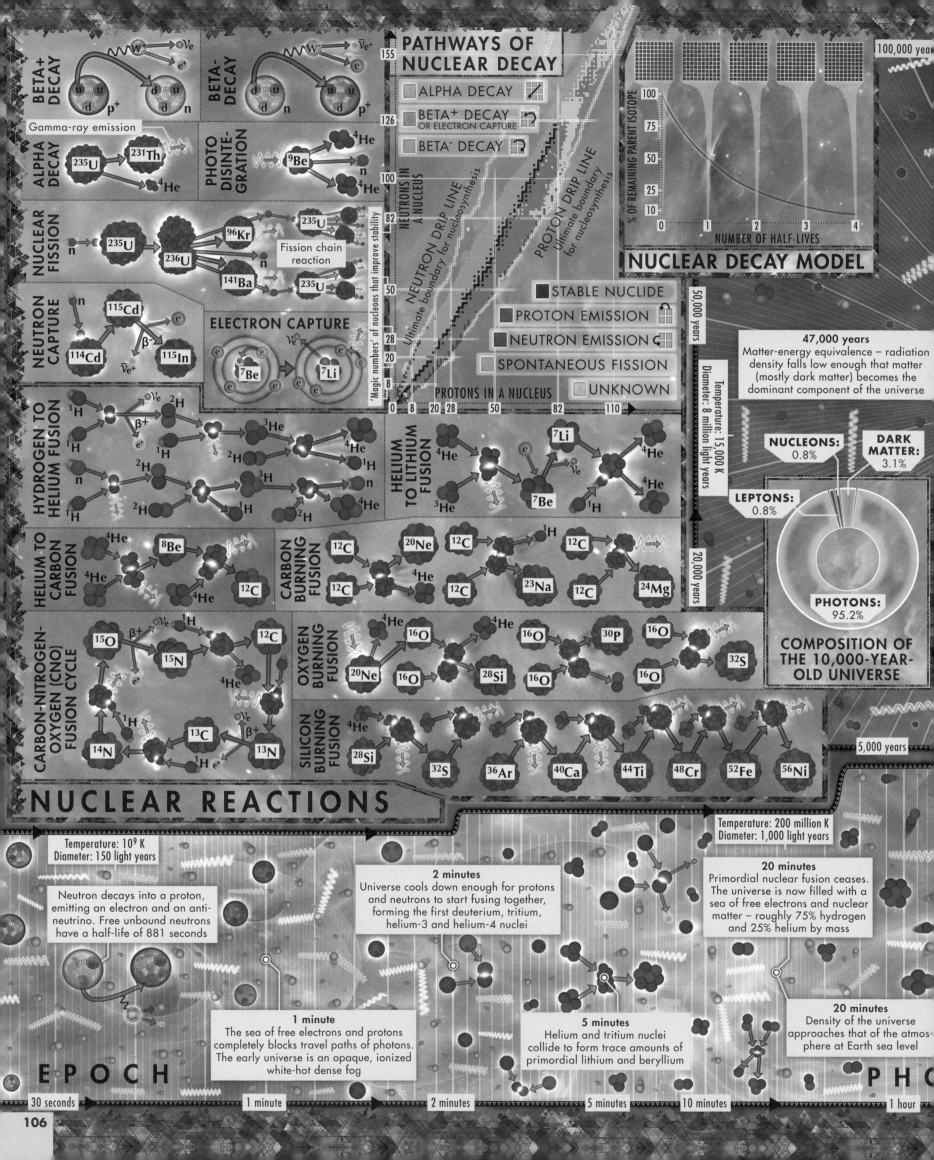

PATHWAYS OF NUCLEAR DECAY

BETA+ DECAY
u u d p⁺ → d u d n
W⁺ → ν_e → e⁺

BETA⁻ DECAY
d u d n → u u d p⁺
W⁻ → $\bar{\nu}_e$ → e⁻

Gamma-ray emission

ALPHA DECAY
²³⁵U → ²³¹Th
⁴He

PHOTO-DISINTEGRATION
⁹Be → ⁴He + n + ⁴He

NUCLEAR FISSION
n → ²³⁵U → ²³⁶U → ⁹⁶Kr → ²³⁵U
n
¹⁴¹Ba → ²³⁵U
Fission chain reaction

NEUTRON CAPTURE
n → ¹¹⁵Cd
¹¹⁴Cd → β⁻ → ¹¹⁵In
$\bar{\nu}_e$⁺
e

ELECTRON CAPTURE
ν_e
e e ⁷Be → ⁷Li e
e e

ALPHA DECAY
BETA⁺ DECAY OR ELECTRON CAPTURE
BETA⁻ DECAY

NEUTRON DRIP LINE — Ultimate boundary for nucleosynthesis
PROTON DRIP LINE — Ultimate boundary for nucleosynthesis
NEUTRONS IN A NUCLEUS
'Magic numbers' of nucleons that improve stability
155 126 100 82 50 28 20 8

STABLE NUCLIDE
PROTON EMISSION
NEUTRON EMISSION
SPONTANEOUS FISSION
UNKNOWN

PROTONS IN A NUCLEUS
0 8 20 28 50 82 110

HYDROGEN TO HELIUM FUSION
¹H ¹H → ν_e ²H
β⁺ e⁺
¹H ¹H → ²H
n
²H → ³He → ⁴He ¹H
²H → ³H n
²H → ¹H ⁴He

HELIUM TO LITHIUM FUSION
⁴He → ⁷Li → ⁴He
e
ν_e
³He → ⁷Be ¹H → ⁴He

HELIUM TO CARBON FUSION
⁴He → ⁸Be → ¹²C
⁴He ⁴He

CARBON BURNING FUSION
¹²C → ²⁰Ne → ¹²C → ¹H → ¹²C
¹²C ⁴He ¹²C ²³Na ¹²C ²⁴Mg

CARBON-NITROGEN-OXYGEN (CNO) FUSION CYCLE
¹⁵O → β⁺ → ν_e ¹H → ¹²C
¹⁵N e⁺ ⁴He
¹⁴N → ¹³C → β⁺ → ν_e
¹H → ¹³N
¹H e⁺

OXYGEN BURNING FUSION
⁴He ⁴He
¹⁶O → ¹⁶O → ³⁰P → ¹⁶O
²⁰Ne ¹⁶O ²⁸Si ¹⁶O ¹⁶O ³²S

SILICON BURNING FUSION
⁴He
²⁸Si → ³²S → ³⁶Ar → ⁴⁰Ca → ⁴⁴Ti → ⁴⁸Cr → ⁵²Fe → ⁵⁶Ni

NUCLEAR REACTIONS

NUCLEAR DECAY MODEL

% OF REMAINING PARENT ISOTOPE
100 75 50 25 10
NUMBER OF HALF-LIVES
0 1 2 3 4

100,000 years

47,000 years
Matter-energy equivalence – radiation density falls low enough that matter (mostly dark matter) becomes the dominant component of the universe

NUCLEONS: 0.8%
DARK MATTER: 3.1%
LEPTONS: 0.8%
PHOTONS: 95.2%

COMPOSITION OF THE 10,000-YEAR-OLD UNIVERSE

Temperature: 15,000 K
Diameter: 8 million light years

50,000 years
20,000 years
5,000 years

EPOCH

Temperature: 10⁹ K
Diameter: 150 light years

Neutron decays into a proton, emitting an electron and an anti-neutrino. Free unbound neutrons have a half-life of 881 seconds

1 minute
The sea of free electrons and protons completely blocks travel paths of photons. The early universe is an opaque, ionized white-hot dense fog

2 minutes
Universe cools down enough for protons and neutrons to start fusing together, forming the first deuterium, tritium, helium-3 and helium-4 nuclei

5 minutes
Helium and tritium nuclei collide to form trace amounts of primordial lithium and beryllium

Temperature: 200 million K
Diameter: 1,000 light years

20 minutes
Primordial nuclear fusion ceases. The universe is now filled with a sea of free electrons and nuclear matter – roughly 75% hydrogen and 25% helium by mass

20 minutes
Density of the universe approaches that of the atmosphere at Earth sea level

PHO

30 seconds | 1 minute | 2 minutes | 5 minutes | 10 minutes | 1 hour

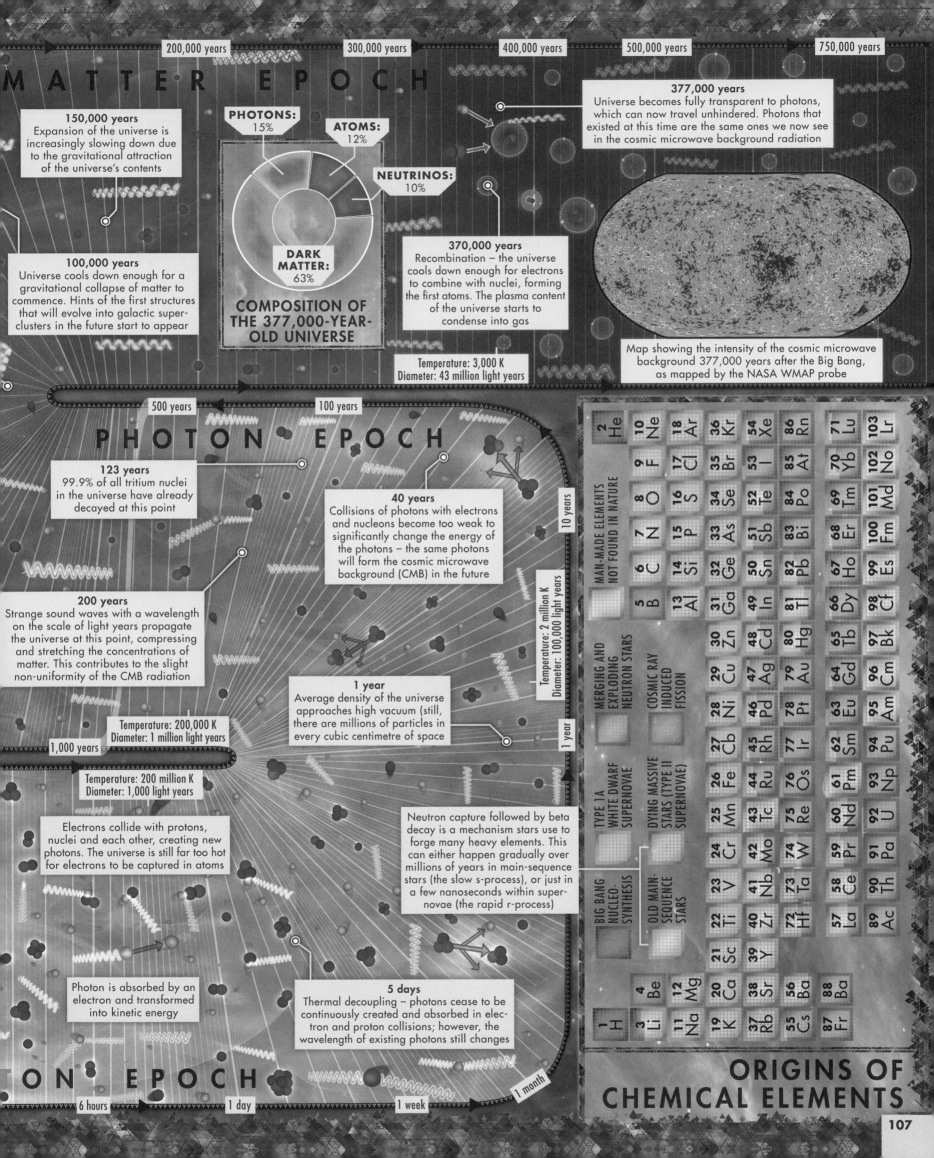

MATTER EPOCH

150,000 years
Expansion of the universe is increasingly slowing down due to the gravitational attraction of the universe's contents

100,000 years
Universe cools down enough for a gravitational collapse of matter to commence. Hints of the first structures that will evolve into galactic super-clusters in the future start to appear

PHOTONS: 15%

ATOMS: 12%

NEUTRINOS: 10%

DARK MATTER: 63%

COMPOSITION OF THE 377,000-YEAR-OLD UNIVERSE

370,000 years
Recombination – the universe cools down enough for electrons to combine with nuclei, forming the first atoms. The plasma content of the universe starts to condense into gas

Temperature: 3,000 K
Diameter: 43 million light years

377,000 years
Universe becomes fully transparent to photons, which can now travel unhindered. Photons that existed at this time are the same ones we now see in the cosmic microwave background radiation

Map showing the intensity of the cosmic microwave background 377,000 years after the Big Bang, as mapped by the NASA WMAP probe

PHOTON EPOCH

123 years
99.9% of all tritium nuclei in the universe have already decayed at this point

200 years
Strange sound waves with a wavelength on the scale of light years propagate the universe at this point, compressing and stretching the concentrations of matter. This contributes to the slight non-uniformity of the CMB radiation

Temperature: 200,000 K
Diameter: 1 million light years

40 years
Collisions of photons with electrons and nucleons become too weak to significantly change the energy of the photons – the same photons will form the cosmic microwave background (CMB) in the future

Temperature: 2 million K
Diameter: 100,000 light years

1 year
Average density of the universe approaches high vacuum (still, there are millions of particles in every cubic centimetre of space)

10 years

1 year

1,000 years

Temperature: 200 million K
Diameter: 1,000 light years

Electrons collide with protons, nuclei and each other, creating new photons. The universe is still far too hot for electrons to be captured in atoms

Neutron capture followed by beta decay is a mechanism stars use to forge many heavy elements. This can either happen gradually over millions of years in main-sequence stars (the slow s-process), or just in a few nanoseconds within super-novae (the rapid r-process)

Photon is absorbed by an electron and transformed into kinetic energy

5 days
Thermal decoupling – photons cease to be continuously created and absorbed in elec-tron and proton collisions; however, the wavelength of existing photons still changes

...ON EPOCH

ORIGINS OF CHEMICAL ELEMENTS

MAN-MADE ELEMENTS NOT FOUND IN NATURE

MERGING AND EXPLODING NEUTRON STARS

COSMIC RAY INDUCED FISSION

TYPE 1A WHITE DWARF SUPERNOVAE

DYING MASSIVE STARS (TYPE II SUPERNOVAE)

BIG BANG NUCLEO-SYNTHESIS

OLD MAIN-SEQUENCE STARS

								2 He
							10 Ne	
							18 Ar	
							36 Kr	
							54 Xe	
							86 Rn	
							71 Lu	
							103 Lr	

9 F, 17 Cl, 35 Br, 53 I, 85 At, 70 Yb, 102 No

8 O, 16 S, 34 Se, 52 Te, 84 Po, 69 Tm, 101 Md

7 N, 15 P, 33 As, 51 Sb, 83 Bi, 68 Er, 100 Fm

6 C, 14 Si, 32 Ge, 50 Sn, 82 Pb, 67 Ho, 99 Es

5 B, 13 Al, 31 Ga, 49 In, 81 Tl, 66 Dy, 98 Cf

30 Zn, 48 Cd, 80 Hg, 65 Tb, 97 Bk

29 Cu, 47 Ag, 79 Au, 64 Gd, 96 Cm

28 Ni, 46 Pd, 78 Pt, 63 Eu, 95 Am

27 Cb, 45 Rh, 77 Ir, 62 Sm, 94 Pu

26 Fe, 44 Ru, 76 Os, 61 Pm, 93 Np

25 Mn, 43 Tc, 75 Re, 60 Nd, 92 U

24 Cr, 42 Mo, 74 W, 59 Pr, 91 Pa

23 V, 41 Nb, 73 Ta, 58 Ce, 90 Th

22 Ti, 40 Zr, 72 Hf, 57 La, 89 Ac

21 Sc, 39 Y

4 Be, 12 Mg, 20 Ca, 38 Sr, 56 Ba, 88 Ba

1 H, 3 Li, 11 Na, 19 K, 37 Rb, 55 Cs, 87 Fr

107

DARK AGES

At this time, photons of the cosmic background are still spectrally red visible light. As space itself expands, however, so does the wavelength of photons travelling across it. By the time those photons reach us, they will be stretched by a factor of over 1,000, putting them into the microwave part of the light spectrum

3 million years
Cosmic background radiation shifts beyond the visible light range, plunging the universe into more than 100 million years of darkness

Spin of the electron flips to cancel out that of the proton

Temperature: 1,300 K
Diameter: 180 million light years

Hydrogen atoms transition into a lower, more stable energy state, emitting a 21 cm radio wavelength photon. This radiation of the so-called 'hydrogen line' permeates the entire universe. Detecting these red-shifted hydrogen line photons is the only way we can explore the 'dark ages' of the cosmos

GALACTIC EVOLUTION

STREAMING GAS MODEL

DARK MATTER HALO MERGER MODEL

Filaments of cold gas are gravitationally pulled into yet indistinct dark matter haloes (top-down galaxy formation)

Distinct bubbles of dark matter (haloes) filled with gas collide and merge with each other (bottom-up galaxy formation)

Temperature: 400 K
Diameter: 660 million light years

Flowing gas heats up, compresses and condenses into the very first stars, as distinct galaxies take shape

Colliding gas clouds collapse into the first stars, as clusters of protogalaxies form

PROTOGALAXIES

Protogalaxies collide with each other, triggering a widespread formation of new stars. Central galactic black holes are also created during this process

A spiral galaxy eventually forms after a number of young galaxies coalesce. Central black holes of those galaxies merge with each other as well, absorb gas and gradually grow to many millions of solar masses

Two galaxies violently interact with each other until they eventually merge. It's highly unlikely that any two stars ever collide during this event, however, the orbits of stars within both galaxies are violently disrupted

'Starburst galaxy' – disturbed clouds of interstellar gas start to collapse and form stars at a much higher rate than normal, rapidly depleting the galaxy of star-forming gas

Large galaxies, such as the Milky Way, often 'cannibalize' smaller galaxies in their neighbourhood. They grow larger, and use the gas to continue new star formation

Active 'radio galaxy' – remaining interstellar gas falls into the quasar in the centre of the galaxy, which emits extremely bright jets

As stars form and die, all galaxies eventually become depleted of star-forming gas. This process usually proceeds from the inside out. The core bulge region of most galaxies already contains little gas and mostly old red stars; and as it gradually expands, new star formation grinds to a halt

Elliptical galaxy depleted of interstellar gas, populated almost entirely by red dwarfs, old yellow stars and stellar remnants

100 million years
First galaxies start to take shape as the dark matter structures along with their gas contract. Heating of this gas produces energetic UV light photons that start to knock electrons out from atoms, causing the gas to transform into plasma. Universe begins to light up again as the millions of years of darkness come to a close

150 million years
Gas clouds within protogalaxies begin to contract into dense, gravitationally bound protostellar globules

D A R K   A G E S

100 million years
120 million years
140 million years
160 million years

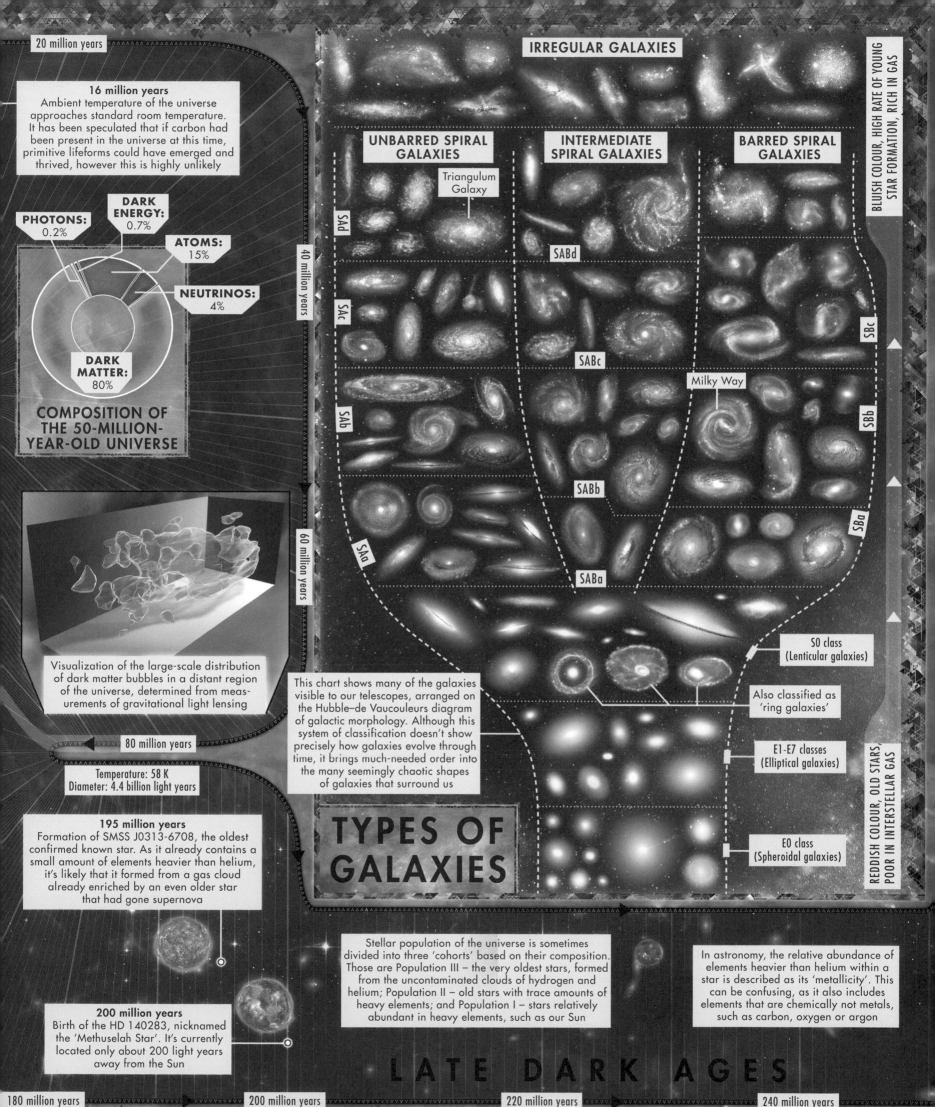

16 million years
Ambient temperature of the universe approaches standard room temperature. It has been speculated that if carbon had been present in the universe at this time, primitive lifeforms could have emerged and thrived, however this is highly unlikely

PHOTONS: 0.2%

DARK ENERGY: 0.7%

ATOMS: 15%

NEUTRINOS: 4%

DARK MATTER: 80%

COMPOSITION OF THE 50-MILLION-YEAR-OLD UNIVERSE

Visualization of the large-scale distribution of dark matter bubbles in a distant region of the universe, determined from measurements of gravitational light lensing

80 million years

Temperature: 58 K
Diameter: 4.4 billion light years

195 million years
Formation of SMSS J0313-6708, the oldest confirmed known star. As it already contains a small amount of elements heavier than helium, it's likely that it formed from a gas cloud already enriched by an even older star that had gone supernova

200 million years
Birth of the HD 140283, nicknamed the 'Methuselah Star'. It's currently located only about 200 light years away from the Sun

IRREGULAR GALAXIES

UNBARRED SPIRAL GALAXIES

Triangulum Galaxy

SAd

SAc

SAb

SAa

INTERMEDIATE SPIRAL GALAXIES

SABd

SABc

SABb

SABa

BARRED SPIRAL GALAXIES

Milky Way

SBc

SBb

SBa

BLUISH COLOUR, HIGH RATE OF YOUNG STAR FORMATION, RICH IN GAS

S0 class (Lenticular galaxies)

Also classified as 'ring galaxies'

E1-E7 classes (Elliptical galaxies)

E0 class (Spheroidal galaxies)

REDDISH COLOUR, OLD STARS, POOR IN INTERSTELLAR GAS

This chart shows many of the galaxies visible to our telescopes, arranged on the Hubble–de Vaucouleurs diagram of galactic morphology. Although this system of classification doesn't show precisely how galaxies evolve through time, it brings much-needed order into the many seemingly chaotic shapes of galaxies that surround us

TYPES OF GALAXIES

Stellar population of the universe is sometimes divided into three 'cohorts' based on their composition. Those are Population III – the very oldest stars, formed from the uncontaminated clouds of hydrogen and helium; Population II – old stars with trace amounts of heavy elements; and Population I – stars relatively abundant in heavy elements, such as our Sun

In astronomy, the relative abundance of elements heavier than helium within a star is described as its 'metallicity'. This can be confusing, as it also includes elements that are chemically not metals, such as carbon, oxygen or argon

LATE DARK AGES

GALACTIC EPOCH

1.5 billion years
Formation of the 'Baby Boom Galaxy', an extremely bright starburst galaxy, with the most intense star production ever recorded. Young stars are formed there at a rate of one every 2.2 hours, while the Milky Way produces just 10 new stars annually

1.7 billion years
Parent galaxy of the Methuselah Star (one of the oldest known stars), is devoured by the young and growing Milky Way

Due to extreme gravitational lensing, a flat accretion disk around a black hole would appear significantly distorted, especially near the event horizon

1.6 billion years
GRB 080916C – the most energetic gamma-ray burst ever recorded, equivalent to 8,000 normal supernovae. It lasted for over 23 minutes (average duration for gamma-ray bursts is only about 2 seconds). The cause of this event remains shrouded in mystery

1.3 billion years
Formation of the planet PSR J1719-1438 b that orbits around a milisecond pulsar. It weighs more than Jupiter, but is composed mostly of carbon, a lot of it in exotic, ultra-dense forms of diamond

'Smiley galaxy', distorted into its unique shape by a gravitational lens

Image of the galaxy as observed on Earth

Distant galaxy

Massive cluster of galaxies

Milky Way

Halo of dark matter

As predicted by Einstein's theory of relativity, massive objects cause space-time around them to curve – and any light passing through that curved region is bent as well. This causes the effect of 'gravitational lensing' that causes objects behind the gravity lens to appear as multiple images, sometimes as so-called 'Einstein rings'

GRAVITATIONAL LENSING

1 billion years
End of the reionization, as the expansion of the universe has caused the matter to become too diffuse. Most of the interstellar plasma starts to convert back into gas, and the universe remains almost completely transparent from this point on

1 billion years
Birth of the oldest known exoplanet, gas giant PSR B1620-26 b, inside the globular cluster Messier 4. It now orbits around a binary star system composed of a neutron star pulsar and a white dwarf

950 million years
Formation of the hyperluminous quasar SDSS J0100+2802 with a central black hole of over 12 billion solar masses. Its light output is over 40,000 times higher than the entire Milky Way galaxy

Hypothetical 'quasi-star' – a massive Population III star drawing its energy not from nuclear fusion, but from the heating of gas falling into the black hole in its core. Such stars would be over 1,000 times more massive than the Sun, and over 60 AU in diameter

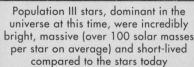

Population III stars, dominant in the universe at this time, were incredibly bright, massive (over 100 solar masses per star on average) and short-lived compared to the stars today

380 million years
Formation of UDFj-39546284, the oldest known quasar (active galactic supermassive black hole surrounded by a luminous accretion disc)

300 million years
It's likely that the first gas giant planets in the universe formed around this time. It's probable that most of them were eventually ejected from their orbits, some of them surviving as frigid rogue planets in the interstellar space to this day

Lifespan of a quasi-star would be limited to only about 7 million years, after which the black hole would have grown to thousands of solar masses and absorbed the star entirely. This has been suggested as the way many of the supermassive black holes in the centres of galaxies originated

2 billion years
SN 1000+0216, the oldest known supernova. Its brightness far exceeded the luminosity of its own galaxy, and was detectable on Earth from 2006 to 2008. The original star was probably up to 250 times more massive than the Sun

2.4 billion years
Omega Centauri forms the largest of the many globular clusters orbiting the Milky Way. It currently contains over 10 million stars, most of them much dimmer and older compared to the Sun

2.6 billion years
Formation of the Kepler-444 binary star system. One of the stars hosts a system of at least five rocky exoplanets, all of them smaller than the Earth. They orbit extremely close to their star, and are far too hot to be habitable

2 billion years
Globular cluster Messier 15, known to have a black hole of over 1,500 solar masses, forms in the Milky Way's dark matter halo

2.2 billion years
Formation of WD 0346+246, the oldest known white dwarf. Originally extremely hot, it has since cooled to less than 3,000 K, making it quite dim and reddish in colour

COMPOSITION OF THE 2.5-BILLION-YEAR-OLD UNIVERSE

DARK ENERGY: 20.5%

DARK MATTER: 67%

FREE HYDROGEN AND HELIUM: 11%

HYDROGEN AND HELIUM IN STARS: 0.5%

PHOTONS: 0.05%

ALL HEAVIER ELEMENTS: 0.01%

NEUTRINOS: 1%

Temperature: 10 K
Diameter: 25 billion light years

900 million years
Large-scale structures such as galactic superclusters and galactic groups begin to appear as galaxies and are attracted together, forming a complex network of nodes and filaments that connect them

800 million years
Metal-enriched Population II stars begin to form at an increased rate, as interstellar star-forming clouds are showered by the ejecta from many consecutive supernovae and neutron star mergers

725 million years
Oldest known globular clusters – gravitationally bound spherical systems of stars located outside of the main disc of their galaxy, in its halo. The present-day Milky Way has over 150 globular clusters orbiting around it – within them, some of the oldest known stars can be found

C O S M I C   R E N A I S S A N C E

700 million years

850 million years
Formation of the Milky Way's galactic halo. However, it will take several billion years for its gas and stars to form into a flattened disk, and even longer for its signature spiral arms to fully develop

760 million years
Starburst galaxy GN-108036 forms. Even though it is over 100 times less massive than the Milky Way, more than 30 times more young stars are born within it every year

670 million years
EGS-zs8-1 – the oldest known starburst galaxy. It probably formed after two smaller young protogalaxies collided

650 million years

Temperature: 21 K
Diameter: 12 billion light years

Temperature: 33 K
Diameter: 7.5 billion light years

650 million years
Very high rate of galaxy formation, as dark matter haloes and gas they contain contract and become distinct objects

400 million years
The oldest and the most distant confirmed galaxy, GN-z11, forms

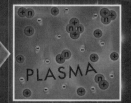

GAS → PLASMA

The Hubble Ultra Deep Field shows the universe stretching back in time as far as just 500 million years after the Big Bang. The area of the sky shown in this image is over 10 times smaller than the full Moon, and every dot represents an individual galaxy

450 million years
Reionization of the universe is complete, as energetic photons from quasars and young stars transform the entirety of interstellar gas into plasma. The universe becomes mostly opaque again

600 million years

600 million years
GRB 090423 – the oldest known gamma-ray burst. These are among the most energetic events in the universe, and are believed to be jets of extremely energetic photons released during a supernova. If such a gamma-ray burst anywhere in the Milky Way pointed directly at the solar system, it would probably cause a mass extinction event

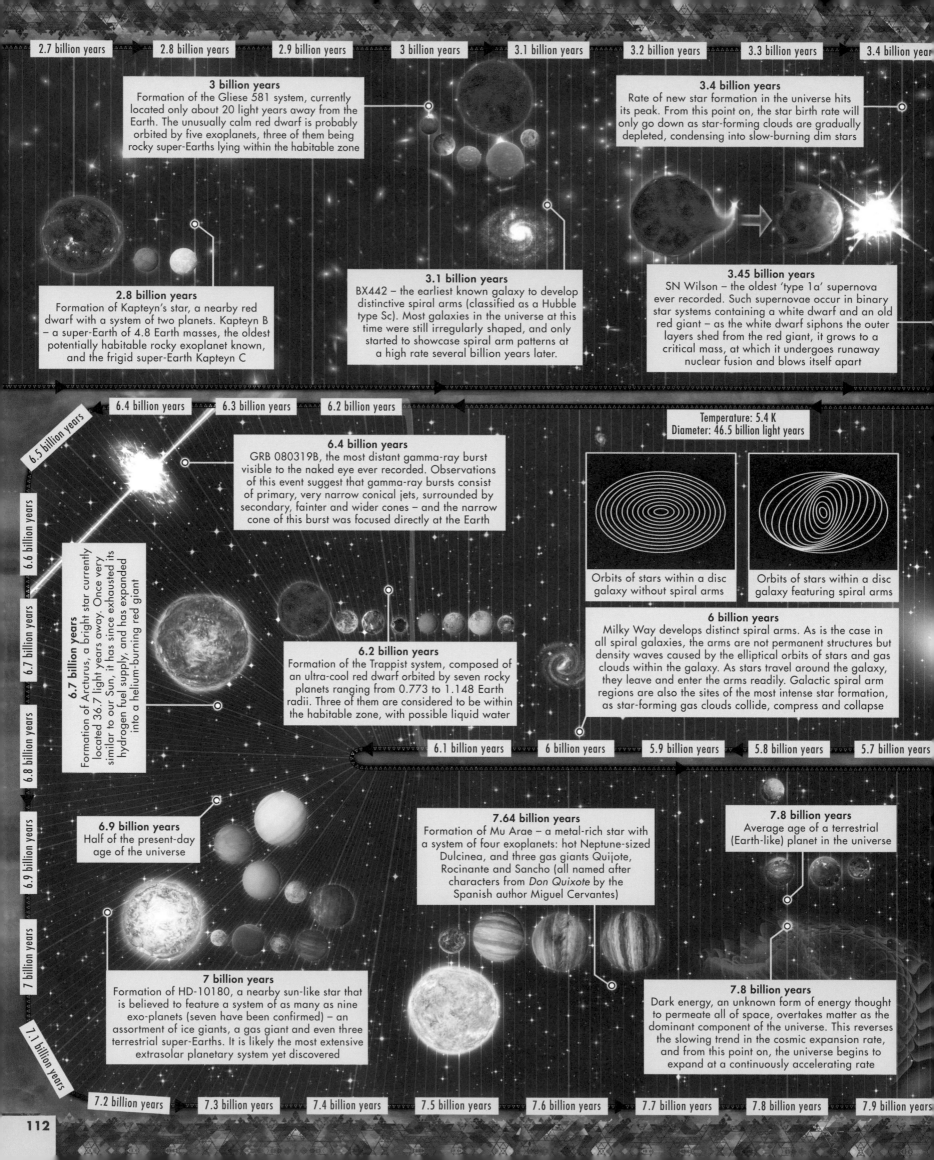

3 billion years
Formation of the Gliese 581 system, currently located only about 20 light years away from the Earth. The unusually calm red dwarf is probably orbited by five exoplanets, three of them being rocky super-Earths lying within the habitable zone

3.4 billion years
Rate of new star formation in the universe hits its peak. From this point on, the star birth rate will only go down as star-forming clouds are gradually depleted, condensing into slow-burning dim stars

2.8 billion years
Formation of Kapteyn's star, a nearby red dwarf with a system of two planets. Kapteyn B – a super-Earth of 4.8 Earth masses, the oldest potentially habitable rocky exoplanet known, and the frigid super-Earth Kapteyn C

3.1 billion years
BX442 – the earliest known galaxy to develop distinctive spiral arms (classified as a Hubble type Sc). Most galaxies in the universe at this time were still irregularly shaped, and only started to showcase spiral arm patterns at a high rate several billion years later.

3.45 billion years
SN Wilson – the oldest 'type 1a' supernova ever recorded. Such supernovae occur in binary star systems containing a white dwarf and an old red giant – as the white dwarf siphons the outer layers shed from the red giant, it grows to a critical mass, at which it undergoes runaway nuclear fusion and blows itself apart

◂ 6.4 billion years ◂ 6.3 billion years ◂ 6.2 billion years
6.5 billion years
6.6 billion years
6.7 billion years
6.8 billion years
6.9 billion years
7 billion years
7.1 billion years

Temperature: 5.4 K
Diameter: 46.5 billion light years

6.4 billion years
GRB 080319B, the most distant gamma-ray burst visible to the naked eye ever recorded. Observations of this event suggest that gamma-ray bursts consist of primary, very narrow conical jets, surrounded by secondary, fainter and wider cones – and the narrow cone of this burst was focused directly at the Earth

Orbits of stars within a disc galaxy without spiral arms

Orbits of stars within a disc galaxy featuring spiral arms

6 billion years
Milky Way develops distinct spiral arms. As is the case in all spiral galaxies, the arms are not permanent structures but density waves caused by the elliptical orbits of stars and gas clouds within the galaxy. As stars travel around the galaxy, they leave and enter the arms readily. Galactic spiral arm regions are also the sites of the most intense star formation, as star-forming gas clouds collide, compress and collapse

6.7 billion years
Formation of Arcturus, a bright star currently located 36.7 light years away. Once very similar to our Sun, it has since exhausted its hydrogen fuel supply, and has expanded into a helium-burning red giant

6.2 billion years
Formation of the Trappist system, composed of an ultra-cool red dwarf orbited by seven rocky planets ranging from 0.773 to 1.148 Earth radii. Three of them are considered to be within the habitable zone, with possible liquid water

6.9 billion years
Half of the present-day age of the universe

7.64 billion years
Formation of Mu Arae – a metal-rich star with a system of four exoplanets: hot Neptune-sized Dulcinea, and three gas giants Quijote, Rocinante and Sancho (all named after characters from *Don Quixote* by the Spanish author Miguel Cervantes)

7.8 billion years
Average age of a terrestrial (Earth-like) planet in the universe

7 billion years
Formation of HD-10180, a nearby sun-like star that is believed to feature a system of as many as nine exo-planets (seven have been confirmed) – an assortment of ice giants, a gas giant and even three terrestrial super-Earths. It is likely the most extensive extrasolar planetary system yet discovered

7.8 billion years
Dark energy, an unknown form of energy thought to permeate all of space, overtakes matter as the dominant component of the universe. This reverses the slowing trend in the cosmic expansion rate, and from this point on, the universe begins to expand at a continuously accelerating rate

7.2 billion years ▸ 7.3 billion years ▸ 7.4 billion years ▸ 7.5 billion years ▸ 7.6 billion years ▸ 7.7 billion years ▸ 7.8 billion years ▸ 7.9 billion years

112

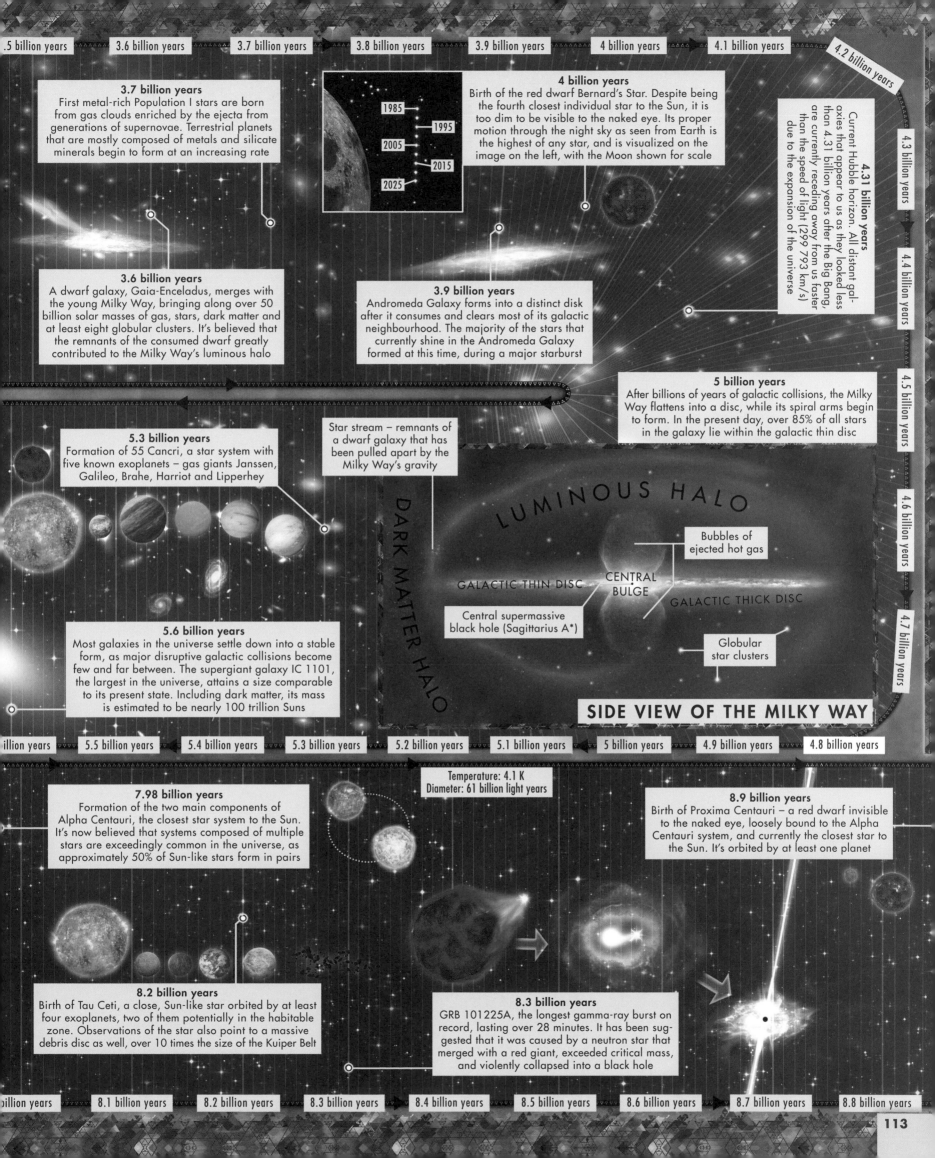

3.7 billion years
First metal-rich Population I stars are born from gas clouds enriched by the ejecta from generations of supernovae. Terrestrial planets that are mostly composed of metals and silicate minerals begin to form at an increasing rate

1985 1995 2005 2015 2025

4 billion years
Birth of the red dwarf Bernard's Star. Despite being the fourth closest individual star to the Sun, it is too dim to be visible to the naked eye. Its proper motion through the night sky as seen from Earth is the highest of any star, and is visualized on the image on the left, with the Moon shown for scale

4.31 billion years
Current Hubble horizon. All distant galaxies that appear to us as they looked less than 4.31 billion years after the Big Bang, are currently receding away from us faster than the speed of light (299 793 km/s) due to the expansion of the universe

3.6 billion years
A dwarf galaxy, Gaia-Enceladus, merges with the young Milky Way, bringing along over 50 billion solar masses of gas, stars, dark matter and at least eight globular clusters. It's believed that the remnants of the consumed dwarf greatly contributed to the Milky Way's luminous halo

3.9 billion years
Andromeda Galaxy forms into a distinct disk after it consumes and clears most of its galactic neighbourhood. The majority of the stars that currently shine in the Andromeda Galaxy formed at this time, during a major starburst

5 billion years
After billions of years of galactic collisions, the Milky Way flattens into a disc, while its spiral arms begin to form. In the present day, over 85% of all stars in the galaxy lie within the galactic thin disc

5.3 billion years
Formation of 55 Cancri, a star system with five known exoplanets – gas giants Janssen, Galileo, Brahe, Harriot and Lipperhey

Star stream – remnants of a dwarf galaxy that has been pulled apart by the Milky Way's gravity

LUMINOUS HALO

DARK MATTER HALO

GALACTIC THIN DISC

CENTRAL BULGE

Bubbles of ejected hot gas

GALACTIC THICK DISC

Central supermassive black hole (Sagittarius A*)

Globular star clusters

SIDE VIEW OF THE MILKY WAY

5.6 billion years
Most galaxies in the universe settle down into a stable form, as major disruptive galactic collisions become few and far between. The supergiant galaxy IC 1101, the largest in the universe, attains a size comparable to its present state. Including dark matter, its mass is estimated to be nearly 100 trillion Suns

7.98 billion years
Formation of the two main components of Alpha Centauri, the closest star system to the Sun. It's now believed that systems composed of multiple stars are exceedingly common in the universe, as approximately 50% of Sun-like stars form in pairs

Temperature: 4.1 K
Diameter: 61 billion light years

8.9 billion years
Birth of Proxima Centauri – a red dwarf invisible to the naked eye, loosely bound to the Alpha Centauri system, and currently the closest star to the Sun. It's orbited by at least one planet

8.2 billion years
Birth of Tau Ceti, a close, Sun-like star orbited by at least four exoplanets, two of them potentially in the habitable zone. Observations of the star also point to a massive debris disc as well, over 10 times the size of the Kuiper Belt

8.3 billion years
GRB 101225A, the longest gamma-ray burst on record, lasting over 28 minutes. It has been suggested that it was caused by a neutron star that merged with a red giant, exceeded critical mass, and violently collapsed into a black hole

FORMATION AND EVOLUTION OF THE SOLAR SYSTEM

CONTRACTING FRAGMENT OF A MOLECULAR CLOUD

PROTOPLANETARY DISC

Proto-Sun begins to form and shine, as the kinetic energy of infalling dust and gas is converted into heat and light

Radiance of the proto-Sun drives gases and volatiles away from the inner solar system, leaving only solid grains behind – the protoplanetary disc differentiates

In this 'T-Tauri' stage, the Sun is over five times more luminous than it is now

4.53 BILLION YEARS AGO

Four rocky planets of the inner solar system form relatively close to their present-day orbits

Mercury
Venus
Earth
Mars

Giant planets clean up their orbital neighbourhoods

4.54 BILLION YEARS AGO

Protoplanets

Accretion discs around the solar system's giant planets will eventually form their many moons

Saturn
Planetesimals
Jupiter
Uranus
Neptune

4.56 BILLION YEARS AGO

Kilometre-sized planetesimals in the inner solar system gradually coalesce into dozens of protoplanets ranging from smaller than the Moon to over one mass of Mars

To-scale orbits of the four giant planets in the solar system

4 BILLION YEARS AGO

Most of the Kuiper Belt's mass is eventually scattered. Giant planets settle into a new orbital configuration, which more or less persists into the present day

The Sun enters the main-sequence stage, as stable nuclear fusion of hydrogen into helium commences within its core

4.5 BILLION YEARS AGO

A 1:2 orbital resonance develops between Jupiter and Saturn, causing Neptune to be ejected outwards. This sends the orbits of most Kuiper Belt planetesimals into chaos, and results in the 'Late Heavy Bombardment' of the inner solar system

3 BILLION YEARS AGO

Gravitational interactions of a supernova disrupt the molecular cloud, causing it to contract

The cloud collapses and fragments, as gas condenses into protostellar globules and new stars are born

Stellar wind and supernovae blow the remaining gas off, stopping the formation of new stars

9.2 billion years
A nearby supernova triggers the collapse of the pre-solar molecular cloud (estimated to be originally 65 light years across). It will eventually condense into hundreds of stars, and give rise to the young solar system

OPEN STAR CLUSTERS

Most stars, including our own Sun, have started their life in a so-called 'open star cluster'. These are relatively dense groups of thousands to tens of thousands of stars that formed from a single giant molecular gas cloud (a nebula dense and cold enough for atoms to combine into molecules). Unlike globular clusters, open clusters are only weakly gravitationally bound, and their stars usually disperse throughout the galaxy after 10–100 million years. Some well-known open clusters include the Pleiades and Hyades. Many bright nebulae, such as the Eagle Nebula and Orion Nebula, are currently undergoing collapse, and will transform into open clusters in a few million years

9.23 billion years
The Sun is born from a contracting fragment of a giant molecular cloud. A rotating protoplanetary disc of dust and gas forms around the bright protostar, which is still fuelled only by the kinetic energy of the infalling matter

8.9 billion years   9 billion years   9.1 billion years   9.15 billion years   9.2 billion years   9.25 billion ye

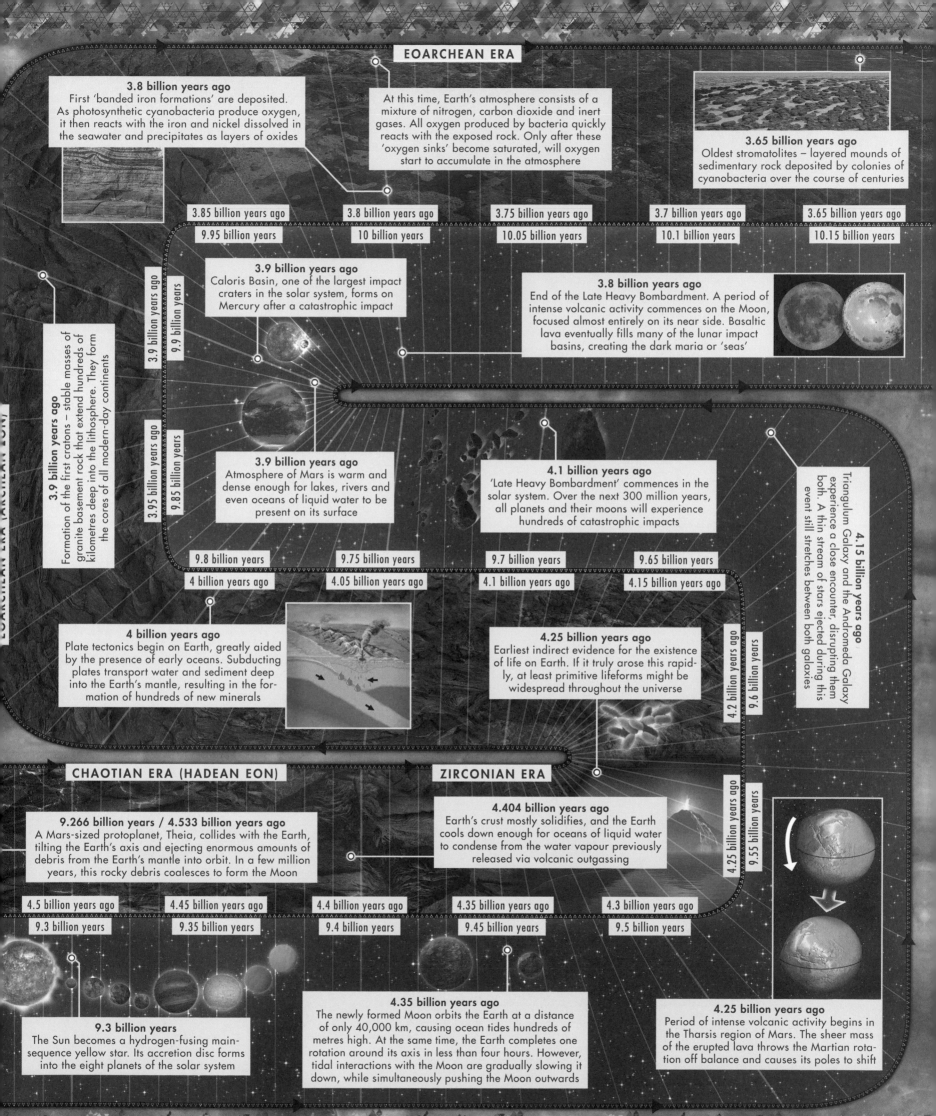

EOARCHEAN ERA

3.8 billion years ago
First 'banded iron formations' are deposited. As photosynthetic cyanobacteria produce oxygen, it then reacts with the iron and nickel dissolved in the seawater and precipitates as layers of oxides

At this time, Earth's atmosphere consists of a mixture of nitrogen, carbon dioxide and inert gases. All oxygen produced by bacteria quickly reacts with the exposed rock. Only after these 'oxygen sinks' become saturated, will oxygen start to accumulate in the atmosphere

3.65 billion years ago
Oldest stromatolites – layered mounds of sedimentary rock deposited by colonies of cyanobacteria over the course of centuries

| 3.85 billion years ago | 3.8 billion years ago | 3.75 billion years ago | 3.7 billion years ago | 3.65 billion years ago |
| 9.95 billion years | 10 billion years | 10.05 billion years | 10.1 billion years | 10.15 billion years |

3.9 billion years ago
Caloris Basin, one of the largest impact craters in the solar system, forms on Mercury after a catastrophic impact

3.8 billion years ago
End of the Late Heavy Bombardment. A period of intense volcanic activity commences on the Moon, focused almost entirely on its near side. Basaltic lava eventually fills many of the lunar impact basins, creating the dark maria or 'seas'

3.9 billion years ago / **9.9 billion years**

3.9 billion years ago
Formation of the first cratons – stable masses of granite basement rock that extend hundreds of kilometres deep into the lithosphere. They form the cores of all modern-day continents

3.95 billion years ago / **9.85 billion years**

3.9 billion years ago
Atmosphere of Mars is warm and dense enough for lakes, rivers and even oceans of liquid water to be present on its surface

4.1 billion years ago
'Late Heavy Bombardment' commences in the solar system. Over the next 300 million years, all planets and their moons will experience hundreds of catastrophic impacts

4.15 billion years ago
Triangulum Galaxy and the Andromeda Galaxy experience a close encounter, disrupting them both. A thin stream of stars ejected during this event still stretches between both galaxies

| 9.8 billion years | 9.75 billion years | 9.7 billion years | 9.65 billion years |
| 4 billion years ago | 4.05 billion years ago | 4.1 billion years ago | 4.15 billion years ago |

4 billion years ago
Plate tectonics begin on Earth, greatly aided by the presence of early oceans. Subducting plates transport water and sediment deep into the Earth's mantle, resulting in the formation of hundreds of new minerals

4.25 billion years ago
Earliest indirect evidence for the existence of life on Earth. If it truly arose this rapidly, at least primitive lifeforms might be widespread throughout the universe

4.2 billion years ago / **9.6 billion years**

CHAOTIAN ERA (HADEAN EON)

ZIRCONIAN ERA

9.266 billion years / 4.533 billion years ago
A Mars-sized protoplanet, Theia, collides with the Earth, tilting the Earth's axis and ejecting enormous amounts of debris from the Earth's mantle into orbit. In a few million years, this rocky debris coalesces to form the Moon

4.404 billion years ago
Earth's crust mostly solidifies, and the Earth cools down enough for oceans of liquid water to condense from the water vapour previously released via volcanic outgassing

4.25 billion years ago / **9.55 billion years**

| 4.5 billion years ago | 4.45 billion years ago | 4.4 billion years ago | 4.35 billion years ago | 4.3 billion years ago |
| 9.3 billion years | 9.35 billion years | 9.4 billion years | 9.45 billion years | 9.5 billion years |

9.3 billion years
The Sun becomes a hydrogen-fusing main-sequence yellow star. Its accretion disc forms into the eight planets of the solar system

4.35 billion years ago
The newly formed Moon orbits the Earth at a distance of only 40,000 km, causing ocean tides hundreds of metres high. At the same time, the Earth completes one rotation around its axis in less than four hours. However, tidal interactions with the Moon are gradually slowing it down, while simultaneously pushing the Moon outwards

4.25 billion years ago
Period of intense volcanic activity begins in the Tharsis region of Mars. The sheer mass of the erupted lava throws the Martian rotation off balance and causes its poles to shift

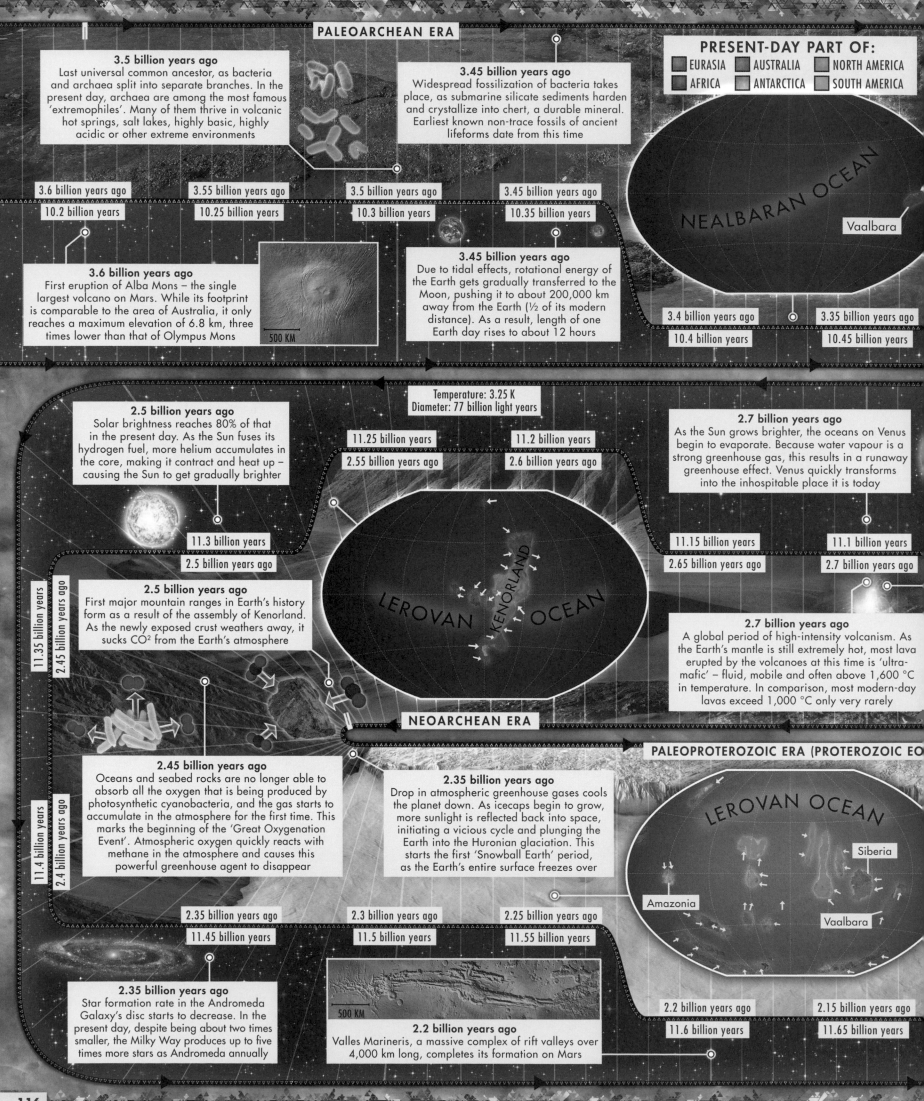

PRESENT-DAY PART OF:
- EURASIA
- AFRICA
- AUSTRALIA
- ANTARCTICA
- NORTH AMERICA
- SOUTH AMERICA

3.5 billion years ago
Last universal common ancestor, as bacteria and archaea split into separate branches. In the present day, archaea are among the most famous 'extremophiles'. Many of them thrive in volcanic hot springs, salt lakes, highly basic, highly acidic or other extreme environments

3.45 billion years ago
Widespread fossilization of bacteria takes place, as submarine silicate sediments harden and crystallize into chert, a durable mineral. Earliest known non-trace fossils of ancient lifeforms date from this time

NEALBARAN OCEAN

Vaalbara

| 3.6 billion years ago | 3.55 billion years ago | 3.5 billion years ago | 3.45 billion years ago |
| 10.2 billion years | 10.25 billion years | 10.3 billion years | 10.35 billion years |

3.6 billion years ago
First eruption of Alba Mons – the single largest volcano on Mars. While its footprint is comparable to the area of Australia, it only reaches a maximum elevation of 6.8 km, three times lower than that of Olympus Mons

500 KM

3.45 billion years ago
Due to tidal effects, rotational energy of the Earth gets gradually transferred to the Moon, pushing it to about 200,000 km away from the Earth (½ of its modern distance). As a result, length of one Earth day rises to about 12 hours

| 3.4 billion years ago | 3.35 billion years ago |
| 10.4 billion years | 10.45 billion years |

Temperature: 3.25 K
Diameter: 77 billion light years

2.5 billion years ago
Solar brightness reaches 80% of that in the present day. As the Sun fuses its hydrogen fuel, more helium accumulates in the core, making it contract and heat up – causing the Sun to get gradually brighter

| 11.25 billion years | 11.2 billion years |
| 2.55 billion years ago | 2.6 billion years ago |

2.7 billion years ago
As the Sun grows brighter, the oceans on Venus begin to evaporate. Because water vapour is a strong greenhouse gas, this results in a runaway greenhouse effect. Venus quickly transforms into the inhospitable place it is today

| 11.3 billion years |
| 2.5 billion years ago |

LEROVAN OCEAN

KENORLAND

| 11.15 billion years | 11.1 billion years |
| 2.65 billion years ago | 2.7 billion years ago |

2.5 billion years ago
First major mountain ranges in Earth's history form as a result of the assembly of Kenorland. As the newly exposed crust weathers away, it sucks CO_2 from the Earth's atmosphere

2.7 billion years ago
A global period of high-intensity volcanism. As the Earth's mantle is still extremely hot, most lava erupted by the volcanoes at this time is 'ultra-mafic' – fluid, mobile and often above 1,600 °C in temperature. In comparison, most modern-day lavas exceed 1,000 °C only very rarely

11.35 billion years
2.45 billion years ago

2.45 billion years ago
Oceans and seabed rocks are no longer able to absorb all the oxygen that is being produced by photosynthetic cyanobacteria, and the gas starts to accumulate in the atmosphere for the first time. This marks the beginning of the 'Great Oxygenation Event'. Atmospheric oxygen quickly reacts with methane in the atmosphere and causes this powerful greenhouse agent to disappear

2.35 billion years ago
Drop in atmospheric greenhouse gases cools the planet down. As icecaps begin to grow, more sunlight is reflected back into space, initiating a vicious cycle and plunging the Earth into the Huronian glaciation. This starts the first 'Snowball Earth' period, as the Earth's entire surface freezes over

LEROVAN OCEAN

Siberia

Amazonia

Vaalbara

11.4 billion years
2.4 billion years ago

| 2.35 billion years ago | 2.3 billion years ago | 2.25 billion years ago |
| 11.45 billion years | 11.5 billion years | 11.55 billion years |

2.35 billion years ago
Star formation rate in the Andromeda Galaxy's disc starts to decrease. In the present day, despite being about two times smaller, the Milky Way produces up to five times more stars as Andromeda annually

500 KM

2.2 billion years ago
Valles Marineris, a massive complex of rift valleys over 4,000 km long, completes its formation on Mars

| 2.2 billion years ago | 2.15 billion years ago |
| 11.6 billion years | 11.65 billion years |

3.33 billion years ago
Oldest major granite deposits form. In the molten basaltic magma of the mantle, lighter minerals slowly rise to the top and eventually crystallize into granite while still underground

3.25 billion years ago
A 58-km asteroid impacts the primordial supercontinent of Vaalbara. It leaves behind the oldest known crater, 400 km in diameter, remnants of which persist to the present day

3.1 billion years ago
As a result of the proximity of the young Moon, and because almost the entirety of the Earth is covered by oceans, any existing landmasses experience daily tides up to 100 m high, and are frequently pummelled by hurricane-force winds

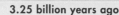

3.3 billion years ago	3.25 billion years ago	3.2 billion years ago	3.15 billion years ago	3.1 billion years ago
10.5 billion years	10.55 billion years	10.6 billion years	10.65 billion years	10.7 billion years

3.3 billion years ago
Intense volcanism on the Moon produces a thin atmosphere of carbon dioxide and water vapour, about as dense as that of modern-day Mars. However, in less than 70 million years, it completely escapes into space

3.1 billion years ago
As a result of the faintness of the young Sun, Venus probably still hosts oceans of liquid water and is potentially habitable. However, no evidence of any possible lifeforms remains, as the entirety of the planet has since undergone volcanic resurfacing several times

3.05 billion years ago

3 billion years ago
Large-scale formation of continental crust takes place. The most ancient parts of Siberia, Canada and the Baltic Shield, as well as the oldest Antarctic rocks, date to this time period. As new land-masses rise above the ocean surface, photosynthesizing cyanobacteria experience a population explosion in the shallow coastal waters

2.85 billion years ago
GW170104, a black hole merger, takes place, producing one of the most distant gravitational wave signals ever detected. During the collision, up to two solar masses were converted to energy, and radiated away as gravitational waves

3 billion years ago
After most of the atmosphere of Mars thins to its present density, its oceans either freeze or dry up, leaving it a cold, barren planet

3 billion years ago
10.8 billion years

11.05 billion years	11 billion years	10.95 billion years	10.8 billion years	10.85 billion years
2.75 billion years ago	2.8 billion years ago	2.85 billion years ago	2.9 billion years ago	2.95 billion years ago

2.707 billion years ago
Eruption of the Blake River Megacaldera, in present-day Ontario and Quebec. It is the oldest known supervolcanic eruption

2.9 billion years ago
Start of the Pongola glaciation – the first known frigid period in Earth's history. It is likely that it was triggered by the population explosion of cyanobacteria, and the subsequent rapid drop in atmospheric CO_2

2.1 billion years ago
Volcanism gradually recovers the levels of greenhouse gases in Earth's atmosphere, and the Huronian glaciation ends. During the subsequent global population explosion, it's believed that the first advanced unicellular organisms – eukaryotes – emerged

1.9 billion years ago
Mitochondria (originally a type of independent bacteria), get incorporated into eukaryotic cells and start a symbiotic relationship that still persists today. First bacteriophages (viruses attacking bacteria) emerge at this time as well

Proto-Laurentia

LEROVAN OCEAN

NENA

COLUMBIAN OCEAN

Vaalbara

Siberia

ATLANTICA

2.1 billion years ago	2.05 billion years ago	2 billion years ago	1.95 billion years ago
11.7 billion years	11.75 billion years	11.8 billion years	11.85 billion years

2.1 billion years ago
The solar system completes its tenth orbit around the centre of the Milky Way. One 'galactic year' – the time it takes for the Sun to orbit once around the galactic centre – is about 230 million years

2 billion years ago
Hoag's Object, a mysterious ring-shaped galaxy, attains its unique configuration, possibly as a result of a galactic collision

1.9 billion years ago	1.85 billion years ago
11.9 billion years	11.95 billion years

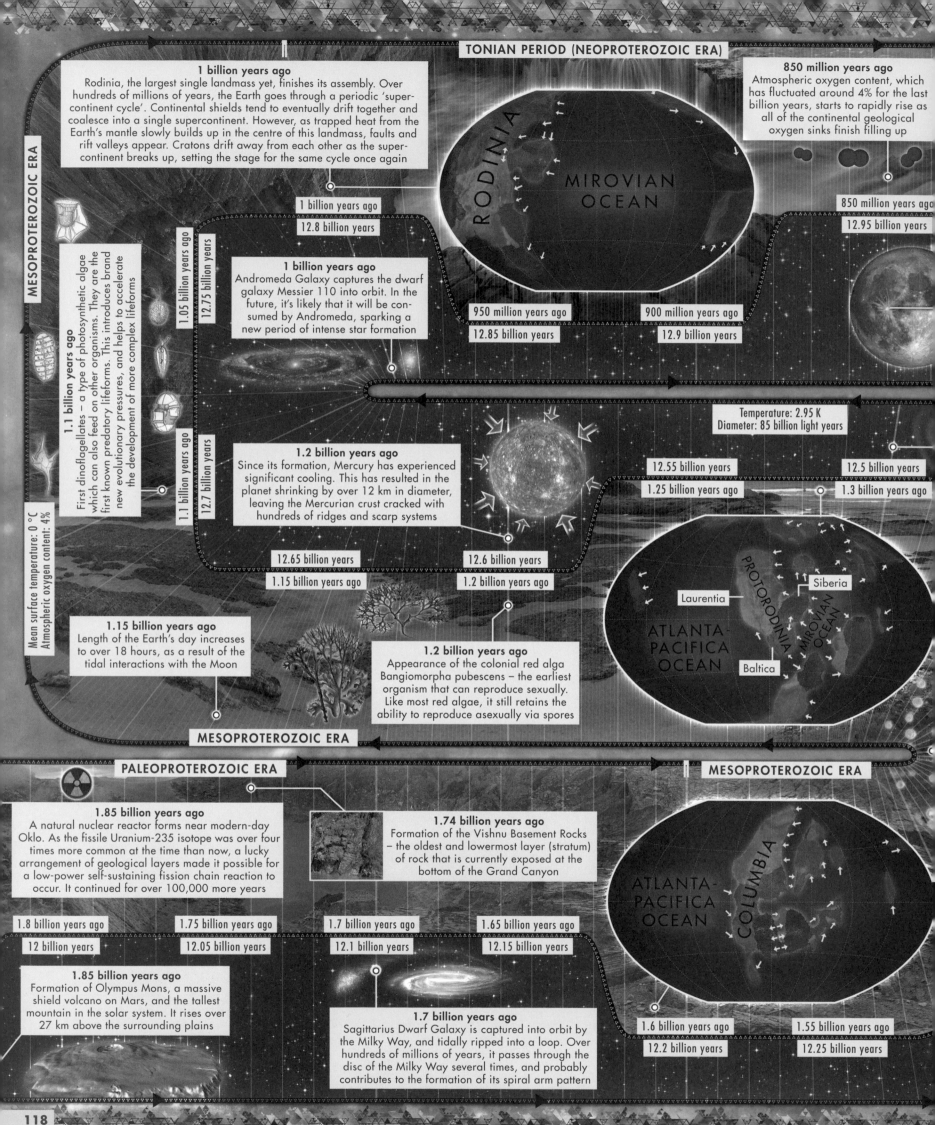

1 billion years ago
Rodinia, the largest single landmass yet, finishes its assembly. Over hundreds of millions of years, the Earth goes through a periodic 'super-continent cycle'. Continental shields tend to eventually drift together and coalesce into a single supercontinent. However, as trapped heat from the Earth's mantle slowly builds up in the centre of this landmass, faults and rift valleys appear. Cratons drift away from each other as the super-continent breaks up, setting the stage for the same cycle once again

850 million years ago
Atmospheric oxygen content, which has fluctuated around 4% for the last billion years, starts to rapidly rise as all of the continental geological oxygen sinks finish filling up

RODINIA

MIROVIAN OCEAN

1 billion years ago
12.8 billion years

850 million years ago
12.95 billion years

MESOPROTEROZOIC ERA

1.05 billion years ago
12.75 billion years

1 billion years ago
Andromeda Galaxy captures the dwarf galaxy Messier 110 into orbit. In the future, it's likely that it will be consumed by Andromeda, sparking a new period of intense star formation

950 million years ago
12.85 billion years

900 million years ago
12.9 billion years

1.1 billion years ago
First dinoflagellates – a type of photosynthetic algae which can also feed on other organisms. They are the first known predatory lifeforms. This introduces brand new evolutionary pressures, and helps to accelerate the development of more complex lifeforms

1.05 billion years ago
12.75 billion years

Temperature: 2.95 K
Diameter: 85 billion light years

1.1 billion years ago
12.7 billion years

1.2 billion years ago
Since its formation, Mercury has experienced significant cooling. This has resulted in the planet shrinking by over 12 km in diameter, leaving the Mercurian crust cracked with hundreds of ridges and scarp systems

12.55 billion years
1.25 billion years ago

12.5 billion years
1.3 billion years ago

Mean surface temperature: 0 °C
Atmospheric oxygen content: 4%

12.65 billion years
1.15 billion years ago

12.6 billion years
1.2 billion years ago

1.15 billion years ago
Length of the Earth's day increases to over 18 hours, as a result of the tidal interactions with the Moon

PROTORODINIA

Siberia

Laurentia

MIROVIAN OCEAN

ATLANTA-PACIFICA OCEAN

Baltica

1.2 billion years ago
Appearance of the colonial red alga Bangiomorpha pubescens – the earliest organism that can reproduce sexually. Like most red algae, it still retains the ability to reproduce asexually via spores

MESOPROTEROZOIC ERA

1.85 billion years ago
A natural nuclear reactor forms near modern-day Oklo. As the fissile Uranium-235 isotope was over four times more common at the time than now, a lucky arrangement of geological layers made it possible for a low-power self-sustaining fission chain reaction to occur. It continued for over 100,000 more years

1.74 billion years ago
Formation of the Vishnu Basement Rocks – the oldest and lowermost layer (stratum) of rock that is currently exposed at the bottom of the Grand Canyon

COLUMBIA

ATLANTA-PACIFICA OCEAN

1.8 billion years ago
12 billion years

1.75 billion years ago
12.05 billion years

1.7 billion years ago
12.1 billion years

1.65 billion years ago
12.15 billion years

1.85 billion years ago
Formation of Olympus Mons, a massive shield volcano on Mars, and the tallest mountain in the solar system. It rises over 27 km above the surrounding plains

1.7 billion years ago
Sagittarius Dwarf Galaxy is captured into orbit by the Milky Way, and tidally ripped into a loop. Over hundreds of millions of years, it passes through the disc of the Milky Way several times, and probably contributes to the formation of its spiral arm pattern

1.6 billion years ago
12.2 billion years

1.55 billion years ago
12.25 billion years

TONIAN PERIOD

760 million years ago
Appearance of the first protozoa – highly advanced single-celled organisms such as amoeba, choanoflagellates and ciliates. Unlike algae and other simpler eukaryotes, they feed on other organisms or biological debris and are often capable of movement. They also lack a rigid cell wall – a property which they share with animals

720 million years ago
Beginning of the aptly named 'Cryogenian Period', during which the Earth freezes over at least twice. Ice sheets up to 1 km thick reach as far as the equator, making this the second 'snowball' period in Earth's history

CRYOGENIAN PERIOD
Mean surface temperature: 2 °C
Atmospheric oxygen content: 12%

PANNOTIA

MIROVIAN OCEAN

PANNOTIA

PANNOTIA

800 million years ago	750 million years ago
13 billion years	13.05 billion years

850 million years ago
An asteroid 6 km in diameter impacts the Moon, leaving behind the highly prominent crater Copernicus, with its bright rayed pattern. These rays of ejecta thrown out during impacts gradually fade away due to space weathering, and can usually be seen only around younger craters

750 million years ago
Formation of the binary brown dwarf system Luhman 16, currently located only 6.5 light years away from the Sun. As brown dwarfs radiate almost no visible light, they can only be spotted by infrared space telescopes

700 million years ago	675 million years ago
13.1 billion years	13.125 billion years

GRAVITATIONAL WAVES

1.3 billion years ago
GW150914, a merger of two black holes of 36 and 29 solar masses, takes place. The gravitational wave signal from this event reached Earth in 2015, and was the first such disturbance ever to be detected by the LIGO detector

According to Albert Einstein's theory of general relativity, matter and energy bend and distort the fabric of spacetime. Gravity is a direct result of these distortions. Acceleration of masses (such as orbital motion) produces changing ripples, 'gravitational waves', in the fabric of space-time, that continuously distort and flex it, and propagate across the universe at the speed of light. Usually, their effect is so weak that it's undetectable, but, in extreme cases, such as in black hole mergers, the entire cosmos quakes, as the sheer energy output in the form of gravitational waves can briefly eclipse the luminosity of all the stars and quasars in the observable universe combined. All orbits in the universe will eventually decay after quadrillions of years, as the kinetic energy of the bodies is slowly converted into gravitational waves

12.45 billion years

1.35 billion years ago

INDIRECT DETECTION

Binary systems of neutron stars that often complete one orbit in less than a couple of minutes emit gravitational waves heavily as they lose momentum and spiral closer towards each other. By detecting this orbital decay in a binary pulsar system, evidence for gravitational waves was first revealed in 1993

ENERGY OUTPUT IN GRAVITATIONAL WAVES

6 W (watts)
(5 phones)

200 W
(4 lightbulbs)

EUROPA ORBITING JUPITER
x110

EARTH ORBITING THE SUN
x33

1.4 billion years ago
Widespread diversification of eukaryotic life takes place. The first eukaryotic algae begin populating the oceans, and many of them aggregate in small cellular colonies. This is thought to be a precursor to the eventual evolution of multicellularity

1.4 billion years ago

12.4 billion years

DIRECT DETECTION

In 2015, the flexing of space-time itself caused by passing gravitational waves was detected for the first time, by LIGO laser interferometers. Using lasers to continuously measure the distance between two points in space, it's possible to detect distortions smaller than 1/1000 the diameter of a single proton. In order to rule out false positives and allow the origin of gravitational waves to be determined, two LIGO detectors have been built in separate locations in the US

680 W
(vacuum cleaner)

VENUS ORBITING THE SUN
x200,000

6700 W
(average household)

JUPITER ORBITING THE SUN
x750,000

150 MW
(passenger jet)

5 GW
(large power plant)

ALPHA CENTAURI A AND B
x660 million

SIRIUS A AND B
x2 trillion

1.45 billion years ago

12.35 billion years

INTERFERENCE PATTERNS

Default (no distortions)

Passing gravitational wave

100 PW
(solar energy striking the face of the Earth every second)

10^{22} PW
(entire global oil reserves burning up every second)

KEPLER-35 A AND B
x100 billion

SPICA A AND B
x10^{27}

2.3 billion years

MIRROR

MIRROR

LIGHT STORAGE ARM (4 KM LONG)

PARTIALLY REFLECTIVE MIRRORS

LIGHT STORAGE ARM (4 KM LONG)

Interferometers such as LIGO superimpose two laser beams of the same frequency in a way that perfectly cancels each other out. If the length of the detector arms shifts even by a tiny amount, the beams can no longer cancel out each other, and a distinctive interference pattern is detected

LASER

LASER BEAM SPLITTER

LIGHT DETECTOR

10^{28} W
(energy output of 25 suns)

HM CANCRI
(BINARY WHITE DWARFS)

10^{49} W
(gravitational wave power output at its peak, higher than the luminosity of the entire known universe)

GW150914 BLACK HOLE MERGER

Mean surface temperature: 17 °C
Atmospheric oxygen content: 8%

660 million years ago
During a brief warm period separating the two Cryogenian 'snowball Earth' glaciations, the first true multicellular animals appear in the form of sea sponges. They still form the most primitive part of the animal kingdom, lacking any distinct organ systems and incapable of voluntary movement

600 million years ago
Supercontinent Pannotia begins to rift apart. Roughly at the same time, the ozone layer starts to take shape in the Earth's stratosphere, absorbing up to 99% of the harmful solar UV-B radiation that would otherwise kill most lifeforms on the surface

PANTHALASSIC OCEAN

Siberia

PANNOTIA

PANNOTIA

650 million years ago	625 million years ago	600 million years ago
13.15 billion years	13.175 billion years	13.2 billion years

625 million years ago
Formation of Capella, one of the brightest stars in the northern night sky. In reality, it's not a single star, but a quadruple star system, composed of two pairs of binary stars that orbit each other in turn

600 million years ago
Hyades, one of the oldest known open star clusters, form after the collapse of a giant molecular cloud. Originally, the Hyades were much brighter and contained many more stars, but, most have since died or have dispersed

580 million years ago	570 million years ago
13.22 billion years	13.23 billion years

13.38 billion years	13.37 billion years	13.36 billion years	13.35 billion years
420 million years ago	430 million years ago	440 million years ago	450 million years ago

13.39 billion years		
410 million years ago		

420 million years ago
First sharks swim in the Silurian seas, feeding on the distant human ancestors in the form of lobe-finned fishes

433 million years ago
Cooksonia – the oldest and simplest vascular land plant. It has a simple branching photo-synthesising stem with vascular tissue inside, and lacks specialized leaves and roots

450 million years ago
Thanks to the ozone layer blocking harmful solar radiation, the first plants (tiny, moss-like liverworts) begin to spread across the Earth's surface. About 20 million years later, they are followed by the first land arthropods, ancestors of the modern scorpions and spiders

410 million years ago
Appearance of the fungal kingdom, as the first yeasts, moulds and mushrooms evolve. The most iconic genus of Devonian fungi was without a doubt Prototaxites. These giant lichens (fungi symbiotic with algae) could reach up to 8 m in height, and formed the world's first forests

445 million years ago
Ordovician-Silurian mass extinction – the first of the five major animal extinction events. Up to 85% of all marine species go extinct. The unusual speed at which this occurred has led to a hypothesis that it was triggered by a nearby gamma-ray burst

460 million years ago
Appearance of the first jawed fish. During the next 30 million years, these lifeforms grow to dominate the global ecosystem, as they diversify into armoured placoderms, cartilaginous sharks, and bony lobe and ray-finned fishes

Mean surface temperature: 18 °C
Atmospheric oxygen content: 14%

SILURIAN PERIOD

13.4 billion years		
400 million years ago		

DEVONIAN PERIOD

Mean surface temperature: 22 °C
Atmospheric oxygen content: 15%

CARBONIFEROUS PERIOD

395 million years ago
First insects appear. They quickly become the most prolific group of animals on the planet, a title they still hold to the present day

375 million years ago
Beginning of the Late Devonian mass extinction, during which 19% of all animal families and 75% of species went extinct. Brachiopods, trilobites and corals were especially hard hit

340 million years ago
Carboniferous wetland forests spread throughout the globe. As trees die and fall into the swamp, they do not decay completely, but transform into carbon-rich peat. After hundreds of millions of years, after being covered by sediment, this peat transforms into the coal deposits of the present

13.41 billion years		
390 million years ago		

Siberia

South China

Laurussia

PROTOTETHYS OCEAN

GONDWANA

360 million years ago
First ferns appear, some reaching the size of trees

365 million years ago
Evolution of Ichthyostega – the first vertebrate with four functional legs. This amphibian possessed both gills and lungs, was probably a carnivore, and inhabited, shallow, swampy environments

330 million years ago
First amniotes (animals that lay eggs with a protective fluid-filled amniotic sac) emerge. They are the first vertebrates adapted to a fully terrestrial lifestyle, and are the ancestors of all modern reptiles and mammals

380 million years ago
13.42 billion years

370 million years ago	360 million years ago	350 million years ago	340 million years ago
13.43 billion years	13.44 billion years	13.45 billion years	13.46 billion years

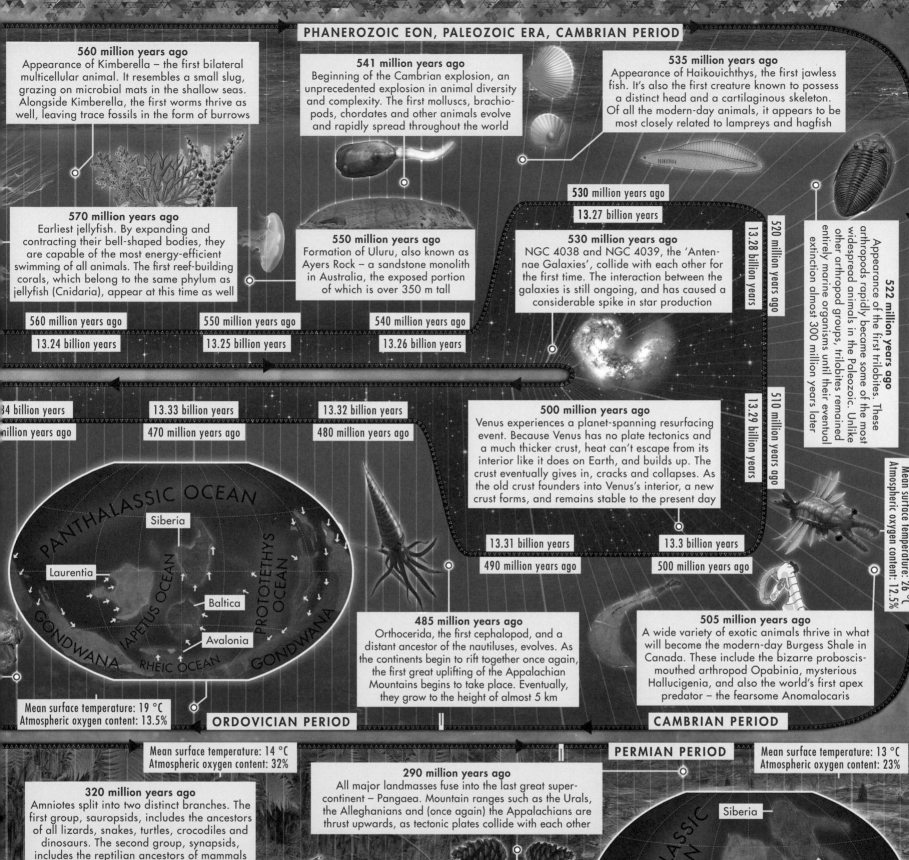

560 million years ago
Appearance of Kimberella – the first bilateral multicellular animal. It resembles a small slug, grazing on microbial mats in the shallow seas. Alongside Kimberella, the first worms thrive as well, leaving trace fossils in the form of burrows

541 million years ago
Beginning of the Cambrian explosion, an unprecedented explosion in animal diversity and complexity. The first molluscs, brachiopods, chordates and other animals evolve and rapidly spread throughout the world

535 million years ago
Appearance of Haikouichthys, the first jawless fish. It's also the first creature known to possess a distinct head and a cartilaginous skeleton. Of all the modern-day animals, it appears to be most closely related to lampreys and hagfish

570 million years ago
Earliest jellyfish. By expanding and contracting their bell-shaped bodies, they are capable of the most energy-efficient swimming of all animals. The first reef-building corals, which belong to the same phylum as jellyfish (Cnidaria), appear at this time as well

550 million years ago
Formation of Uluru, also known as Ayers Rock – a sandstone monolith in Australia, the exposed portion of which is over 350 m tall

530 million years ago
13.27 billion years

530 million years ago
NGC 4038 and NGC 4039, the 'Antennae Galaxies', collide with each other for the first time. The interaction between the galaxies is still ongoing, and has caused a considerable spike in star production

520 million years ago

13.28 billion years

522 million years ago
Appearance of the first trilobites. These arthropods rapidly became some of the most widespread animals in the Paleozoic. Unlike other arthropod groups, trilobites remained entirely marine organisms until their eventual extinction almost 300 million years later

| 560 million years ago | 550 million years ago | 540 million years ago |
| 13.24 billion years | 13.25 billion years | 13.26 billion years |

34 billion years | 13.33 billion years | 13.32 billion years

million years ago | 470 million years ago | 480 million years ago

500 million years ago
Venus experiences a planet-spanning resurfacing event. Because Venus has no plate tectonics and a much thicker crust, heat can't escape from its interior like it does on Earth, and builds up. The crust eventually gives in, cracks and collapses. As the old crust founders into Venus's interior, a new crust forms, and remains stable to the present day

510 million years ago

13.29 billion years

Mean surface temperature: 26 °C
Atmospheric oxygen content: 12.5%

PANTHALASSIC OCEAN
Siberia
Laurentia
Baltica
Avalonia
IAPETUS OCEAN
PROTOTETHYS OCEAN
GONDWANA
RHEIC OCEAN
GONDWANA

| 13.31 billion years | 13.3 billion years |
| 490 million years ago | 500 million years ago |

485 million years ago
Orthocerida, the first cephalopod, and a distant ancestor of the nautiluses, evolves. As the continents begin to rift together once again, the first great uplifting of the Appalachian Mountains begins to take place. Eventually, they grow to the height of almost 5 km

505 million years ago
A wide variety of exotic animals thrive in what will become the modern-day Burgess Shale in Canada. These include the bizarre proboscis-mouthed arthropod Opabinia, mysterious Hallucigenia, and also the world's first apex predator – the fearsome Anomalocaris

Mean surface temperature: 19 °C
Atmospheric oxygen content: 13.5%

ORDOVICIAN PERIOD

CAMBRIAN PERIOD

Mean surface temperature: 14 °C
Atmospheric oxygen content: 32%

PERMIAN PERIOD

Mean surface temperature: 13 °C
Atmospheric oxygen content: 23%

290 million years ago
All major landmasses fuse into the last great supercontinent – Pangaea. Mountain ranges such as the Urals, the Alleghanians and (once again) the Appalachians are thrust upwards, as tectonic plates collide with each other

320 million years ago
Amniotes split into two distinct branches. The first group, sauropsids, includes the ancestors of all lizards, snakes, turtles, crocodiles and dinosaurs. The second group, synapsids, includes the reptilian ancestors of mammals

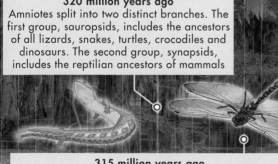

300 million years ago
Emergence of the first conifers. Adaptations such as pollen, cones and seeds allow them to quickly supplant the giant lycopods and tree ferns, and become the dominant plant group

PANTHALASSIC OCEAN
Siberia
PANGAEA
PALEO-TETHYS OCEAN

315 million years ago
The high oxygen content of the carboniferous atmosphere allows arthropods to grow to unprecedented sizes. Giant predatory dragonfly Meganeura dominates the skies, while the 2 m-long giant millipede Arthopleura peacefully grazes upon the plants on the forest floor

295 million years ago
The first beetles appear. Synapsids grow dominant, as the iconic dimetrodon, carrying a skin-covered bony sail on its back, becomes the apex predator of the early Permian

| n years ago | 320 million years ago | 310 million years ago | 300 million years ago | 290 million years ago | 280 million years ago |
| lion years | 13.48 billion years | 13.49 billion years | 13.5 billion years | 13.51 billion years | 13.52 billion years |

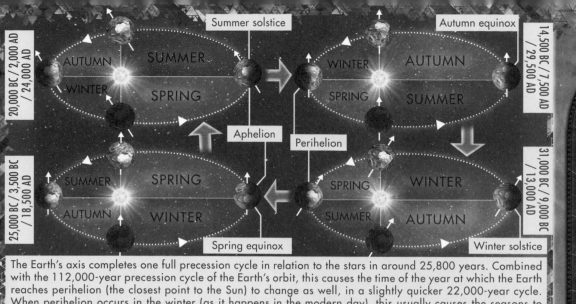

Summer solstice | Autumn equinox

AUTUMN | SUMMER | WINTER | AUTUMN
WINTER | SPRING | SPRING | SUMMER

20,000 BC / 2,000 AD / 24,000 AD

14,500 BC / 7,500 AD / 29,500 AD

Aphelion | Perihelion

25,000 BC / 3,500 AD / 18,500 BC

SUMMER | SPRING | SPRING | WINTER
AUTUMN | WINTER | SUMMER | AUTUMN

31,000 BC / 9,000 AD / 13,000 AD

Spring equinox | Winter solstice

The Earth's axis completes one full precession cycle in relation to the stars in around 25,800 years. Combined with the 112,000-year precession cycle of the Earth's orbit, this causes the time of the year at which the Earth reaches perihelion (the closest point to the Sun) to change as well, in a slightly quicker 22,000-year cycle. When perihelion occurs in the winter (as it happens in the modern day), this usually causes the seasons to be less extreme, and makes winters shorter and more moderate. When the Earth is closest to the Sun in the summer, however, the seasonal effects are amplified, winters are longer and colder. This has a cyclical effect on Earth's climate, and is believed to be responsible for the periodic dry and humid periods of the Sahara Desert. The illustration above shows the long-term precession cycle of the northern hemisphere seasons

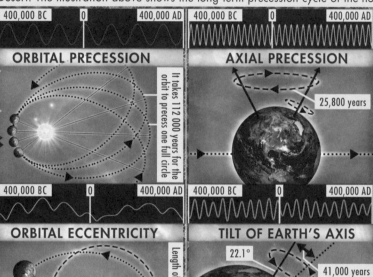

ORBITAL PRECESSION

400,000 BC | 0 | 400,000 AD

It takes 112 000 years for the orbit to precess one full circle

AXIAL PRECESSION

400,000 BC | 0 | 400,000 AD

25,800 years

400,000 BC | 0 | 400,000 AD

ORBITAL ECCENTRICITY

Length of the year is unchanged

TILT OF EARTH'S AXIS

400,000 BC | 0 | 400,000 AD

22.1°
41,000 years
24.5°

110,000 years

MOVEMENT OF THE CELESTIAL NORTH POLE DUE TO THE PRECESSION OF EARTH'S AXIS

Polaris
2,000
2,000 BC
Thuban
Kochab
6,000
5,000 BC
10,000
8,000 BC
14,000
Vega

Over the course of millennia, the Earth goes through periodic cyclical changes in its movement around the Sun and its rotational axis. This happens due to gravitational interactions with other bodies in the solar system. Variations in axial tilt and especially in Earth's orbital eccentricity have a major influence on the Earth's climate. Evidence from deep arctic Greenland ice cores show that in the last 2.5 million years, the cyclical glacial periods ('ice ages') and warm interglacial periods have closely followed either the 41,000-year-long axial tilt cycle or (especially the most recent ice ages) the 110-000-year old orbital eccentricity cycle.

MILANKOVITCH CYCLES

260 million years ago
Appearance of the first cynodonts – synapsid ancestors of the mammals. It's likely that many of them already possess rudimentary hair and fur

250 million years ago
Massive volcanic eruptions of the Siberian Traps trigger the Permian mass extinction, which results in the disappearance of up to 96% of all marine and 70% of all land vertebrate species. All remaining trilobites and sea scorpions go extinct

233 million years ago
First dinosaurs appear. They can be divided into two distinct groups – saurischia (lizard-hipped), which includes the giant herbivorous sauropods and bipedal theropods such as T-Rex, and ornithischia (bird-hipped), which includes the stegosaurs, ankylosaurs and horned ceratopsians

265 million years ago
Length of the Earth's day increases to over 22 hours, as evidenced by the growth patterns of fossilized coral reefs

245 million years ago
First ichthyosaurs – marine carnivorous reptiles, resembling modern-day dolphins and whales

240 million years ago
13.56 billion years

235 million years ago
Formation of the binary system of Sirius. Originally the smaller of the two stars, Sirius A appears as the brightest star in the present night sky

270 million years ago
13.53 billion years

260 million years ago
13.54 billion years

250 million years ago
13.55 billion years

230 million years ago
13.57 billion years

66.04 million years ago
A giant asteroid impacts modern-day Chicxulub, Mexico. This triggers the last great mass extinction, which wipes out 75% of all the Earth's species. No land vertebrates weighing more than 25 kg survive, and all non-avian dinosaurs go extinct

68 million years ago
Appearance of the two most iconic dinosaur species – the massive bipedal predator Tyrannosaurus Rex, and the rhinoceros-like Triceratops, with its large bony frill and three horns

70 million years ago
13.73 billion years

70 million years ago
Formation of Polaris – a yellow supergiant, orbited by two smaller Sun-like stars. Currently, it is the closest bright star to the Earth's celestial north pole

80 million years ago
Evolution of the first colonial ants. In the present day, all the ant species form up to 20% of all animal biomass

80 million years ago
13.72 billion years

13.71 billion years
90 million years ago

LAURASIA
ATLANTIC OCEAN
SOUTH AMERICA
AFRICA
TETHYS OCEAN
PACIFIC OCEAN

CRETACEOUS PERIOD

Mean surface temperature: 22 °C
Atmospheric oxygen content: 15%

MESOZOIC ERA, TRIASS

220 million years ago
Appearance of the first flies and crocodiles in Triassic swamps

225 million years ago
First true mammals evolve from cynodonts. While they are warm-blooded and nourish their young with breast milk, all early mammals still lay eggs

220 million years
13.58 billion years

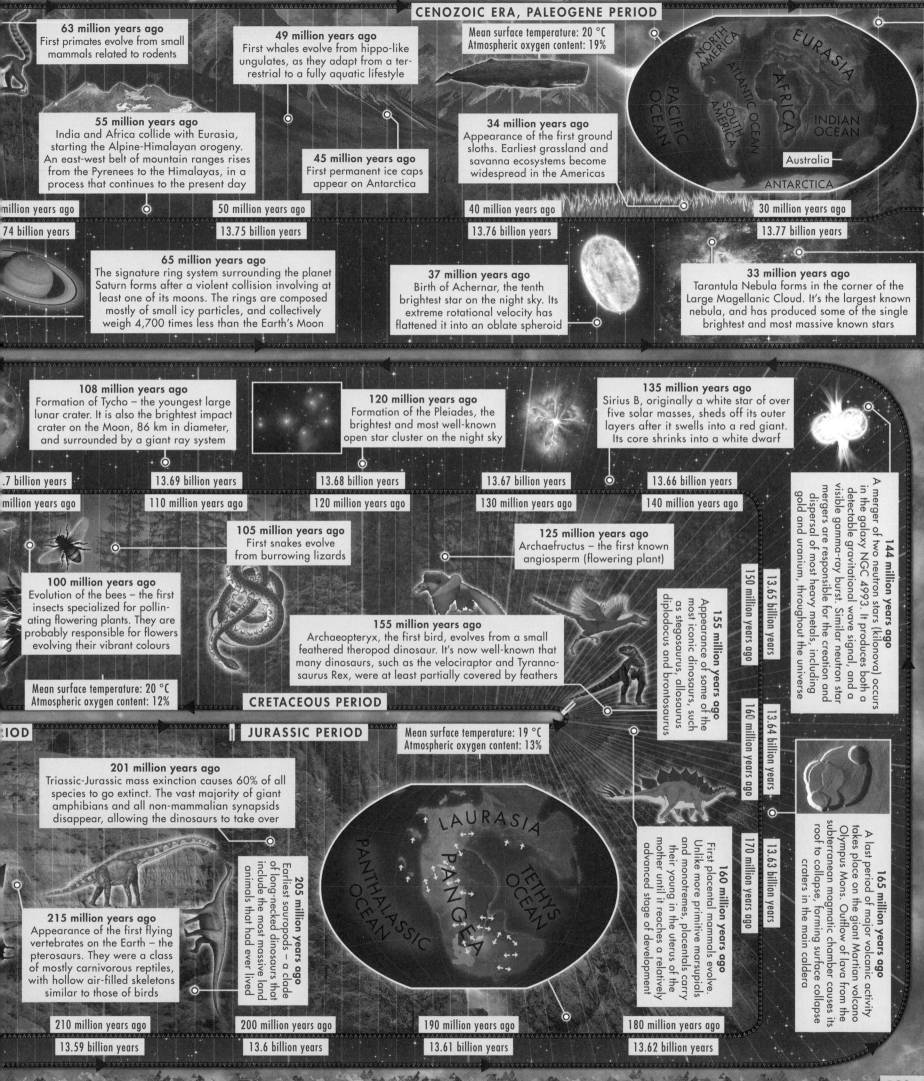

Mean surface temperature: 20 °C
Atmospheric oxygen content: 19%

63 million years ago
First primates evolve from small mammals related to rodents

49 million years ago
First whales evolve from hippo-like ungulates, as they adapt from a terrestrial to a fully aquatic lifestyle

55 million years ago
India and Africa collide with Eurasia, starting the Alpine-Himalayan orogeny. An east-west belt of mountain ranges rises from the Pyrenees to the Himalayas, in a process that continues to the present day

45 million years ago
First permanent ice caps appear on Antarctica

34 million years ago
Appearance of the first ground sloths. Earliest grassland and savanna ecosystems become widespread in the Americas

NORTH AMERICA · EURASIA · PACIFIC OCEAN · ATLANTIC OCEAN · AFRICA · SOUTH AMERICA · INDIAN OCEAN · Australia · ANTARCTICA

million years ago | 50 million years ago | 40 million years ago | 30 million years ago

74 billion years | 13.75 billion years | 13.76 billion years | 13.77 billion years

65 million years ago
The signature ring system surrounding the planet Saturn forms after a violent collision involving at least one of its moons. The rings are composed mostly of small icy particles, and collectively weigh 4,700 times less than the Earth's Moon

37 million years ago
Birth of Achernar, the tenth brightest star on the night sky. Its extreme rotational velocity has flattened it into an oblate spheroid

33 million years ago
Tarantula Nebula forms in the corner of the Large Magellanic Cloud. It's the largest known nebula, and has produced some of the single brightest and most massive known stars

108 million years ago
Formation of Tycho – the youngest large lunar crater. It is also the brightest impact crater on the Moon, 86 km in diameter, and surrounded by a giant ray system

120 million years ago
Formation of the Pleiades, the brightest and most well-known open star cluster on the night sky

135 million years ago
Sirius B, originally a white star of over five solar masses, sheds off its outer layers after it swells into a red giant. Its core shrinks into a white dwarf

.7 billion years | 13.69 billion years | 13.68 billion years | 13.67 billion years | 13.66 billion years

million years ago | 110 million years ago | 120 million years ago | 130 million years ago | 140 million years ago

105 million years ago
First snakes evolve from burrowing lizards

125 million years ago
Archaefructus – the first known angiosperm (flowering plant)

100 million years ago
Evolution of the bees – the first insects specialized for pollinating flowering plants. They are probably responsible for flowers evolving their vibrant colours

Mean surface temperature: 20 °C
Atmospheric oxygen content: 12%

155 million years ago
Archaeopteryx, the first bird, evolves from a small feathered theropod dinosaur. It's now well-known that many dinosaurs, such as the velociraptor and Tyrannosaurus Rex, were at least partially covered by feathers

CRETACEOUS PERIOD

144 million years ago
A merger of two neutron stars (kilonova) occurs in the galaxy NGC 4993. It produces both a detectable gravitational wave signal, and a visible gamma-ray burst. Similar neutron star mergers are responsible for the creation and dispersal of most heavy metals, including gold and uranium, throughout the universe

13.65 billion years

150 million years ago

13.64 billion years

160 million years ago

155 million years ago
Appearance of some of the most iconic dinosaurs, such as stegosaurus, allosaurus, diplodocus and brontosaurus

JURASSIC PERIOD

Mean surface temperature: 19 °C
Atmospheric oxygen content: 13%

201 million years ago
Triassic-Jurassic mass extinction causes 60% of all species to go extinct. The vast majority of giant amphibians and all non-mammalian synapsids disappear, allowing the dinosaurs to take over

160 million years ago
First placental mammals evolve. Unlike more primitive marsupials and monotremes, placentals carry their young in the uterus of the mother until it reaches a relatively advanced stage of development

13.63 billion years

170 million years ago

165 million years ago
A last period of major volcanic activity takes place on the giant Martian volcano Olympus Mons. Outflow of lava from the subterranean magmatic chamber causes its roof to collapse, forming surface collapse craters in the main caldera

LAURASIA · PANTHALASSIC OCEAN · PANGEA · TETHYS OCEAN

215 million years ago
Appearance of the first flying vertebrates on the Earth – the pterosaurs. They were a class of mostly carnivorous reptiles, with hollow air-filled skeletons similar to those of birds

205 million years ago
Earliest sauropods – a clade of long-necked dinosaurs that include the most massive land animals that had ever lived

210 million years ago | 200 million years ago | 190 million years ago | 180 million years ago

13.59 billion years | 13.6 billion years | 13.61 billion years | 13.62 billion years

PERIOD

NEOGENE PERIOD, MIOCENE EPOCH

28 million years ago
Evolution of Paraceratherium – a massive hornless rhinoceros, and the largest land mammal to have ever lived. Some specimens might have exceeded 25 tons

14 million years ago
Common ancestor of humans and the orangutans

12 million years ago
Common ancestor of humans and the gorillas

26 million years ago
Appearance of deinotherium – the earliest elephant

20 million years ago
Emergence of many modern mammal groups, including but not limited to big cats, giraffes, anteaters, deer, bovids, kangaroos and seals

15 million years ago
Great apes (hominids) split from the ancestors of the gibbons (lesser apes). Unlike other primates, great apes lack tails, and have highly developed brains

25 million years ago
13.775 billion years

20 million years ago
13.78 billion years

15 million years ago
13.785 billion years

10 million years ago
13.79 billion years

30 million years ago
LAMOST-HVS1, the closest hypervelocity star to the solar system, is accelerated to an escape trajectory out of the Milky Way. This can happen when a binary star system gets too close to a massive black hole – one of the stars gets gravitationally captured by the black hole, while the other is ejected at a speed exceeding 500 km/s

17 million years ago
Most recent known lunar volcanic activity. Under its surface, the Moon still hosts many empty lava tubes. Some of them are accessible via collapsed skylights, and could be one of the best sites for a future lunar base

9.9 million years ago
Formation of Mu Cephei, a red supergiant. Placed in the Sun's position, it would extend between the orbits of Jupiter and Saturn

1.7 million years ago
Eruption of the youngest lava flows from the flanks of Olympus Mons. By counting the amount of craters over an area, it's possible to quite accurately estimate the age of any surface on Mars

2 million years ago
Origin of the V830 Tauri b – the youngest known exoplanet. It's a 'hot Jupiter' that orbits close to a contracting Sun-like protostar

2.3 million years ago
A supernova explodes less than 300 light years away from Earth. It showers our planet with dust containing a short-lived radioactive isotope iron-60, trace amounts of which can still be found in sediments on the ocean floor

13.798 billion years
1.6 million years ago

13.798 billion years
1.8 million years ago

13.798 billion years
2 million years ago

13.797 billion years
2.5 million years ago

1.9 million years ago
Appearance of Homo erectus, the first human species to migrate outside Africa and spread throughout Europe and Asia

QUATERNARY PERIOD, PLEISTOCENE EPOCH

2.58 million years ago
Beginning of the Quaternary glaciation, the most recent and still ongoing ice age, characterized by a cycle of cold glacial and hot interglacial periods

1.16 million years ago
A sunlike star, 2MASS J0610-246, with its companion red dwarf, passes closer than 0.76 light years from the Sun, disturbing the orbits of comets in the Oort cloud

13.798 billion years
1.4 million years ago

13.798 billion years
1.2 million years ago

Earliest evidence of the controlled use
of fire by Homo erectus. The invention of cooking enabled the energy-hungry human brains to further grow in size
(1 million years ago)

2.2 million years ago
Evolution of Paranthropus boisei, nicknamed 'Nutcracker Man' due to his unusually large teeth and jawbone

2.8 million years ago
Isthmus of Panama bridges North and South America. This shifts the ocean currents and causes the northern hemisphere to cool down significantly. Animals native to North America begin to migrate south and vice versa, sparking the Great American Interchange

PLEISTOCENE EPOCH

700,000 years ago
Most recent reversal of the Earth's magnetic field. Such events happen every 200–300,000 years on average, meaning we are now overdue for one

180,000 years ago
Homo sapiens first steps out of Africa and starts to slowly spread throughout Eurasia

800,000 years ago
Birth of the blue hypergiant R136a1, the brightest and the most massive known star in the universe, in the heart of the Tarantula Nebula

13.799 billion years
1 million years ago

640,000 years ago
Most recent eruption of the Yellowstone supervolcano, located over a hotspot in Wyoming. More than 1,000 km³ of lava and volcanic ash is ejected over the course of this catastrophic event

315,000 years ago
First modern humans (Homo sapiens) appear, likely having evolved from Homo erectus in North Africa

150,000 years ago
Mitochondrial Eve – the most recent common matrilinear ancestor of all living humans

800,000 years ago

600,000 years ago

400,000 years ago

300,000 years ago

200,000 years ago

560,000 years ago
Comet West, an extremely long period comet with an aphelion of over one light year, passes through the inner solar system. It returns again in 1976 as one of the brightest comets of the last century, visible even in daylight

300,000 years ago
Origin of the Trifid Nebula, currently the youngest star-forming region in the Milky Way. It is gradually transforming into an open star cluster, and already contains thousands of newborn stars

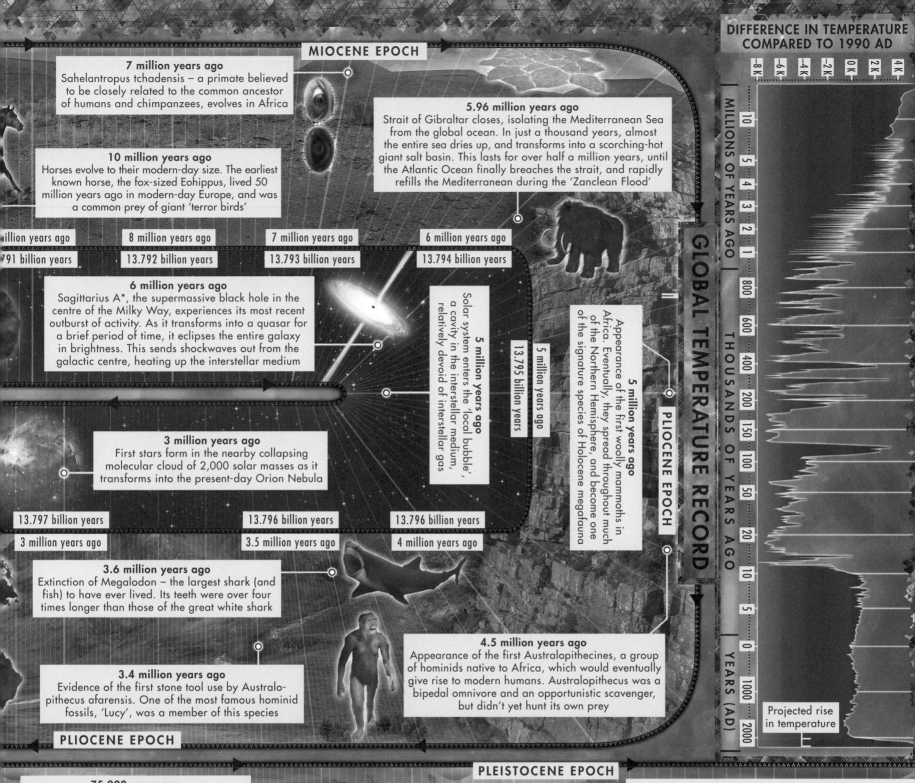

7 million years ago
Sahelantropus tchadensis – a primate believed to be closely related to the common ancestor of humans and chimpanzees, evolves in Africa

10 million years ago
Horses evolve to their modern-day size. The earliest known horse, the fox-sized Eohippus, lived 50 million years ago in modern-day Europe, and was a common prey of giant 'terror birds'

5.96 million years ago
Strait of Gibraltar closes, isolating the Mediterranean Sea from the global ocean. In just a thousand years, almost the entire sea dries up, and transforms into a scorching-hot giant salt basin. This lasts for over half a million years, until the Atlantic Ocean finally breaches the strait, and rapidly refills the Mediterranean during the 'Zanclean Flood'

...illion years ago	8 million years ago	7 million years ago	6 million years ago
...791 billion years	13.792 billion years	13.793 billion years	13.794 billion years

6 million years ago
Sagittarius A*, the supermassive black hole in the centre of the Milky Way, experiences its most recent outburst of activity. As it transforms into a quasar for a brief period of time, it eclipses the entire galaxy in brightness. This sends shockwaves out from the galactic centre, heating up the interstellar medium

Solar system enters the 'local bubble', a cavity in the interstellar medium, relatively devoid of interstellar gas

13.795 billion years

5 million years ago

5 million years ago
Appearance of the first woolly mammoths in Africa. Eventually, they spread throughout much of the Northern Hemisphere, and become one of the signature species of Holocene megafauna

PLIOCENE EPOCH

3 million years ago
First stars form in the nearby collapsing molecular cloud of 2,000 solar masses as it transforms into the present-day Orion Nebula

13.797 billion years	13.796 billion years	13.796 billion years
3 million years ago	3.5 million years ago	4 million years ago

3.6 million years ago
Extinction of Megalodon – the largest shark (and fish) to have ever lived. Its teeth were over four times longer than those of the great white shark

4.5 million years ago
Appearance of the first Australopithecines, a group of hominids native to Africa, which would eventually give rise to modern humans. Australopithecus was a bipedal omnivore and an opportunistic scavenger, but didn't yet hunt its own prey

3.4 million years ago
Evidence of the first stone tool use by Australopithecus afarensis. One of the most famous hominid fossils, 'Lucy', was a member of this species

PLIOCENE EPOCH

GLOBAL TEMPERATURE RECORD

-8K -6K -4K -2K 0K 2K 4K

MILLIONS OF YEARS AGO
10 5 4 3 2 1

THOUSANDS OF YEARS AGO
800 600 400 200 150 100 50 20 10 5

YEARS (AD)
0 1000 2000

Projected rise in temperature

PLEISTOCENE EPOCH

75,000 years ago
Toba supervolcano in Indonesia erupts in the most destructive volcanic event of the last 25 million years. The global cooling that follows kills off the majority of humans on Earth

70,000 years ago
First modern humans move to Europe and start interbreeding with the local population of Neanderthals. In the present day, the measurable Neanderthal ancestry in some western European populations (especially Basques) can exceed 5%

20,000 years ago
End of the 'Last Glacial Maximum', and the beginning of the global deglaciation. Ice sheets covering large swathes of Eurasia and North America start to melt, eventually raising the global sea levels by more than 120 m

100,000 years ago
Extinction of Gigantopithecus, the largest primate to have ever lived (weighing over three times more than a modern gorilla). It was closely related to orangutans

50,000 years ago
An iron-nickel meteorite impacts modern-day Arizona with a force of 10 megatons, excavating the 1.1 km wide 'Barringer Crater'. It's one of the few impact craters on Earth that has not eroded away yet

100,000 years ago	80,000 years ago	60,000 years ago	40,000 years ago	30,000 years ago	25,000 years ago

150,000 years ago
Formation of the oldest surviving planetary nebula. They are produced by dying stars of between 0.8 and eight solar masses, as they cast off their outer layers into deep space, while their core transforms into a white dwarf

70,000 years ago
Scholz's Star, a dim red dwarf with a brown dwarf companion, passes through the outer Oort Cloud

15,000 years ago
Origin of the neutron star SGR 1806-20. It's a so-called 'magnetar' – a young, active neutron star with an extremely strong magnetic field. If a magnetar were to be placed halfway from the Earth to the Moon, it's magnetic field would wipe out the information on all the world's credit cards. At a distance of 1,000 km, it would stretch the very atoms into thin spindles, rendering the chemistry of life impossible

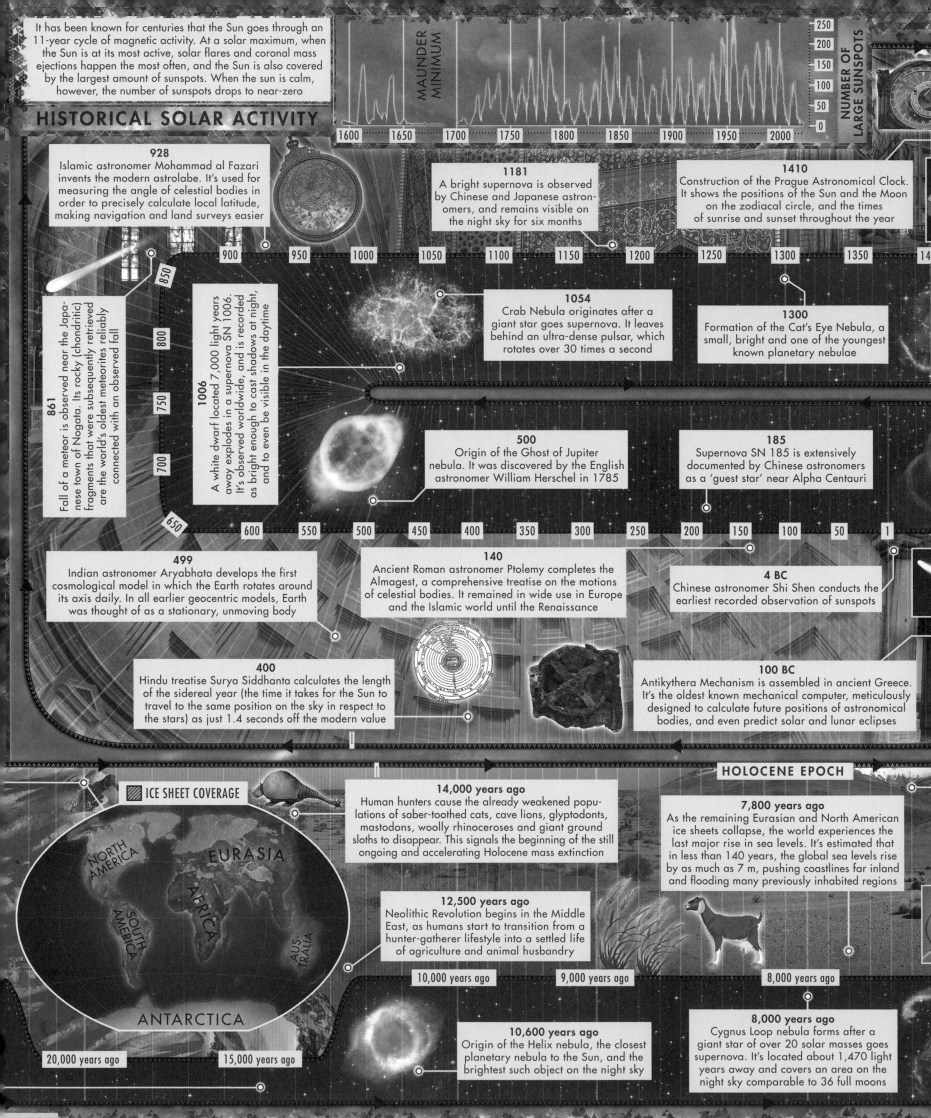

It has been known for centuries that the Sun goes through an 11-year cycle of magnetic activity. At a solar maximum, when the Sun is at its most active, solar flares and coronal mass ejections happen the most often, and the Sun is also covered by the largest amount of sunspots. When the sun is calm, however, the number of sunspots drops to near-zero

MAUNDER MINIMUM

NUMBER OF LARGE SUNSPOTS

250 200 150 100 50 0

1600 1650 1700 1750 1800 1850 1900 1950 2000

HISTORICAL SOLAR ACTIVITY

928
Islamic astronomer Mohammad al Fazari invents the modern astrolabe. It's used for measuring the angle of celestial bodies in order to precisely calculate local latitude, making navigation and land surveys easier

1181
A bright supernova is observed by Chinese and Japanese astronomers, and remains visible on the night sky for six months

1410
Construction of the Prague Astronomical Clock. It shows the positions of the Sun and the Moon on the zodiacal circle, and the times of sunrise and sunset throughout the year

900 950 1000 1050 1100 1150 1200 1250 1300 1350 140

850

861
Fall of a meteor is observed near the Japanese town of Nogata. Its rocky (chondritic) fragments that were subsequently retrieved are the world's oldest meteorites reliably connected with an observed fall

800 750 700

1006
A white dwarf located 7,000 light years away explodes in a supernova SN 1006. It's observed worldwide, and is recorded as bright enough to cast shadows at night, and to even be visible in the daytime

1054
Crab Nebula originates after a giant star goes supernova. It leaves behind an ultra-dense pulsar, which rotates over 30 times a second

1300
Formation of the Cat's Eye Nebula, a small, bright and one of the youngest known planetary nebulae

500
Origin of the Ghost of Jupiter nebula. It was discovered by the English astronomer William Herschel in 1785

185
Supernova SN 185 is extensively documented by Chinese astronomers as a 'guest star' near Alpha Centauri

650

600 550 500 450 400 350 300 250 200 150 100 50 1

499
Indian astronomer Aryabhata develops the first cosmological model in which the Earth rotates around its axis daily. In all earlier geocentric models, Earth was thought of as a stationary, unmoving body

140
Ancient Roman astronomer Ptolemy completes the Almagest, a comprehensive treatise on the motions of celestial bodies. It remained in wide use in Europe and the Islamic world until the Renaissance

4 BC
Chinese astronomer Shi Shen conducts the earliest recorded observation of sunspots

400
Hindu treatise Surya Siddhanta calculates the length of the sidereal year (the time it takes for the Sun to travel to the same position on the sky in respect to the stars) as just 1.4 seconds off the modern value

100 BC
Antikythera Mechanism is assembled in ancient Greece. It's the oldest known mechanical computer, meticulously designed to calculate future positions of astronomical bodies, and even predict solar and lunar eclipses

HOLOCENE EPOCH

▨ **ICE SHEET COVERAGE**

14,000 years ago
Human hunters cause the already weakened populations of saber-toothed cats, cave lions, glyptodonts, mastodons, woolly rhinoceroses and giant ground sloths to disappear. This signals the beginning of the still ongoing and accelerating Holocene mass extinction

7,800 years ago
As the remaining Eurasian and North American ice sheets collapse, the world experiences the last major rise in sea levels. It's estimated that in less than 140 years, the global sea levels rise by as much as 7 m, pushing coastlines far inland and flooding many previously inhabited regions

NORTH AMERICA
EURASIA
AFRICA
SOUTH AMERICA
AUSTRALIA
ANTARCTICA

12,500 years ago
Neolithic Revolution begins in the Middle East, as humans start to transition from a hunter-gatherer lifestyle into a settled life of agriculture and animal husbandry

10,000 years ago   9,000 years ago   8,000 years ago

10,600 years ago
Origin of the Helix nebula, the closest planetary nebula to the Sun, and the brightest such object on the night sky

8,000 years ago
Cygnus Loop nebula forms after a giant star of over 20 solar masses goes supernova. It's located about 1,470 light years away and covers an area on the night sky comparable to 36 full moons

20,000 years ago   15,000 years ago

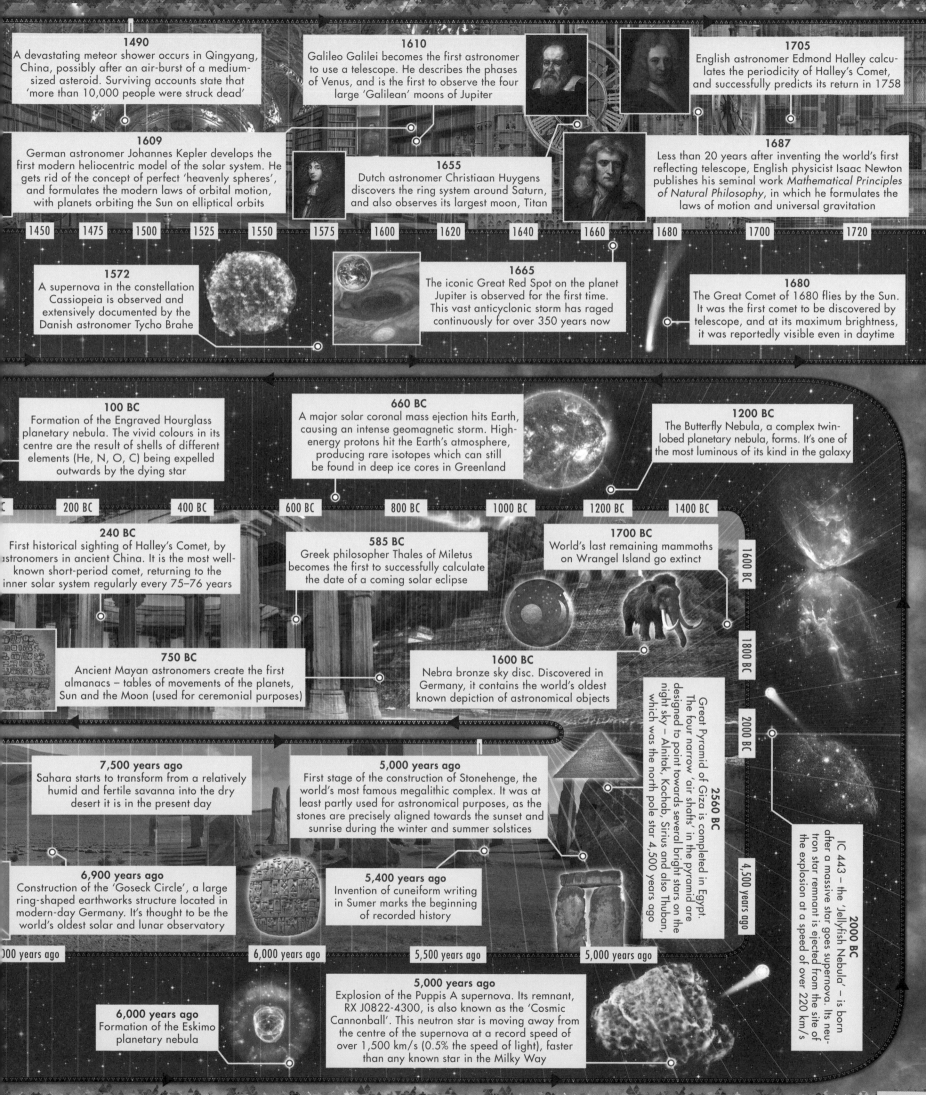

1490
A devastating meteor shower occurs in Qingyang, China, possibly after an air-burst of a medium-sized asteroid. Surviving accounts state that 'more than 10,000 people were struck dead'

1610
Galileo Galilei becomes the first astronomer to use a telescope. He describes the phases of Venus, and is the first to observe the four large 'Galilean' moons of Jupiter

1705
English astronomer Edmond Halley calculates the periodicity of Halley's Comet, and successfully predicts its return in 1758

1609
German astronomer Johannes Kepler develops the first modern heliocentric model of the solar system. He gets rid of the concept of perfect 'heavenly spheres', and formulates the modern laws of orbital motion, with planets orbiting the Sun on elliptical orbits

1655
Dutch astronomer Christiaan Huygens discovers the ring system around Saturn, and also observes its largest moon, Titan

1687
Less than 20 years after inventing the world's first reflecting telescope, English physicist Isaac Newton publishes his seminal work *Mathematical Principles of Natural Philosophy*, in which he formulates the laws of motion and universal gravitation

| 1450 | 1475 | 1500 | 1525 | 1550 | 1575 | 1600 | 1620 | 1640 | 1660 | 1680 | 1700 | 1720 |

1572
A supernova in the constellation Cassiopeia is observed and extensively documented by the Danish astronomer Tycho Brahe

1665
The iconic Great Red Spot on the planet Jupiter is observed for the first time. This vast anticyclonic storm has raged continuously for over 350 years now

1680
The Great Comet of 1680 flies by the Sun. It was the first comet to be discovered by telescope, and at its maximum brightness, it was reportedly visible even in daytime

100 BC
Formation of the Engraved Hourglass planetary nebula. The vivid colours in its centre are the result of shells of different elements (He, N, O, C) being expelled outwards by the dying star

660 BC
A major solar coronal mass ejection hits Earth, causing an intense geomagnetic storm. High-energy protons hit the Earth's atmosphere, producing rare isotopes which can still be found in deep ice cores in Greenland

1200 BC
The Butterfly Nebula, a complex twin-lobed planetary nebula, forms. It's one of the most luminous of its kind in the galaxy

| 200 BC | 400 BC | 600 BC | 800 BC | 1000 BC | 1200 BC | 1400 BC |

240 BC
First historical sighting of Halley's Comet, by astronomers in ancient China. It is the most well-known short-period comet, returning to the inner solar system regularly every 75–76 years

585 BC
Greek philosopher Thales of Miletus becomes the first to successfully calculate the date of a coming solar eclipse

1700 BC
World's last remaining mammoths on Wrangel Island go extinct

| 1600 BC | 1800 BC | 2000 BC | 4,500 years ago |

750 BC
Ancient Mayan astronomers create the first almanacs – tables of movements of the planets, Sun and the Moon (used for ceremonial purposes)

1600 BC
Nebra bronze sky disc. Discovered in Germany, it contains the world's oldest known depiction of astronomical objects

2560 BC
Great Pyramid of Giza is completed in Egypt. The four narrow 'air shafts' in the pyramid are designed to point towards several bright stars on the night sky – Alnitak, Kochab, Sirius and also Thuban, which was the north pole star 4,500 years ago

7,500 years ago
Sahara starts to transform from a relatively humid and fertile savanna into the dry desert it is in the present day

5,000 years ago
First stage of the construction of Stonehenge, the world's most famous megalithic complex. It was at least partly used for astronomical purposes, as the stones are precisely aligned towards the sunset and sunrise during the winter and summer solstices

2000 BC
IC 443 – the 'Jellyfish Nebula' – is born after a massive star goes supernova. Its neutron star remnant is ejected from the site of the explosion at a speed of over 220 km/s

6,900 years ago
Construction of the 'Goseck Circle', a large ring-shaped earthworks structure located in modern-day Germany. It's thought to be the world's oldest solar and lunar observatory

5,400 years ago
Invention of cuneiform writing in Sumer marks the beginning of recorded history

| 000 years ago | 6,000 years ago | 5,500 years ago | 5,000 years ago |

6,000 years ago
Formation of the Eskimo planetary nebula

5,000 years ago
Explosion of the Puppis A supernova. Its remnant, RX J0822-4300, is also known as the 'Cosmic Cannonball'. This neutron star is moving away from the centre of the supernova at a record speed of over 1,500 km/s (0.5% the speed of light), faster than any known star in the Milky Way

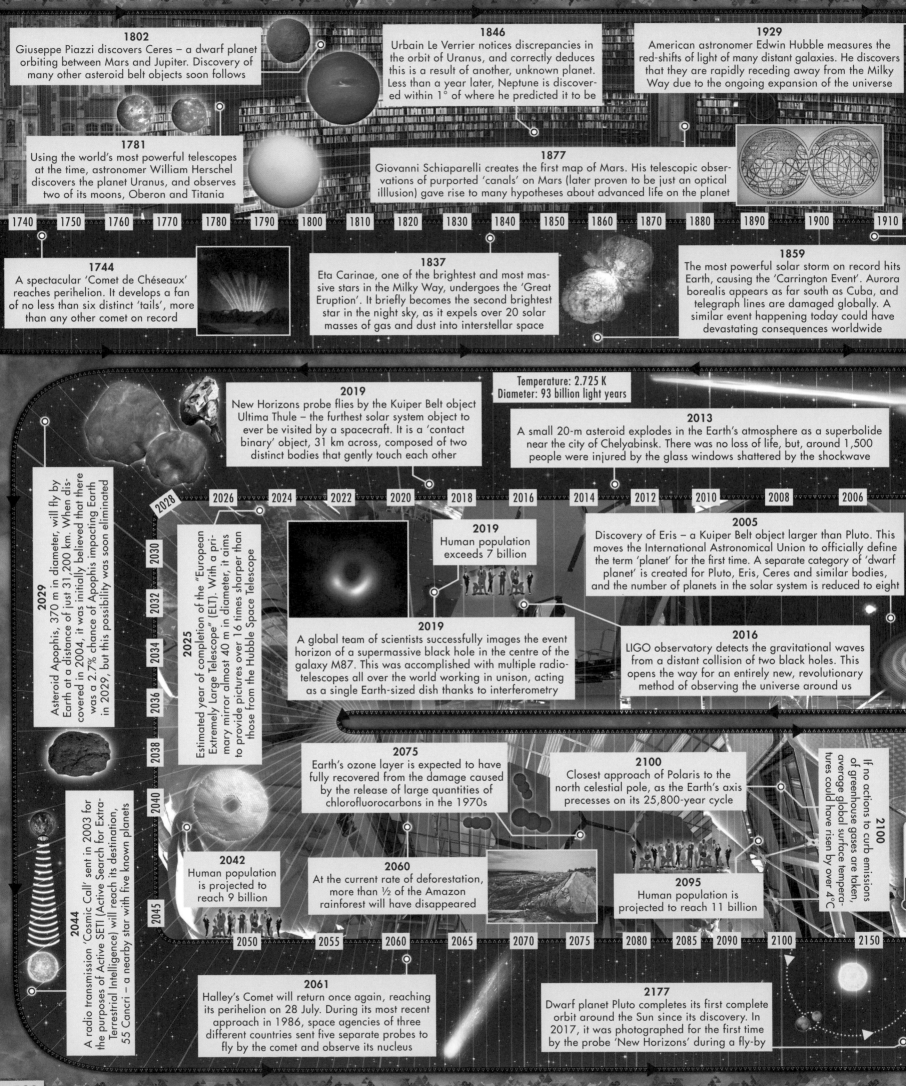

1802
Giuseppe Piazzi discovers Ceres – a dwarf planet orbiting between Mars and Jupiter. Discovery of many other asteroid belt objects soon follows

1846
Urbain Le Verrier notices discrepancies in the orbit of Uranus, and correctly deduces this is a result of another, unknown planet. Less than a year later, Neptune is discovered within 1° of where he predicted it to be

1929
American astronomer Edwin Hubble measures the red-shifts of light of many distant galaxies. He discovers that they are rapidly receding away from the Milky Way due to the ongoing expansion of the universe

1781
Using the world's most powerful telescopes at the time, astronomer William Herschel discovers the planet Uranus, and observes two of its moons, Oberon and Titania

1877
Giovanni Schiaparelli creates the first map of Mars. His telescopic observations of purported 'canals' on Mars (later proven to be just an optical illusion) gave rise to many hypotheses about advanced life on the planet

| 1740 | 1750 | 1760 | 1770 | 1780 | 1790 | 1800 | 1810 | 1820 | 1830 | 1840 | 1850 | 1860 | 1870 | 1880 | 1890 | 1900 | 1910 |

1744
A spectacular 'Comet de Chéseaux' reaches perihelion. It develops a fan of no less than six distinct 'tails', more than any other comet on record

1837
Eta Carinae, one of the brightest and most massive stars in the Milky Way, undergoes the 'Great Eruption'. It briefly becomes the second brightest star in the night sky, as it expels over 20 solar masses of gas and dust into interstellar space

1859
The most powerful solar storm on record hits Earth, causing the 'Carrington Event'. Aurora borealis appears as far south as Cuba, and telegraph lines are damaged globally. A similar event happening today could have devastating consequences worldwide

2019
New Horizons probe flies by the Kuiper Belt object Ultima Thule – the furthest solar system object to ever be visited by a spacecraft. It is a 'contact binary' object, 31 km across, composed of two distinct bodies that gently touch each other

Temperature: 2.725 K
Diameter: 93 billion light years

2013
A small 20-m asteroid explodes in the Earth's atmosphere as a superbolide near the city of Chelyabinsk. There was no loss of life, but, around 1,500 people were injured by the glass windows shattered by the shockwave

| 2028 | 2026 | 2024 | 2022 | 2020 | 2018 | 2016 | 2014 | 2012 | 2010 | 2008 | 2006 |

2029
Asteroid Apophis, 370 m in diameter, will fly by Earth at a distance of just 31,200 km. When discovered in 2004, it was initially believed that there was a 2.7% chance of Apophis impacting Earth in 2029, but this possibility was soon eliminated

2025
Estimated year of completion of the "European Extremely Large Telescope" (ELT). With a primary mirror almost 40 m in diameter, it aims to provide pictures over 16 times sharper than those from the Hubble Space Telescope

2019
Human population exceeds 7 billion

2005
Discovery of Eris – a Kuiper Belt object larger than Pluto. This moves the International Astronomical Union to officially define the term 'planet' for the first time. A separate category of 'dwarf planet' is created for Pluto, Eris, Ceres and similar bodies, and the number of planets in the solar system is reduced to eight

2019
A global team of scientists successfully images the event horizon of a supermassive black hole in the centre of the galaxy M87. This was accomplished with multiple radio-telescopes all over the world working in unison, acting as a single Earth-sized dish thanks to interferometry

2016
LIGO observatory detects the gravitational waves from a distant collision of two black holes. This opens the way for an entirely new, revolutionary method of observing the universe around us

| 2030 | 2032 | 2034 | 2036 | 2038 | 2040 | 2045 |

2075
Earth's ozone layer is expected to have fully recovered from the damage caused by the release of large quantities of chlorofluorocarbons in the 1970s

2100
Closest approach of Polaris to the north celestial pole, as the Earth's axis precesses on its 25,800-year cycle

If no actions to curb emissions of greenhouse gases are taken, average global surface temperatures could have risen by over 4°C **2100**

2044
A radio transmission 'Cosmic Call' sent in 2003 for the purposes of Active SETI (Active Search for Extra-Terrestrial Intelligence) will reach its destination, 55 Cancri – a nearby star with five known planets

2042
Human population is projected to reach 9 billion

2060
At the current rate of deforestation, more than ½ of the Amazon rainforest will have disappeared

2095
Human population is projected to reach 11 billion

| 2050 | 2055 | 2060 | 2065 | 2070 | 2075 | 2080 | 2085 | 2090 | 2100 | 2150 |

2061
Halley's Comet will return once again, reaching its perihelion on 28 July. During its most recent approach in 1986, space agencies of three different countries sent five separate probes to fly by the comet and observe its nucleus

2177
Dwarf planet Pluto completes its first complete orbit around the Sun since its discovery. In 2017, it was photographed for the first time by the probe 'New Horizons' during a fly-by

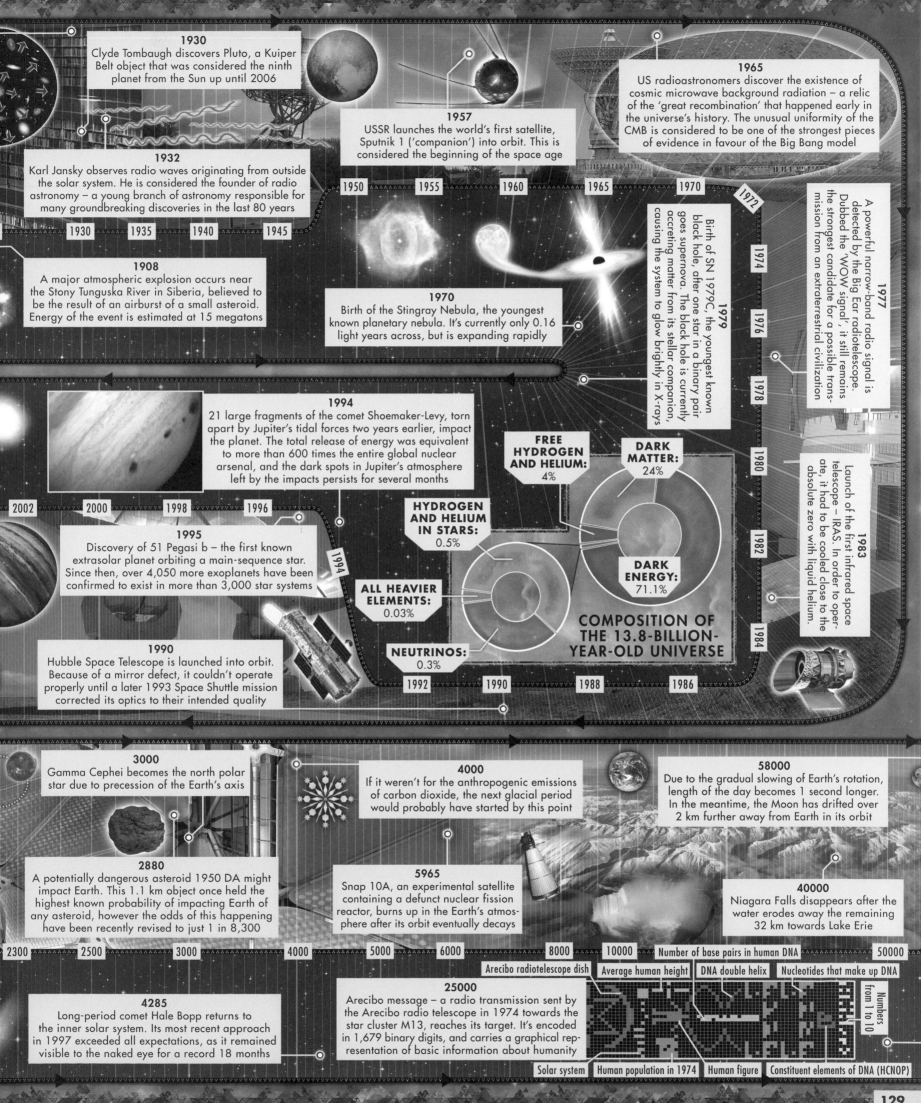

1930
Clyde Tombaugh discovers Pluto, a Kuiper Belt object that was considered the ninth planet from the Sun up until 2006

1957
USSR launches the world's first satellite, Sputnik 1 ('companion') into orbit. This is considered the beginning of the space age

1965
US radioastronomers discover the existence of cosmic microwave background radiation – a relic of the 'great recombination' that happened early in the universe's history. The unusual uniformity of the CMB is considered to be one of the strongest pieces of evidence in favour of the Big Bang model

1932
Karl Jansky observes radio waves originating from outside the solar system. He is considered the founder of radio astronomy – a young branch of astronomy responsible for many groundbreaking discoveries in the last 80 years

1950 | 1955 | 1960 | 1965 | 1970 | 1972

1930 | 1935 | 1940 | 1945

1977
A powerful narrow-band radio signal is detected by the Big Ear radiotelescope. Dubbed the 'WOW signal', it still remains the strongest candidate for a possible transmission from an extraterrestrial civilization

1974 | 1976 | 1978

1908
A major atmospheric explosion occurs near the Stony Tunguska River in Siberia, believed to be the result of an airburst of a small asteroid. Energy of the event is estimated at 15 megatons

1970
Birth of the Stingray Nebula, the youngest known planetary nebula. It's currently only 0.16 light years across, but is expanding rapidly

1979
Birth of SN 1979C, the youngest known black hole, after one star in a binary pair goes supernova. The black hole is currently accreting matter from its stellar companion, causing the system to glow brightly in X-rays

1980

1982

1983
Launch of the first infrared space telescope – IRAS. In order to operate, it had to be cooled close to the absolute zero with liquid helium.

1994
21 large fragments of the comet Shoemaker-Levy, torn apart by Jupiter's tidal forces two years earlier, impact the planet. The total release of energy was equivalent to more than 600 times the entire global nuclear arsenal, and the dark spots in Jupiter's atmosphere left by the impacts persists for several months

FREE HYDROGEN AND HELIUM: 4%

DARK MATTER: 24%

HYDROGEN AND HELIUM IN STARS: 0.5%

DARK ENERGY: 71.1%

2002 | 2000 | 1998 | 1996

1995
Discovery of 51 Pegasi b – the first known extrasolar planet orbiting a main-sequence star. Since then, over 4,050 more exoplanets have been confirmed to exist in more than 3,000 star systems

1994

ALL HEAVIER ELEMENTS: 0.03%

COMPOSITION OF THE 13.8-BILLION-YEAR-OLD UNIVERSE

1984

1990
Hubble Space Telescope is launched into orbit. Because of a mirror defect, it couldn't operate properly until a later 1993 Space Shuttle mission corrected its optics to their intended quality

NEUTRINOS: 0.3%

1992 | 1990 | 1988 | 1986

3000
Gamma Cephei becomes the north polar star due to precession of the Earth's axis

4000
If it weren't for the anthropogenic emissions of carbon dioxide, the next glacial period would probably have started by this point

58000
Due to the gradual slowing of Earth's rotation, length of the day becomes 1 second longer. In the meantime, the Moon has drifted over 2 km further away from Earth in its orbit

2880
A potentially dangerous asteroid 1950 DA might impact Earth. This 1.1 km object once held the highest known probability of impacting Earth of any asteroid, however the odds of this happening have been recently revised to just 1 in 8,300

5965
Snap 10A, an experimental satellite containing a defunct nuclear fission reactor, burns up in the Earth's atmosphere after its orbit eventually decays

40000
Niagara Falls disappears after the water erodes away the remaining 32 km towards Lake Erie

2300 | 2500 | 3000 | 4000 | 5000 | 6000 | 8000 | 10000 | Number of base pairs in human DNA | 50000

Arecibo radiotelescope dish | Average human height | DNA double helix | Nucleotides that make up DNA

4285
Long-period comet Hale Bopp returns to the inner solar system. Its most recent approach in 1997 exceeded all expectations, as it remained visible to the naked eye for a record 18 months

25000
Arecibo message – a radio transmission sent by the Arecibo radio telescope in 1974 towards the star cluster M13, reaches its target. It's encoded in 1,679 binary digits, and carries a graphical representation of basic information about humanity

Numbers from 1 to 10

Solar system | Human population in 1974 | Human figure | Constituent elements of DNA (HCNOP)

CONSTELLATIONS THROUGHOUT THE AGES

	50 000 BC	2000	50 000	100 000
BIG DIPPER				
ORION				
SOUTHERN CROSS				
LEO				
CASSIOPEIA				
LYRA				

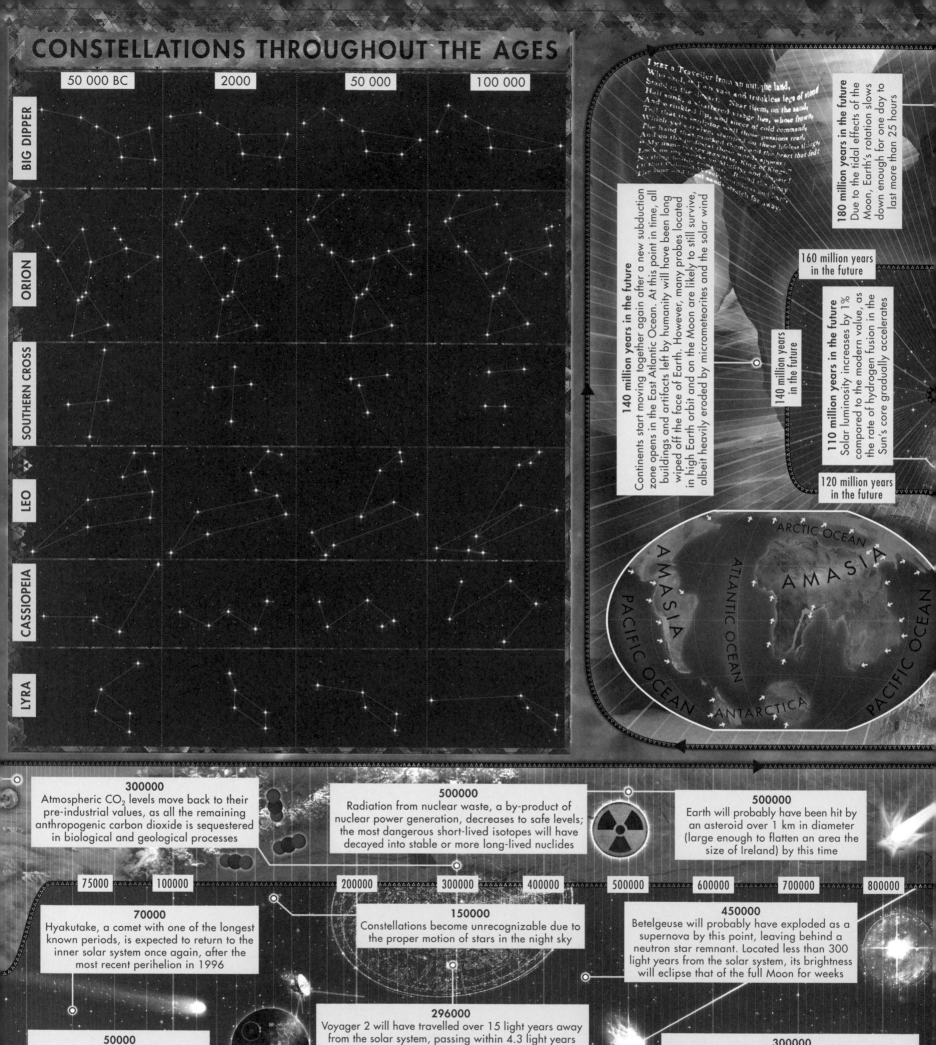

180 million years in the future
Due to the tidal effects of the Moon, Earth's rotation slows down enough for one day to last more than 25 hours

160 million years in the future

140 million years in the future
Solar luminosity increases by 1% compared to the modern value, as the rate of hydrogen fusion in the Sun's core gradually accelerates

140 million years in the future

140 million years in the future
Continents start moving together again after a new subduction zone opens in the East Atlantic Ocean. At this point in time, all buildings and artifacts left by humanity will have been long wiped off the face of Earth. However, many probes located in high Earth orbit and on the Moon are likely to still survive, albeit heavily eroded by micrometeorites and the solar wind

120 million years in the future

ARCTIC OCEAN

A M A S I A

AMASIA

PACIFIC OCEAN

ATLANTIC OCEAN

ANTARCTICA

PACIFIC OCEAN

300000
Atmospheric CO_2 levels move back to their pre-industrial values, as all the remaining anthropogenic carbon dioxide is sequestered in biological and geological processes

500000
Radiation from nuclear waste, a by-product of nuclear power generation, decreases to safe levels; the most dangerous short-lived isotopes will have decayed into stable or more long-lived nuclides

500000
Earth will probably have been hit by an asteroid over 1 km in diameter (large enough to flatten an area the size of Ireland) by this time

75000	100000	200000	300000	400000	500000	600000	700000	800000

70000
Hyakutake, a comet with one of the longest known periods, is expected to return to the inner solar system once again, after the most recent perihelion in 1996

150000
Constellations become unrecognizable due to the proper motion of stars in the night sky

450000
Betelgeuse will probably have exploded as a supernova by this point, leaving behind a neutron star remnant. Located less than 300 light years from the solar system, its brightness will eclipse that of the full Moon for weeks

296000
Voyager 2 will have travelled over 15 light years away from the solar system, passing within 4.3 light years of the Sirius system. The probe carries the Voyager Golden Record, which contains sounds and images portraying the diversity of Earth's life and culture

50000
Antares, a bright red supergiant already nearing the end of its life, will have exploded as a supernova

300000
WR-104, a Wolf-Rayet star 7,500 light years away, is expected to explode as a supernova, potentially producing a destructive gamma-ray burst

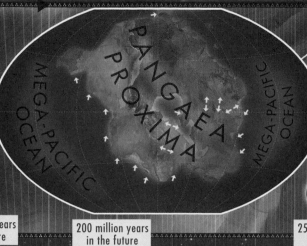

PANGAEA PROXIMA

MEGA-PACIFIC OCEAN

MEGA-PACIFIC OCEAN

...years ...re

250 million years in the future
All the Earth's major landmasses converge once again into a supercontinent – Pangaea Proxima. The interior of this supercontinent will probably be an extreme desert, with daytime temperatures routinely exceeding 70 °C

400 million years in the future

500 million years in the future

400 million years in the future
Interacting 'Antennae Galaxies' conclude their merging process, as they coalesce into a single massive elliptical galaxy

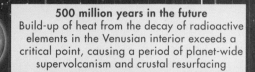

300 million years in the future
Earth's underground reserves of coal, natural gas and petroleum naturally restore to their pre-industrial levels

200 million years in the future

250 million years in the future

300 million years in the future

230 million years in the future
Solar system completes one full orbit around the centre of the Milky Way galaxy from its present-day position

330 million years in the future
Luminosity of the Sun rises by 3% compared to the modern day. This causes the polar dry ice deposits on Mars to fully evaporate, doubling the density of the planet's carbon dioxide-rich atmosphere

500 million years in the future
Build-up of heat from the decay of radioactive elements in the Venusian interior exceeds a critical point, causing a period of planet-wide supervolcanism and crustal resurfacing

100 million years in the future
Tarantula Nebula, a massive star-forming region in the Large Magellanic Cloud, transforms into a tightly bound globular star cluster

80 million years in the future
Estimated lifespan of the ring system around Saturn. Every second, Saturn's rings lose several tons of small ice particles due to atmospheric drag

40 million years in the future

40 million years in the future
Earth's climate warms enough for the Antarctic ice sheet to fully melt down

30 million years in the future

...00 million years in the future

80 million years in the future

60 million years in the future

100 million years in the future
Earth will probably have been hit by an asteroid over 10 km in diameter (comparable to the Chicxulub impactor which wiped out the dinosaurs) by this time

50 million years in the future
Movement of the tectonic plates along the San Andreas Fault puts Los Angeles and San Francisco right next to each other

80 million years in the future
Big Island becomes the last of the present-day Hawaiian Islands to disappear under the ocean. However, as the Pacific plate moves above the local hotspot, an entire new chain of islands will have replaced Hawaii in the mean time

40 million years in the future
Africa's collision with Eurasia results in uplifting of a major mountain range, rivalling the modern Himalayas in height

15 million years in the future
The widening East African Rift becomes completely flooded by the Indian ocean, causing the Somalian plate to fully separate from the African mainland

20 million years in the future

10 million years in the future
Estimated time until the Earth's animal and plant biodiversity recovers from the ongoing Holocene mass extinction

950000
Barringer meteor crater in the desert of Arizona erodes into unrecognizability

1.5 million years in the future
Earth will probably have undergone another catastrophic supervolcanic eruption (comparable to the Lake Toba eruption 75,000 years ago) by this time

1 million years in the future

1.5 million years in the future

1 million
Desdemona and Cressida, two small moons of Uranus, are expected to have collided with each other by this point. This might result in a bright ring system forming around the planet

4 million years in the future
Northwards movement of Africa seals the Strait of Gibraltar permanently. Mediterranean Sea rapidly dries up, transforming into a giant salt basin

10 million years in the future

8.4 million years in the future
Mount Rushmore, a granite mountain that hosts the giant sculpted faces of four US presidents, erodes away into unrecognizability

2 million years in the future

5 million years in the future

1.28 million years in the future
Orange dwarf star Gliese 710 will approach the Sun as close as 0.22 light years. For several millennia, the frequency of comet arrivals visible to the naked eye is expected to increase more than a hundredfold

2.5 million years in the future
Galactic shockwaves of ionized hot gas from the most recent active phase of the Milky Way's central black hole reach the solar system

7.6 million years in the future
Orbit of the Martian moon Phobos decays from 6,000 km in the present to less than 4,000 km above the Martian surface. As it crosses its 'roche limit', tidal forces of Mars tear Phobos apart, forming a short-lived ring system around the planet

550 million years in the future
As the Moon moves outwards and the Sun slowly increases in size, complete solar eclipses on Earth become impossible. The length of one Earth day is now almost at 27 hours

800 million years in the future
Average temperature on Earth exceeds 40 °C. Poles and oceans become the only places where animal and plant life can still survive, as most of the planet becomes an inhospitable wasteland

600 million years in the future
Rate at which silicate minerals react with carbon dioxide increases as a result of the rising solar luminosity, to the point that atmospheric CO_2 levels drop below 50 ppm. This makes C3 photosynthesis, the pathway most plants rely on to survive, effectively impossible. They are completely replaced by plants that rely on more complex C4 photosynthesis (like many grass species), that can function with CO_2 levels as low as 10 ppm

1 billion years in the future
27% of all water in the modern-day oceans will have been subducted deep into the Earth's mantle by this point

600 million years in the future	700 million years in the future	800 million years in the future	900 million years in the future	1 billion years in the future

550 million years in the future
Our planet will probably have been hit by a devastating gamma-ray burst from a nearby supernova by this time

750 million years in the future
Sirius A, the brightest star in the modern night sky, expands into a red giant, as it enters the last stage of its lifespan

1.05 billion years in the future
Luminosity of the Sun rises by 10% compared to the modern day, pushing Earth towards the edge of the circumstellar habitable zone

Temperature: 2.61 K
Diameter: 97 billion light years

3.3 billion years in the future
There is a 1% chance that gravitational effects of Jupiter will have caused Mercury to collide with Venus by this time. Other possible scenarios include Venus colliding with Earth, Mercury plunging into the Sun, or Mars being ejected from the solar system

2.75 billion years in the future
As the Moon gradually drifts away from the Earth, its stabilizing effect on our planet's rotational axis lessens. Earth's axial tilt becomes chaotic, poles begin to shift dramatically on the order of millions of years

3.5 billion years in the future	3 billion years in the future		2.5 billion years in the future

3.6 billion years in the future
Neptune's largest moon Triton (believed to be a captured dwarf planet from the Kuiper Belt) disintegrates after it passes through the planet's Roche limit. An extensive ring system forms around Neptune from the ice and rock debris

3 billion years in the future
Average surface temperature on Earth exceeds 200 °C, even at the poles. The heat eventually conducts its way deep underground, and even the sturdiest of extremophiles are destroyed, wiping the Earth clean of all life

4.5 billion years in the future
After billions of years of erosion and volcanic resurfacing, all direct geological evidence of life on Earth is wiped out. Space satellites and probes, the last remnants of human civilization, will have long been disintegrated into fine dust by micrometeorites and cosmic rays

2.8 billion years in the future
Rising solar heat combined with volcanic activity initiates a runaway outgassing of water vapour and carbon dioxide from the Earth's crust into the atmosphere. Conditions on the surface of our planet start to resemble those on modern-day Venus, but are still nowhere as extreme

4 billion years in the future		

5.5 billion years in the future
Length of one Earth's day slows down to over 50 hours. By this point, tidal effects have also pushed the Moon so far from our planet that it appears two times smaller than the Sun in the sky

6.75 billion years in the future
Earth becomes tidally locked to the Sun. Temperatures on the illuminated side exceed 1,500 °C due to the extreme solar irradiance, causing silicate rock to melt into oceans of lava

4.5 billion years in the future	5 billion years in the future	6 billion years in the future

4 billion years in the future
Milky Way and the Andromeda Galaxy collide with each other, and begin to gradually coalesce. Over the course of this galactic merger, collision between any two stars is highly unlikely, but many, including the Sun, could get ejected into intergalactic space

4.4 billion years in the future
Mars is now well within the habitable zone, its surface temperatures comparable with modern-day Earth. It will stay like this for roughly 2 billion more years

5.4 billion years in the future
The Sun runs out of the hydrogen supply in its core, and leaves the main sequence. After hydrogen fusion moves outwards to the shell surrounding its core, our star starts to rapidly expand, transforming into a red giant

4.6 billion years in the future
Milky Way and Andromeda complete their merging process and reform into a single giant elliptical galaxy, dubbed 'Milkomeda'. Eventually, the two supermassive black holes spiral towards each other and merge as well

6.5 billion years in the future
As the Sun expands, the icy moons of Jupiter melt and experience a brief window of habitability. Soon, this also becomes the case for Saturn's hydrocarbon-rich moon Titan, and finally for Kuiper Belt objects such as Eris and Pluto

Temperature: 2.02 K
Diameter: 125 billion light years

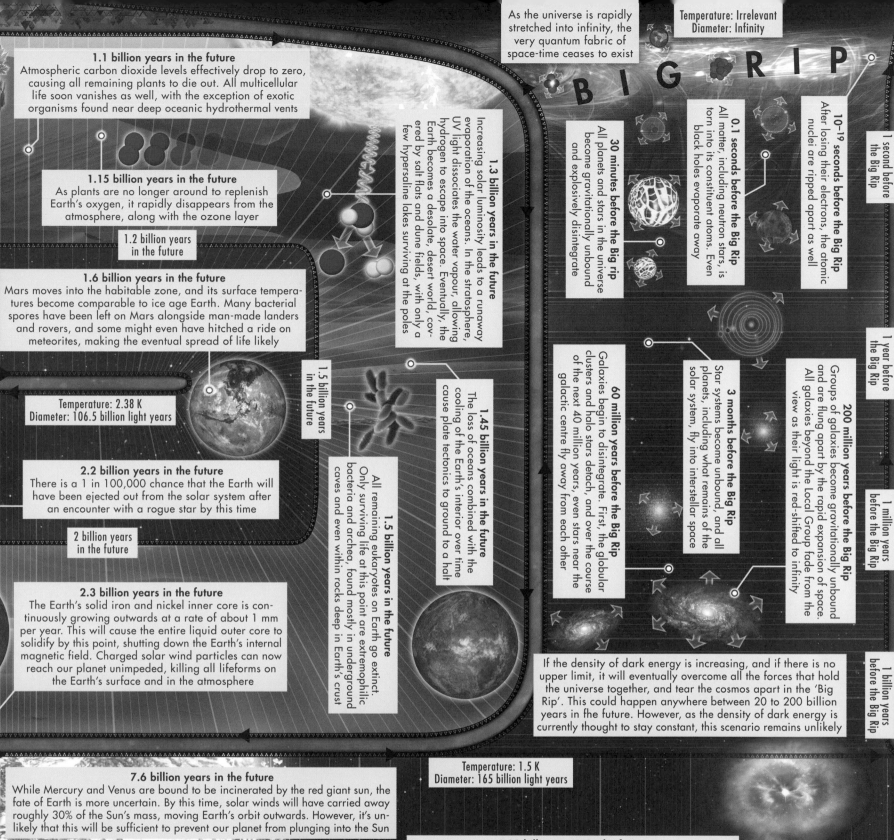

1.1 billion years in the future
Atmospheric carbon dioxide levels effectively drop to zero, causing all remaining plants to die out. All multicellular life soon vanishes as well, with the exception of exotic organisms found near deep oceanic hydrothermal vents

1.15 billion years in the future
As plants are no longer around to replenish Earth's oxygen, it rapidly disappears from the atmosphere, along with the ozone layer

1.2 billion years in the future

1.6 billion years in the future
Mars moves into the habitable zone, and its surface temperatures become comparable to ice age Earth. Many bacterial spores have been left on Mars alongside man-made landers and rovers, and some might even have hitched a ride on meteorites, making the eventual spread of life likely

Temperature: 2.38 K
Diameter: 106.5 billion light years

2.2 billion years in the future
There is a 1 in 100,000 chance that the Earth will have been ejected out from the solar system after an encounter with a rogue star by this time

2 billion years in the future

2.3 billion years in the future
The Earth's solid iron and nickel inner core is continuously growing outwards at a rate of about 1 mm per year. This will cause the entire liquid outer core to solidify by this point, shutting down the Earth's internal magnetic field. Charged solar wind particles can now reach our planet unimpeded, killing all lifeforms on the Earth's surface and in the atmosphere

1.3 billion years in the future
Increasing solar luminosity leads to a runaway evaporation of the oceans. In the stratosphere, UV light dissociates the water vapour, allowing hydrogen to escape into space. Eventually, the Earth becomes a desolate, desert world, covered by salt flats and dune fields, with only a few hypersaline lakes surviving at the poles

1.5 billion years in the future

1.45 billion years in the future
The loss of oceans combined with the cooling of the Earth's interior over time cause plate tectonics to ground to a halt

1.5 billion years in the future
All remaining eukaryotes on Earth go extinct. Only surviving life at this point are extremophilic bacteria and archea, found mostly in underground caves and even within rocks deep in Earth's crust

As the universe is rapidly stretched into infinity, the very quantum fabric of space-time ceases to exist

Temperature: Irrelevant
Diameter: Infinity

BIG RIP

10–19 seconds before the Big Rip
After losing their electrons, the atomic nuclei are ripped apart as well

0.1 seconds before the Big Rip
All matter, including neutron stars, is torn into its constituent atoms. Even black holes evaporate away

30 minutes before the Big Rip
All planets and stars in the universe become gravitationally unbound and explosively disintegrate

3 months before the Big Rip
Star systems become unbound, and all planets, including what remains of the solar system, fly into interstellar space

60 million years before the Big Rip
Galaxies begin to disintegrate. First, the globular clusters and halo stars detach, and over the course of the next 40 million years, even stars near the galactic centre fly away from each other

200 million years before the Big Rip
Groups of galaxies become gravitationally unbound and are flung apart by the rapid expansion of space. All galaxies beyond the Local Group fade from the view as their light is red-shifted to infinity

1 second before the Big Rip

1 year before the Big Rip

1 million years before the Big Rip

1 billion years before the Big Rip

If the density of dark energy is increasing, and if there is no upper limit, it will eventually overcome all the forces that hold the universe together, and tear the cosmos apart in the 'Big Rip'. This could happen anywhere between 20 to 200 billion years in the future. However, as the density of dark energy is currently thought to stay constant, this scenario remains unlikely

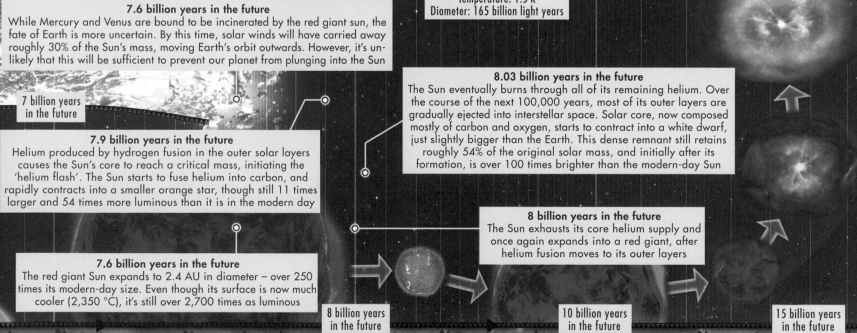

Temperature: 1.5 K
Diameter: 165 billion light years

7.6 billion years in the future
While Mercury and Venus are bound to be incinerated by the red giant sun, the fate of Earth is more uncertain. By this time, solar winds will have carried away roughly 30% of the Sun's mass, moving Earth's orbit outwards. However, it's unlikely that this will be sufficient to prevent our planet from plunging into the Sun

7 billion years in the future

7.9 billion years in the future
Helium produced by hydrogen fusion in the outer solar layers causes the Sun's core to reach a critical mass, initiating the 'helium flash'. The Sun starts to fuse helium into carbon, and rapidly contracts into a smaller orange star, though still 11 times larger and 54 times more luminous than it is in the modern day

7.6 billion years in the future
The red giant Sun expands to 2.4 AU in diameter – over 250 times its modern-day size. Even though its surface is now much cooler (2,350 °C), it's still over 2,700 times as luminous

8.03 billion years in the future
The Sun eventually burns through all of its remaining helium. Over the course of the next 100,000 years, most of its outer layers are gradually ejected into interstellar space. Solar core, now composed mostly of carbon and oxygen, starts to contract into a white dwarf, just slightly bigger than the Earth. This dense remnant still retains roughly 54% of the original solar mass, and initially after its formation, is over 100 times brighter than the modern-day Sun

8 billion years in the future
The Sun exhausts its core helium supply and once again expands into a red giant, after helium fusion moves to its outer layers

8 billion years in the future

10 billion years in the future

15 billion years in the future

In many cosmological models, the Big Crunch would be immediately followed by a rebirth of the universe in a new Big Bang, as a part of an eternal cosmic cycle with neither a beginning nor an end

BIG BOUNCE

Everything in the universe collapses into an infinitely dense singularity

BIG CRUNCH

2×10^{106} years in the future
Even the largest supermassive black holes ultimately decay via the emission of Hawking radiation. This marks the end of the black hole era, and the universe enters the final phase of its existence, as it marches towards the inevitable heat death

10^{1500} years in the future
If protons don't decay, all remaining nucleons in the universe will have transformed into iron-56 (the lowest possible energy state of nuclear matter) via quantum tunnelling by this time. All matter that still exists in the universe assumes the form of cold iron spheres, isolated from each other by an endless void

Moments before the Big Crunch
The extreme temperatures and pressure gradually break down all matter into elementary particles, repeating the numerous 'condensations' that took place after the Big Bang, but in reverse order

10,000 years before the Big Crunch

BLACK

5×10^{68} years in the future
First stellar-mass black holes evaporate away via the emission of 'Hawking radiation', produced by quantum effects at the event horizon. As black holes get smaller, the faster they evaporate and the more Hawking radiation they emit. Eventually, they will become anything but black, as they end their lifespan with a burst of ultra-high-energy gamma rays

10,000 years before the Big Crunch
The increasing heat of cosmic background 'cooks' the remaining stars from outside in, eventually causing them to explode and disperse into clouds of plasma

10^{100} (googol) years in the future

10^{16} years in the future

DEGENERATE

10 billion years before the Big Crunch
As galaxies hurtle towards each other, their light becomes increasingly blue-shifted, making the deep sky far brighter and more vivid

10 million years before the Big Crunch

10^{19} years in the future
After many encounters with each other, objects in the galaxy gradually exchange their kinetic energy in a way that favours lighter objects gaining speed, and heavier objects losing it. This will eventually cause around 90-99% of stellar remnants and brown dwarfs in Milkomeda and other galaxies to gain escape velocity and drift into intergalactic space

100 million years before the Big Crunch
As galaxies hurtle towards each other, they begin to merge into giant clumps

1 million years before the Big Crunch
Collisions between stars become frequent as boundaries between distinct galaxies disappear. Cosmic background radiation rapidly heats up, causing the sky everywhere to glow red

10 billion years before the Big Crunch

10^{15} (1 quadrillion) years in the future

10^{15} (1 quadrillion) years in the future
All planets in the solar system will probably have been ejected from their orbits by this point, after numerous close encounters with other stellar remnants. Temperature of the black dwarf Sun is now only 5 degrees above the absolute

10^{18} years in the future

200 trillion years in the future
Even though the era of star formation has long ended, occasional collisions between brown and black dwarfs help to maintain a relatively stable population of several dozen dim red dwarfs in the entire galaxy

If the influence of dark energy is not powerful enough, or if it reverses in the future, the mutual gravitational attraction of all matter will eventually slow the expansion of the universe down to a halt. This would probably happen anywhere between 50 to 500 billion years in the future. Afterwards, the universe would start contracting, eventually ending in a 'Big Crunch'. From what we know about dark energy, this scenario is considered unlikely

500 trillion years in the future

200 trillion years in the future

100 trillion years in the future

Temperature: 0.8 K
Diameter: 320 billion light years

Temperature: 0.0075 K
Diameter: 33 trillion light years

20 billion years in the future
Triangulum Galaxy, currently the third largest galaxy in the Local Group, merges with Milkomeda after orbiting it for several billion years

100 billion years in the future
Over 99% of stars in Milkomeda are now red dwarfs. As interstellar gas becomes increasingly depleted, rate of new star formation drops to less than one per century (compared to seven per year just in the present-day Milky Way galaxy)

180 billion years in the future
Gliese 581, a red dwarf orbited by five known exoplanets, reaches the end of its lifespan, as it gradually contracts into a low-mass white dwarf

65 billion years in the future
A nearby red dwarf, Lalande 21185, reaches the end of its lifespan

50 billion years in the future
Assuming Earth has survived the death of the Sun, it becomes tidally locked with the Moon, both bodies permanently showing only one face to each other. However, as tidal effects of the white dwarf Sun gradually siphon angular momentum from this system, the lunar orbit will eventually decay, and the Moon will end up disintegrating and colliding with Earth in the far future

150 billion years in the future
All galaxies outside the Local Group disappear beyond the cosmic light horizon, the boundary of the observable universe, as they red-shift to infinity. Messier 81, presently one of the closest galaxies to the Local Group at 11.4 million light years away, will be over 200 trillion light years away by this time, receding at over 10,000 times the speed of light

OBSERVABLE UNIVERSE

20 billion years in the future

50 billion years in the future

100 billion years in the future

200 billion years in the future

$10^{10^{100}}$ (googolplex) years in the future

$10^{10^{26}}$ years in the future
All remaining matter in the universe collapses into black holes via the effects of quantum tunnelling, and subsequently evaporates away into photons and leptons. As all particles drift away from each other for all eternity, the universe enters its final energy state. This scenario, also known as the 'Big Freeze', or the 'heat death', is currently thought of as the most likely eventual fate of the universe

It's theorized that long after the universe has reached its final energy state, a random quantum fluctuation could cause a new Big Bang to commence. This new universe would probably bear little resemblance to the one we live in, with dramatically different laws of physics and fundamental constants. Many scientists believe that our universe is indeed only one of many (possibly infinite) universes in a 'multiverse'. Out of those countless universes, only a few would happen to have the right combination of laws of physics for conscious life to develop – this would explain why our universe appears to many to be finely tuned for life

10^{65} years in the future
All objects in the universe become spherical. On long-enough timescales, even the most rigid bodies of matter behave like a liquid, as their constituent atoms continuously rearrange themselves and diffuse around via quantum tunnelling

10^{43} years in the future
If protons are not stable, all matter in the universe outside of black holes will have decayed by this point. Although proton decay has never been observed, many theoretical models suggest that it does happen, just on very long timescales. It's thought that the decay of protons would take place via the same process that originally caused ordinary matter to predominate over antimatter

10^{35} years in the future

10^{50} years in the future

10^{45} years in the future

10^{40} years in the future

ERA

10^{30} years in the future

10^{20} years in the future
If Earth miraculously survives to this point without getting ejected from the solar system, its orbit will gradually decay via the emission of gravitational waves, causing what remains of our planet to inevitably impact the black dwarf Sun

2×10^{24} years in the future
Half of all tellurium-128, the most stable known radioisotope in the universe, will have decayed

10^{30} years in the future
All rogue planets, stellar black holes, neutron stars and brown and black dwarfs not previously ejected from their galaxies fall into supermassive black holes. By this time, the universe has expanded so vastly in size that all remaining large bodies are now completely isolated within their own observable universe bubbles, incapable of ever interacting with each other in the future in any way

10^{20} years in the future

10^{22} years in the future

10^{24} years in the future

10^{26} years in the future

10^{28} years in the future

30 trillion years in the future

STELLIFEROUS ERA

60 trillion years in the future
All stars in the universe born from interstellar gas clouds will have exhausted their fuel reserves and transformed into white and black dwarfs

40 trillion years in the future
At this point, almost all ordinary matter in the universe has condensed into black holes, neutron stars, planets and black and brown dwarfs. As galaxies run out of interstellar gas and dust, star formation grounds to a halt

30 trillion years in the future
By this time, the black dwarf Sun will likely have undergone a sufficiently close encounter with another star or a stellar remnant to significantly disturb the orbits of the remaining planets

80 trillion years in the future

60 trillion years in the future

40 trillion years in the future

200 billion years in the future
Cosmic microwave background, the oldest light in the universe, cools down to less than 0.0001 K above the absolute zero, rendering it for all intents and purposes undetectable. All astronomical evidence of the Big Bang disappears for ever

Temperature: 10^{-63} K
Diameter: 10^{74} light years

18 trillion years in the future
LHS 2924, the dimmest, coldest and the least massive known star still capable of fusing hydrogen in its core, reaches the end of its lifespan. It's extremely unlikely that any star in the universe could continuously shine for longer than 20 trillion years

20 trillion years in the future

800 billion years in the future
Milkomeda Galaxy begins to markedly decrease in brightness as most red dwarfs have passed through their final 'blue dwarf' stage of peak luminosity by this time. Dead, gradually dimming white dwarfs now greatly outnumber all other stars combined

8 trillion years in the future
Ultra-dim red dwarf TRAPPIST-1, with a system of seven exoplanets, reaches the end of its lifespan

450 billion years in the future
All galaxies in the Local Group will have merged with Milkomeda by this point. The universe is now composed of billions of completely isolated pockets, each hosting one single gigantic galaxy in its centre

4 trillion years in the future
Proxima Centauri reaches the end of its lifespan

10 trillion years in the future
The white dwarf Sun cools down enough that it no longer emits any visible light – turning into a solid 'black dwarf' – the ultimate fate of most stars in the universe

500 billion years in the future

1 trillion years in the future

5 trillion years in the future

10 trillion years in the future

STARS SETTING OVER A GRATEFUL UNIVERSE
(SUN REPLACED WITH OTHER STARS)

THE SUN
(Yellow main sequence star)
1 solar mass

KEPLER-35
(Binary yellow main sequence stars)
0.887 and 0.809 solar masses

EPSILON ERIDANI
(Yellow main sequence star)
0.82 solar masses

V830 TAURI
(Young T-Tauri star)
1.1 solar masses

GLIESE 581
(Red dwarf star)
0.31 solar masses

BARNARD'S STAR
(Red dwarf star)
0.144 solar masses

TRAPPIST-1
(Red dwarf star with four Earth-sized planets within its habitable zone)
0.081 solar masses

S1326447-4728037
(Red dwarf star in a globular cluster)
0.13 solar masses

HD 188753
(Triple star system)
*One star with 1.06 solar masses, and a pair
of stars with 0.96 and 0.67 solar masses*

POLLUX
(Orange giant star)
2.04 solar masses

RZ GRUIS
(Cataclysmic variable double star system)
*White dwarf accreting mass from a yellow main sequence star.
Potential candidate for a type 1a supernova*

ARCTURUS
(Red giant star)
1.14 solar masses

XI TAURI
(Complex quadruple star system)
Exact star masses unknown

PROCYON
(White main sequence star)
1.42 solar masses

HEN 2-428
(Binary white dwarf system)
0.85 and 0.75 solar masses

SIRIUS A
(White main sequence star)
2.02 solar masses

BIBLIOGRAPHY

Wikipedia: The Free Encyclopedia. Wikimedia Foundation Inc. *https://en.wikipedia.org*
Encyclopedia Britannica Online. Encyclopedia Britannica Inc. *https://www.britannica.com/*
Encyclopedia.com, *https://www.encyclopedia.com/*
National Aeronautics and Space Administration. NASA, *https://www.nasa.gov*
European Space Agency. ESA, *https://www.esa.int/ESA*
Aerospace. The Aerospace Corporation, *https://aerospace.org/about*
Krebs, Gunter Dirk. *Gunter's Space Page*, *https://space.skyrocket.de/*
Stathopoulos, Vic. *Aerospaceguide*, *http://www.aerospaceguide.net/*
The Satellite Encyclopedia. TBS Internet, *https://www.tbs-satellite.com/tse/*

Cosmologies throughout history:
Cavendish, Richard. *Mythology – An Illustrated Encyclopedia of the Principal Myths and Religions of the World*. Barnes and Noble, 2003
Dasa, Sadaputa. 'The Universe of the Vedas'. *Krishna.com*. The Bhaktivedanta Book Trust International, Inc. *http://www.krishna.com/universe-vedas*
N. F. Gier. *God, Reason, and the Evangelicals*. University Press of America, 1987
Van Helden, Albert. *The Galileo Project*, *http://galileo.rice.edu/*
Schombert, James. 'Ancient Cosmology'. *University of Oregon*, *http://abyss.uoregon.edu/~js/ast123/lectures/lec01.html*
'Historical Models in the European and Chinese Contexts.' *APFSLT*. Education University of Hong Kong, *https://www.eduhk.hk/apfslt/v6_issue2/liusc/liusc4.htm*
O'Dubhain, Searles. 'Shimmering Lights and Strange Music' *Summerlands*, *http://www.summerlands.com/crossroads/library/otherwor.htm*
Carlson, John B. and Dallison, Ken. 'Ancient Skywatchers', *National Geographic Magazine*. National Geographic Society, March 1990
Grimm, Jacob. 'Images of Old Norse Cosmology' *Germanic Mythology*, *http://www.germanicmythology.com/original/cosmology3.html*
Sermoneta, M. Caetani. *La materia della Divina commedia di Dante Alighieri dichiarata in VI tavole*. G. C. Sansoni, 1821
'Finding Our Place in the Cosmos: From Galileo to Sagan and Beyond'. Library of Congress, *https://www.loc.gov/collections/finding-our-place-in-the-cosmos-with-carl-sagan/*
Kollerstrom, N. *The Hollow World of Edmond Halley*. Science History Publications, 1992
The Flat Earth Society. The Flat Earth Society. *https://www.tfes.org/*
Murtaugh, Lindsey. 'Common Elements in Creation Myths'. Williams College, *https://www.cs.williams.edu/~lindsey/myths/myths.html*

Journey into outer space:
National Weather Service. National Oceanic and Atmospheric Administration (NOAA), *https://www.weather.gov*
US Standard Atmosphere. NOAA & NASA, 1976
Glasstone, Samuel and Dolan, Philip J. *The Effects of Nuclear Weapons*. United States Department of Defense and the Energy Research and Development Administration, 1977
Palt, Karsten. *Flugzeuginfo*, *http://www.flugzeuginfo.net/*
National Aeronautic Association. NAA, *https://naa.aero/*
FAI, Fédération Aéronautique Internationale/World Air Sports Federation, *https://fai.org/*
Nave, R. 'Boiling Point Variation'. *Hyperphysics*. Georgia State University, *http://hyperphysics.phy-astr.gsu.edu/hbase/Kinetic/vappre.html*
Guinness World Records. Guinness World Records Ltd. *https://www.guinnessworld records.com/*
'High Altitude Airship (HAA)'. Lockheed Martin, *http://www.lockheedmartin.com/products/HighAltitudeAirship/index.html*
'Red Sprites, Blue Jets and Elves'. *University at Albany, SUNY*, *https://www.albany.edu/faculty/rgk/atm101/sprite.htm*
Lefebvre, Jean-Luc. *Space Strategy*. Wiley-ISTE, 2017
Szondy, David. 'Artificial meteor shower planned for Japan in 2019'. *New Atlas*, Gizmag Pty. Ltd. *https://newatlas.com/sky-canvas-artificial-meteor-shower-japan/52295/*
Wade, Mark. *Encyclopedia Astronautica*. *http://www.astronautix.com/*
Wall, Mike. 'Monkeys in Space: A Brief Spaceflight History'. *Space.com*. Future US Inc. *https://www.space.com/19505-space-monkeys-chimps-history.html*
Darling, David. 'Encyclopedia of Science'. David Darling, *http://www.daviddarling.info/encyclopedia1.html*
Zak, Anatoly. *Russian Space Web*, *http://www.russianspaceweb.com/*
Earth Observation Portal. ESA, *https://directory.eoportal.org/web/eoportal/home*
Spaceflight 101, Space News and Beyond. Spacecraft & Satellites, *http://spaceflight101.com/*

Scale of the universe:
Australia Telescope National Facility. CSIRO, *http://www.atnf.csiro.au/*
Simberg, Rand. 'Bigelow Aerospace Shows Off Bigger, Badder Space Real Estate'. *Popular Mechanics*. Hearst Magazine Media Inc. *https://www.popularmechanics.com/space/a6247/bigelow-aerospace-ba2100-hotel/*
Huang, Cary and Huang, Michael. 'The Scale of the Universe 2'. Htwins, *https://htwins.net/scale2/*
'Schwarzschild radius of a given mass'. WolframAlpha, *https://www.wolframalpha.com/widgets/view.jsp?id=15c7a7eb32c8610b005811b8640ebc1*
Brown, Mike. 'The Dwarf Planets'. Caltech Division of Geological and Planetary Sciences. California Institute of Technology, *http://web.gps.caltech.edu/~mbrown/dwarfplanets/*
'JPL Small-Body Database Browser'. *Solar System Dynamics*. California Institute of Technology, *https://ssd.jpl.nasa.gov/sbdb.cgi?sstr=226*
'PDS Asteroid/Dust Archive'. The Planetary Data System. *NASA*, *https://sbn.psi.edu/pds/*
Exoplanets.org. The Pennsylvania State University, *http://exoplanets.org/*
The Extrasolar Planet Encyclopaedia. *http://exoplanet.eu/*
NASA Exoplanet Archive. IPAC, *https://exoplanetarchive.ipac.caltech.edu/*
The SAO/NASA Astrophysics Data System. Harvard-Smithsonian Center for Astrophysics, http://adsabs.harvard.edu/
SIMBAD Astronomical Database. Université de Strasbourg, *http://simbad.u-strasbg.fr/simbad/*
Extrasolar Planets Catalogue. Kyoto University, *http://www.exoplanetkyoto.org/*
Planetary Habitability Laboratory. University of Puerto Rico at Arecibo, *http://phl.upr.edu/projects/habitable-exoplanets-catalog*
Open Exoplanets Catalogue, *http://www.openexoplanetcatalogue.com/*
Sudarsky, D., Burrows, A. and Pinto, P. *Albedo and Reflection Spectra of Extrasolar Giant Planets*. The Astrophysical Journal, 2000
Batygin, Konstantin and Brown, Michael E. *Evidence for a Distant Giant Planet in the Solar System*. The Astronomical Journal, 2016
IAU. International Astronomical Union, *https://www.iau.org/*
Raymond, Sean. 'Forget "Earth-Like" – We'll First Find Aliens on Eyeball Planets'. *Nautilus*. NautilusThink Inc. *http://nautil.us/blog/forget-earth_likewell-first-find-aliens-on-eyeball-planets*
'Lecture 5: Overview of the Solar System, Matter in Thermodynamic Equilibrium'. The University of Arizona, *http://ircamera.as.arizona.edu/astr_250/Lectures/Lec_05sml.htm*
von Boetticher, Alexander. *A Saturn-size low-mass star at the hydrogen-burning limit*. Astronomy & Astrophysics, 2017
'Abundance in the Universe of the elements'. *Periodic Table*, *https://periodictable.com/Properties/A/UniverseAbundance.html*
Inglis-Arkell, Esther. 'The Mysterious Phenomenon Of "Gravity Darkening"'. *IO9*. Gizmodo Media Group, *https://io9.gizmodo.com/the-mysterious-phenomenon-of-gravity-darkening-1638441719*
Whitworth, N. John. *Universe Guide*, *https://www.universeguide.com/*
COSMOS – The SAO Encyclopedia of Astronomy. Swinburne University of Technology, *http://astronomy.swin.edu.au/cosmos/*
Philips, J. P. *The distances of highly evolved planetary nebulae*. Monthly Notices of the Royal Astronomical Society, 2005
Hubble Space Telescope. ESA, *https://www.spacetelescope.org/*
'The one hundred nearest star systems'. Georgia State University, *http://www.astro.gsu.edu/RECONS/TOP100.posted.htm*
The Messier Catalog. SEDS USA, *http://www.messier.seds.org/*
NASA/IPAC Extragalactic Database. California Institute of Technology, *http://ned.ipac.caltech.edu/*
'Basic plan of the Milky Way'. *Galaxy Map*, *http://galaxymap.org/drupal/node/171*
Powell, Richard. *An Atlas of the Universe*. *http://www.atlasoftheuniverse.com/*

The night sky:
Strobel, Nick. 'Magnitude System'. *Astronomy Notes*, *http://www.astronomynotes.com/starprop/s4.htm*
Stellarium Astronomy Software, *https://stellarium.org/*
'Astronomic Terms and Constants'. *Department of Astrophysical Sciences*, Princeton University, *https://www.astro.princeton.edu/~gk/A403/constants.pdf*
'The Constellations'. IAU, International Astronomical Union, *https://www.iau.org/public/themes/constellations/*

Exploring the cosmos:
'Solar System Exploration'. NASA, *https://solarsystem.nasa.gov/missions/*
'NASA Space Science Data Coordinated Archive'. NASA, *https://nssdc.gsfc.nasa.gov/*
Hamilton, Calvin J. 'Chronology of Space Exploration'. *Views of the Solar System*, *http://solarviews.com/eng/craft1.htm*
Braeunig, Robert A. *Rocket & Space Technology*, *http://www.braeunig.us/space/index.htm*
Hambleton, Kathryn. 'Deep Space Gateway to Open Opportunities for Distant Destinations'. *NASA*, *https://www.nasa.gov/feature/deep-space-gateway-to-open-opportunities-for-distant-destinations*
'Advanced Technology Large-Aperture Space Telescope'. STScI. Space Telescope Science Institute, *http://www.stsci.edu/atlast*
Kyle, Ed. *Space Launch Report*, *http://www.spacelaunchreport.com/*
Alway, Peter. *Rockets of the World*. Saturn Press, 1995
SpaceX. Space Exploration Technologies Corp. *https://www.spacex.com/*
Blue Origin. Blue Origin LLC, *https://www.blueorigin.com/*
Energia. S.P. Korolev Rocket and Space Corporation, *https://www.energia.ru/en*
Brugge, Norbert. 'Space launch vehicles all of the world'. *http://www.b14643.de/Spacerockets/index.htm*
Jones, Andrew. 'China reveals details for super-heavy-lift Long March 9 and reusable Long March 8 rockets'. *SpaceNews*, *https://spacenews.com/china-reveals-details-for-super-heavy-lift-long-march-9-and-reusable-long-march-8-rockets/*
JAXA. Japan Aerospace Exploration Agency, *https://global.jaxa.jp/*
Hill, John M. 'List of large reflecting telescope'. The University of Arizona, *http://abell.as.arizona.edu/~hill/list/bigtel99.htm*
Arnett, Bill. 'The World's Largest Optical Telescopes'. Nine Planets, *http://astro.nineplanets.org/bigeyes.html*

Maps of the inner solar system:
'Gazetteer of Planetary Nomenclature'. IAU. International Astrononomical Union, *https://planetarynames.wr.usgs.gov/*
U.S. Geological Survey. USGS, *https://www.usgs.gov*
Mahoney, T. J. *Mercury*. Springer, 2014
Digital Museum of Planetary Mapping, *https://planetarymapping.wordpress.com/*
Earth Impact Database. PASSC, *http://www.passc.net/EarthImpactDatabase/*
Lunar and Planetary Institute. Universities Space Research Association, *https://www.lpi.usra.edu*
Prado, Mark. 'The Apollo and Luna Samples'. *Permanent*, *https://www.permanent.com/l-apollo.htm*

Timeline of the universe:
Terzic, Balša. 'Lecture 13: History of the Very Early Universe'. Northern Illinois University, *http://www.nicadd.niu.edu/~bterzic/PHYS652/Lecture_13.pdf*
Seagrave, Wyken. *History of the Universe*, *https://www.historyoftheuniverse.com/index.php*
Gnedin, Nick. 'Cosmological Calculator for the Flat Universe'. Fermi Research Alliance LLC, *http://home.fnal.gov/~gnedin/cc/*
Wright, Ned. 'Javascript Cosmology Calculator'. *http://www.astro.ucla.edu/~wright/CosmoCalc.html*
Semenov, Dmitry A. 'From the Big Bang to the First Molecules'. Max Planck Institute for Astronomy, *http://www.mpia.de/homes/semenov/Lectures/Heidelberg_Uni_2012/lecture4-first-molecules.pdf*
Sonzogni, Alejandro. 'Interactive Chart of Nuclides'. National Nuclear Data Center. Brookhaven National Laboratory, *https://www.nndc.bnl.gov/nudat2/*
Brau, Jim. 'The Early Universe, Toward the Beginning of Time'. University of Oregon, *https://pages.uoregon.edu/jimbrau/astr123/Notes/Chapter27.html*
'Wilkinson Microwave Anistropy Probe'. NASA, *https://wmap.gsfc.nasa.gov/*
Johnson, Jennifer. *'Origin of Elements in the Solar System'. Science blog from the SDSS*. Sloan Digital Sky Surveys, *http://blog.sdss.org/2017/01/09/origin-of-the-elements-in-the-solar-system/*
Byrd, Deborah. 'Peering toward the Cosmic Dark Ages'. *EarthSky*. EarthSky Communications Inc. *https://earthsky.org/space/cosmic-dark-ages-lyman-alpha-galaxies-lager*
Massey, R. 'Three-dimensional distribution of dark matter in the Universe'. *Hubble Space Telescope*, *https://www.spacetelescope.org/images/heic0701a/*
Gordon, Karl. 'The Spitzer Infrared Nearby Galaxies Surves (SINGS) Hubble Tuning Fork'. Spitzer Science Center, *http://www.spitzer.caltech.edu/images/2095-sig07-025-Lifestyles-of-the-Galaxies-Next-Door*
Battersby, Stephen. 'Biggest black holes may grow inside quasistars'. *New Scientist*. New Scientist Ltd. *https://www.newscientist.com/article/dn12982-biggest-black-holes-may-grow-inside-quasistars/#.Ut1FCftwHs0*
Schombert, James. 'Stellar Populations'. University of Oregon, *http://abyss.uoregon.edu/~js/ast122/lectures/lec26.html*
Hazen, Robert M. 'Mineral Evolution'. *Carnegie Science*, *https://hazen.carnegiescience.edu/research/mineral-evolution*
Scotese, Christopher R. *Paleomap Project*, *http://www.scotese.com/*
'Continental Drift'. Algol, *https://www.youtube.com/watch?v=ovT90wYrVk4*
'Gravitational Wave Luminosity Calculator'. EasyCalculation.com, *https://www.easycalculation.com/physics/astrodynamics/gravitational-waves-luminosity.php*
Commissariat, Tushna. 'LIGO detects first ever gravitational waves – from two merging black holes'. *Physicsworld*, *https://physicsworld.com/a/ligo-detects-first-ever-gravitational-waves-from-two-merging-black-holes/*
Geologic Data Scale, *https://sites.google.com/site/geologicdatascale/*
Byrd, Deborah. 'A quasar Milky Way six million years ago?' *EarthSky*. EarthSky Communications Inc. *https://earthsky.org/space/a-quasar-milky-way-six-million-years-ago*
Tomonori, Nomoto. *HippLiner*, *http://t.nomoto.org/HippLiner/index-e.html*
Strom, Robert G.; Schaber, Gerald G. and Dawson, Douglas D. 'The global resurfacing of Venus'. *Journal of Geophysical Research*, 1994
Meadows, A. J. *The Future of the Universe*. Springer, 2007
May, Brian; Moore, Patrick and Lintott, Chris. *Bang!: The Complete History of the Universe*. Johns Hopkins University Press, 2008
'NASA's Hubble Shows Milky Way is Destined for Head-On Collision'. NASA, *https://www.nasa.gov/mission_pages/hubble/science/milky-way-collide.html*
Heath, Martin J. and Doyle, Laurance R. 'Circumstellar Habitable Zones to Ecodynamic Domains: A Preliminary Review and Suggested Future Directions'. 2009
Adams, Fred C.; Laughlin, Gregory and Graves, Genevieve J. M. 'Red dwarfs and the end of the main sequence'. RevMexAA, 2004
Physics of the Universe, *https://www.physicsoftheuniverse.com/*
Bryner, Jeanna. 'Long Shot: Planet Could Hit Earth in Distant Future'. *Space.com*. Future US, Inc. *https://www.space.com/6824-long-shot-planet-hit-earth-distant-future.html*
'What is the Ultimate Fate of the Universe?' NASA, *https://map.gsfc.nasa.gov/universe/uni_fate.html*
Carroll, Sean M. and Chen, Jennifer. 'Spontaneous Inflation and the Origin of the Arrow of Time'. 2004
Davies, Paul. *The Last Three Minutes: Conjectures About The Ultimate Fate Of The Universe*. Basic Books, 1997

PICTURE SOURCES

Unless noted, all pictorial content in this book has been solely created by the author, Martin Vargic, all exceptions are listed below. Grateful thanks to all those mentioned.

pp. 8-9
'Orion Nebula'. NASA, ESA, M. Robberto. https://en.wikipedia.org/wiki/Orion_Nebula#/media/File:Orion_Nebula_-_Hubble_2006_mosaic_18000.jpg Also used on pages 28-29, 30-31, 32-33, 34-35, 36-37, 38-39, 40-41, 42-43, 44-45, 46-47, 48-49, 50-51, 54-55, 56-57, 58-59, 60-61, 62-63, 64-65, 66-67, 68-69, 82-83, 84-85, 86-87, 88-89, 90-91, 92-93, 94-95, 96-97, 98-99, 100-101, 102-103, 104-105, 106-107,108-109, 110-111, 112-113, 114-115, 116-117, 118-119, 120-121, 122-123, 124-125, 126-127, 128-129, 130-131, 132-133, 134-135

pp. 10-11
Untitled. Hans. *https://pixabay.com/photos/wave-sea-surf-swell-foam-spray-384385/*
'Blue marble'. NASA. *https://www.nasa.gov/image-feature/running-a-real-time-simulation-of-go-no-go-for-apollo-17/* Also used on pages 12-13, 20-21, 22-23, 24-25, 34-35, 36-37, 38-39, 40-41, 42-43, 86-87, 96-97, 116-117, 118-119, 122-123, 124-125, 126-127, 130-131
'Blue marble'. NASA. *https://visibleearth.nasa.gov/view.php?id=74518* Also used on pages 12-13, 20-21, 22-23, 24-25, 34-35, 36-37, 38-39, 40-41, 42-43, 86-87, 96-97, 116-117, 118-119, 122-123, 124-125, 126-127, 130-131
'Blue Marble: Clouds'. NASA. *https://visibleearth.nasa.gov/view.php?id=57747* Also used on pages 12-13, 20-21, 22-23, 24-25, 34-35, 36-37, 38-39, 40-41, 42-43, 86-87, 116-117, 118-119, 122-123, 124-125, 126-127, 130-131
Untitled. Publicdomainpictures. *https://cdn.pixabay.com/photo/2012/03/03/23/06/backdrop-21534_960_720.jpg*
Untitled. RefusedMind. *https://cdn.pixabay.com/photo/2018/02/03/18/59/gateway-3128289_960_720.png*
Untitled. AlexMercier. *https://pixabay.com/photos/rock-texture-stone-surface-2147359/* Also used on pages 12-13, 26-27, 34-35, 36-37, 38-39, 40-41, 42-43, 92-93, 94-95, 96-97, 98-99, 100-101, 110-111, 112-113, 114-115, 132-133
Untitled. TinyRacket. *https://pixabay.com/photos/texture-rock-rough-background-1441967/* Also used on pages 12-13, 26-27, 34-35, 36-37, 38-39, 40-41, 42-43, 92-93, 94-95, 96-97, 98-99, 100-101, 110-111, 112-113, 114-115, 132-133
Untitled. Engin_Akyurt *https://cdn.pixabay.com/photo/2016/11/19/06/21/rock-1838128_960_720.jpg* Also used on pages 12-13, 26-27, 34-35, 36-37, 38-39, 40-41, 42-43, 92-93, 94-95, 96-97, 98-99, 100-101, 110-111, 112-113, 114-115, 132-133
Untitled. sipa. *https://cdn.pixabay.com/photo/2015/06/16/06/50/stone-810912_1280.jpg* Also used on pages 12-13, 26-27, 34-35, 36-37, 38-39, 40-41, 42-43, 92-93, 94-95, 96-97, 98-99, 100-101, 110-111, 112-113, 114-115, 132-133
Untitled. geralt. *https://pixabay.com/photos/star-constellation-universe-bull-2630050/*
'NASA's SDO Sees Sun Emit Mid-Level Flare'. NASA. *https://www.nasa.gov/feature/goddard/nasas-sdo-sees-sun-emit-mid-level-flare-oct-1/* Also used on pages 12-13, 86-87, 110-111, 112-113, 132-133
Untitled. skeeze. https://pixabay.com/photos/cavern-speleothems-cave-underground-554374/
Untitled. werner22brigitte. *https://pixabay.com/photos/smoke-background-artwork-swirl-69124/*
Untitled. HG-fotografie. *https://cdn.pixabay.com/photo/2016/12/11/17/03/fire-1899824__340.jpg* Also used on pages 86-87, 92-93
Untitled. andreas160578 *https://pixabay.com/photos/paper-handmade-handmade-paper-1332008/*
Untitled. ractapopulous. *https://pixabay.com/illustrations/snake-dragon-head-reptile-3d-1940343/*

Untitled. geralt. *https://pixabay.com/photos/star-constellation-universe-twins-2633893/*
Untitled. Pxhere. *https://pxhere.com/en/photo/626294*
'Mercury'. *https://www.solarsystemscope.com/textures/* (under CC BY 4.0) Also used on pages 118-119, 132-133
'Jupiter and the Moon'. Hastings-Trew, James. *http://planetpixelemporium.com/planets.html* Also used on pages 32-33
Untitled. Photos public domain. *http://www.photos-public-domain.com/2011/11/29/black-schist-rock-texture-with-diagonal-bands/* Also used on pages 12-13, 92-93, 94-95, 96-97, 98-99, 100-101
Untitled. Free-photos. *https://pixabay.com/photos/rocky-mountain-cliff-landscape-1149298/* Also used on pages 12-13, 92-93, 94-95, 96-97, 98-99, 100-101
Untitled. stocksnap. *https://pixabay.com/photos/nature-sky-night-stars-2609647*
'Starry Night Sky'. Holly Ireland. *https://pixy.org/369620/* Also used on pages 12-13, 86-87, 94-95, 108-109
'Rosette Nebula'. NASA. *https://www.nasa.gov/multimedia/imagegallery/image_feature_1760.html* Also used on pages 12-13
'Heart and Soul nebula'. NASA. *https://www.nasa.gov/mission_pages/WISE/multimedia/wiseimage20100524.html* Also used on pages 12-13
'Carina nebula'. NASA. *https://www.nasa.gov/multimedia/imagegallery/image_feature_1146.html* Also used on pages 12-13
Untitled. EliasSch. *https://pixabay.com/photos/water-texture-lake-sea-wave-2072218/* Also used on pages 12-13
Untitled. 12019. *https://pixabay.com/photos/yemeni-lava-hot-steam-fire-81175/* Also used on pages 12-13, 40-41, 42-43, 92-93, 96-97, 114-115, 116-117, 132-133
Untitled. skeeze. *https://pixabay.com/photos/hawaii-volcanoes-national-park-518763/* Also used on pages 12-13, 40-41, 42-43, 92-93, 96-97, 114-115, 116-117, 132-133
'Pahoehoe lava'. USGS. *http://www.copyrightfreephotos.hq101.com/v/nature/mountains_volcanoes/Pahoehoe_-_Lava_-_Hawaii_-_Volcanoes_National_Park.jpg.html* Also used on pages 12-13, 40-41, 42-43, 92-93, 96-97, 114-115, 116-117, 132-133
Untitled. Geralt. *https://pixabay.com/illustrations/heart-love-flame-fire-brand-burn-1137257/* Also used on pages 94-95, 96-97, 114-115, 116-117, 132-133
'Light and shadow in the Carina Nebula'. NASA, 92-93 *https://www.nasa.gov/multimedia/imagegallery/image_feature_1063.html*
Untitled. Tykejones. *https://pixabay.com/photos/brick-wall-texture-background-old-738387/*
Untitled. susannp4. *https://pixabay.com/illustrations/snake-reptile-animal-vintage-2082037/*
Untitled. susannp4. *https://pixabay.com/illustrations/snake-reptile-background-isolated-3066876/*
Untitled. Jjfgg. *https://pixabay.com/illustrations/quetzal-bird-animal-cute-drawing-1088589/*
Untitled. NASA. *https://www.jpl.nasa.gov/news/news.php?feature=4234* Also used on pages 12-13
'Earth'. NASA. *https://www.nasa.gov/press-release/nasa-satellite-camera-provides-epic-view-of-earth/* Also used on pages 12-13, 122-123
'Ghostly Face in South Sahara Desert'. NASA. *https://earthobservatory.nasa.gov/images/7076/ghostly-face-in-south-australian-desert*
'Morocco'. NASA. *https://visibleearth.nasa.gov/view.php?id=69844*
Untitled. JuergenPM. *https://pixabay.com/photos/tree-field-cornfield-nature-247122/* Also used on pages 12-13
Untitled. Free_Photos. *https://pixabay.com/photos/sky-ocean-water-blue-sea-waves-918667/* Also used on pages 12-13
'Cloud'. Pngtree. *https://pngtree.com/freepng/cloud_71798.html* Also used on pages 12-13, 16-17, 94-95
'Cloud'. Pngtree. *https://pngtree.com/freepng/clouds--clear--sky--cloudy_506704.html* Also used on pages 12-13, 16-17, 94-95
'Cloud'. Pngtree. *https://pngtree.com/freepng/fog_130914.html*
Also used on pages 12-13, 16-17, 94-95
Untitled. Ridderhof. *https://pixabay.com/photos/background-concrete-stone-grey-old-1725520/* Also used on pages 12-13
Untitled. _Marion. *https://pixabay.com/photos/concrete-wall-wall-concrete-331294/* Also used on pages 12-13
Untitled. RitaE. *https://pixabay.com/photos/background-panorama-sunset-dawn-3104413/*
Untitled. Liggraphy. *https://pixabay.com/photos/olive-tree-tree-olivier-old-nature-3579922/* Also used on pages 12-13
Untitled. Regalshave. *https://pixabay.com/photos/oak-tree-tree-huge-old-charleston-2018822/* Also used on pages 12-13
Untitled. Transparentpng. *http://www.transparentpng.com/details/tree-images-picture_6387.html* (under CC BY 4.0) Also used on pages 12-13
Untitled. Transparentpng. *http://www.transparentpng.com/details/green-tree-images-pictures-download-_6395.html* (under CC BY 4.0) Also used on pages 12-13
'The Moon, Mercury, Jupiter, Venus'. Solarsystemscope. *https://www.solarsystemscope.com/textures/* (under CC BY 4.0) Also used on pages 12-13
'Saturn'. NASA. *https://nssdc.gsfc.nasa.gov/planetary/image/saturn.jpg* Also used on pages 12-13, 34-35, 40-41, 122-123
'Venus'. NASA. *https://nssdc.gsfc.nasa.gov/photo_gallery/photogallery-venus.html* Also used on pages 12-13, 34-35, 36-37, 38-39, 40-41, 42-43, 94-95, 132-133
'Mars'. NASA. *https://mars.nasa.gov/system/site_config_values/meta_share_images/1_142497main_PIA03154-200.jpg* Also used on pages 12-13, 36-37, 100-101, 114-115
Untitled. TheDigitalArtist. *https://pixabay.com/illustrations/whale-ocean-sea-nature-mammal-1696051/*
Untitled. Pngtree. *https://pngtree.com/freepng/birds_905741.html*
'Ireland'. NASA. *https://web.archive.org/web/20030210160415/http://earthobservatory.nasa.gov/Newsroom/NewImages/images.php3?img_id=10889*

pp. 12-13
Untitled. Buddy_nath. *https://pixabay.com/illustrations/space-stars-comet-astronomy-1486556/* Also used on pages 126-127, 130-131
Untitled. Couleur. *https://pixabay.com/photos/giant-eagle-adler-bird-bird-of-prey-2700371/*
Untitled. Connie_sf. *https://pixabay.com/photos/raven-crow-bird-black-flying-2162966/*
Untitled. jLasWilson. *https://pixabay.com/photos/squirrel-grey-brown-fur-cute-1401509/*
Untitled. pen_ash. *https://pixabay.com/photos/hogwarts-castle-harry-potter-osaka-2404482/*
Untitled. zoosnow. *https://pixabay.com/photos/corn-snake-spotted-elaphe-2034552/*
'Watching the Rivers Flow on Greenland'. NASA. *https://earthobservatory.nasa.gov/images/86508/watching-the-rivers-flow-on-greenland*
'Breakup Continues on the Wilkins Ice Shelf'. NASA. *https://earthobservatory.nasa.gov/images/81174/breakup-continues-on-the-wilkins-ice-shelf*
'Mt. Everest from space'. NASA. *https://www.nasa.gov/multimedia/imagegallery/image_feature_152.html*
Untitled. doctor_a. *https://pixabay.com/sk/photos/sopka-hawaii-lava-cloud-popol-2262295/*
Untitled. Kanenori. *https://pixabay.com/photos/mountainous-landscape-sunset-autumn-2402590/*
Untitled. Wikiimages. *https://pixabay.com/photos/mount-st-helens-volcanic-eruption-164848/*
Untitled. Donations_are_apprecated. *https://pixabay.com/photos/desert-mountain-landscape-rock-2646209/*
Untitled. Jplenio. *https://pixabay.com/photos/hintersee-bergsee-mountains-ramsau-3601004/*
Untitled. skeeze. *https://pixabay.com/photos/coyote-wildlife-nature-snow-1901990/*
Untitled. mysticsartdesign. *https://pixabay.com/illustrations/dragon-say-make-fairytale-hour-1814054/*
'Sahara'. NASA. *https://visibleearth.nasa.gov/view.php?id=57230*
Untitled. skeeze. *https://pixabay.com/photos/monument-valley-rock-formations-752454/*
Untitled. SilviaP_Design. *https://pixabay.com/illustrations/fuchs-red-fox-isolated-figure-1121652/*
'Himalayas'. NASA. *https://visibleearth.nasa.gov/view.php?id=63013*
Untitled. ideasam. *https://pixabay.com/photos/cusco-peru-inca-ruin-agriculture-609473/*
Untitled. skeeze. *https://pixabay.com/photos/arch-stone-night-stars-starry-896900/* Also used on ages 108-109, 110-111, 112-113
Untitled. Mr_Murdoch *https://pixabay.com/photos/clouds-space-milky-way-atmosphere-2295189/*
Untitled. publicdomainvectors. *https://publicdomainvectors.org/en/free-clipart/Stripped-snake/54736.html*
Untitled. BarbaraALane. *https://pixabay.com/illustrations/fractal-spikes-gears-abstract-1765218/*
'Landsat image mosaic of antarctica'. NASA. *https://visibleearth.nasa.gov/view.php?id=78592*
Untitled. Simon. *https://pixabay.com/photos/mount-everest-himalayas-nuptse-276995/*
Untitled. Simon. *https://pixabay.com/photos/rough-horn-alpine-2146181/*
Untitled. Simon. *https://pixabay.com/photos/mountain-peak-summit-altitude-top-690104/*
'Callisto'. NASA. *https://photojournal.jpl.nasa.gov/catalog/PIA03456* Also used on pages 32-33, 34-35, 36-37, 38-39, 40-41, 42-43, 114-115
'Ganymede'. USGS. *https://astrogeology.usgs.gov/search/map/Ganymede/Voyager-Galileo/Ganymede_Voyager_GalileoSSI_Global_ClrMosaic_1435m* Also used on pages 32-33, 34-35, 36-37, 38-39, 40-41, 42-43, 114-115
'Europa'. NASA. *https://photojournal.jpl.nasa.gov/catalog/PIA00502* Also used on pages 32-33, 34-35, 36-37, 38-39, 40-41, 42-43, 78-79, 110-111, 112-113, 114-115, 118-119
'Io'. USGS. *https://astrogeology.usgs.gov/Projects/JupiterSatellites/io/Io_SSI_VGR_color_merge_SIMP0_med.cub.jpg* Also used on pages 32-33, 34-35, 36-37, 38-39, 40-41, 42-43, 114-115, 132-133
Untitled. Simon. *https://pixabay.com/photos/mountain-peak-summit-altitude-top-690104/*
Untitled. Free_photos. *https://pixabay.com/photos/aurora-borealis-northern-lights-731456/*
Untitled. Pexels. *https://pixabay.com/photos/animal-turtle-coral-reef-ocean-sea-1868046/*
Untitled. Skeeze. *https://pixabay.com/photos/landscape-panorama-mountains-scenic-2268775/*
Untitled. jplenio. *https://cdn.pixabay.com/photo/2018/08/12/15/29/hintersee-3601004_1280.jpg*
Untitled. PIRO4D. *https://pixabay.com/photos/schrecksee-bergsee-allg%C3%A4u-2534484/*

pp. 14-15
'Eagle Nebula from ESO'. ESO (under CC BY 4.0) https://en.wikipedia.org/wiki/Eagle_Nebula#/media/File:Eagle_Nebula_from_ESO.jpg
'Noctilucent clouds'. Space Exploration Technologies Corporation *https://en.wikipedia.org/wiki/Noctilucent_cloud#/media/File:SpaceX_Noctilucent_Cloud.jpg* Also used on pages 20-21
'First photo of Earth from space'. U.S. Army - White Sands Missile Range/Applied Physics Laboratory. *https://en.wikipedia.org/wiki/V-2_No._13#/media/File:First_photo_from_space.jpg*
'Dark Shuttle Approaching'. NASA/ISS Expedition 22. *https://en.wikipedia.org/wiki/V-2_No._13#/media/File:First_photo_from_space.jpg*
'Aurora Australis from the ISS'. NASA. *https://en.wikipedia.org/wiki/Aurora#/media/File:Aurora_australis_ISS.jpg*
'Ozone hole in 1978 and 2008'. NASA. *https://www.nist.gov/blogs/taking-measure/refrigerants-rescue-plugging-ozone-hole*
'The ISS as it transits the sun during an eclipse'. NASA. *https://en.wikipedia.org/wiki/International_Space_Station#/media/File:2017_Total_Solar_Eclipse_-_ISS_Transit_(NHQ201708210304).jpg*
'Debris plot by NASA'. NASA Orbital Debris Program Office. *https://en.wikipedia.org/wiki/Space_debris#/media/File:Debris-GEO1280.jpg*
'Kinetic Energy Weapon test'. US Department of Defense. *https://en.wikipedia.org/wiki/Light-gas_gun#/media/File:SDIO_KEW_Lexan_projectile.jpg*
'Cumulonimbus cloud over Africa'. NASA. *https://climate.nasa.gov/climate_resources/124/cumulonimbus-cloud-over-africa/*
'SR-71 Blackbird'. Brohmer, Judson. USAF, *https://en.wikipedia.org/wiki/Lockheed_SR-71_Blackbird* Also used on pages 18-19
'Blue jet as seen from the summit of Mauna Kea, Hawaii'. Gemini Observatory/AURA. *https://en.wikipedia.org/wiki/Upper-atmospheric_lightning#/media/File:BlueJet.jpg*
'Kittinger's record-breaking skydive from Excelsior III' Wentzel, Volkmar. USAF, *https://en.wikipedia.org/wiki/File:Kittinger-jump.jpg* Also used on pages 18-19
'Picture taken at approx. 100,000 feet above Oregon'. Hamel, Justin. *https://commons.wikimedia.org/wiki/File:Picture_taken_at_aprox._100,000_feet_above_Oregon_by_Justin_Hamel_and_Chris_Thompson.jpg*

pp. 16-17
Untitled. Skeeze. *https://pixabay.com/photos/ayers-rock-uluru-australia-landmark-1538576/*
Untitled. 8moments. *https://pixabay.com/photos/desert-camel-sand-pyramid-dry-3217765/*
Untitled. nkoks. *https://pixabay.com/photos/paris-france-eiffel-eiffel-tower-1175022/*
'Ryugyong hotel'. Ferris, Joseph. *https://en.wikipedia.org/wiki/Ryugyong_Hotel#/media/File:Ryugyong_Hotel_-_August_27,_2011_(Cropped).jpg/* (Under CC 2.0)
Untitled. terimakasih0. *https://pixabay.com/photos/petronas-towers-twin-towers-malaysia-1445879/*
Untitled. Sputnik72. *https://pixabay.com/photos/toronto-canada-skyline-cn-tower-277693/*
Untitled. cmmellow. *https://pixabay.com/photos/lotte-world-tower-seoul-1791802/*
Untitled. slotfast. *https://pixabay.com/photos/nature-of-tokyo-skytree-tokyo-2496153/*
Untitled. steven_yu. *https://pixabay.com/photos/shanghai-bund-china-city-1484452/*
Untitled. didiwo. *https://pixabay.com/photos/burj-khalifa-building-dubai-city-1096446/*
Untitled. Open clip art library. *http://www.publicdomainfiles.com/show_file.php?id=13944828016016*
Untitled. *https://pxhere.com/en/photo/1355388*
Untitled. piro4d. *https://pixabay.com/sk/photos/bal%C3%B3n-ballonfahrt-maj%C3%A1k-obloha-2331488/*
Untitled. Beaufort, Jean. *https://www.publicdomainpictures.net/en/view-image.phpw?image=223064&picture=mount-rainier*
'Mount Fuji'. Young, Robert. *https://commons.wikimedia.org/wiki/File:Mount_Fuji_Sanroku.jpg* (Under CC 2.0)
'Mauna Kea'. NASA. *https://eoimages.gsfc.nasa.gov/images/imagerecords/92000/92630/HISEAS_pho_2015_March10_web.lrg.jpg*
Untitled. Violetta. *https://pixabay.com/sk/photos/matterhorn-zermatt-%C5%A1vaj%C4%8Diarsko-sneh-425134/*
'Kilimanjaro'. *https://wallpaperbro.com/img/342658.jpg*
'Mount Everest from Rongbuk'. Ocrambo *https://en.wikipedia.org/wiki/File:Mount_Everest_from_Rongbuk_may_2005.*
Untitled. clker. *https://pixabay.com/vectors/balloons-red-blue-yellow-shiny-25737/*
'Skull and crossbones'. *https://en.wikipedia.org/wiki/Skull_and_crossbones_(symbol)#/media/File:Skull_and_Crossbones.svg* Also used on pages 18-19, 132-133
'DG 1000 glider' Hailday, Paul. *https://en.wikipedia.org/wiki/DG_Flugzeugbau_DG-1000#/media/File:DG1000_glider*
Untitled. GDJ. *https://pixabay.com/vectors/paragliding-parachute-silhouette-4037231/*
Untitled. Hodan, George. *https://www.publicdomainpictures.net/en/view-image.php?image=162152&picture=helicopter*
'B2-Spirit'. USAF. *https://commons.wikimedia.org/wiki/File:B-2_Spirit_original.jpg*
Untitled. Wikimediaimages. *https://pixabay.com/photos/boeing-747-air-china-cargo-jumbo-jet-884427/*
Untitled. froehlich-gera. *https://pixabay.com/photos/concorde-aircraft-supersonic-fighter-203813/*
Untitled. terimakasih0. *https://pixabay.com/photos/copper-kettle-kettle-copper-metal-1276542/* Also used on pages 16-17 (Journey into outer space, second spread)
Untitled. skeeze. *https://pixabay.com/photos/yellow-banded-bumblebee-insect-macro-1829241/*
Untitled. OpenClipart-Vector *https://pixabay.com/vectors/atomic-bomb-bomb-nuclear-war-2026117/* Also used on pages 18-19, 20-21, 22-23
Untitled. skeeze. *https://pixabay.com/photos/turkey-vulture-bird-wildlife-nature-1107362/*
'Challenger explosion'. NASA. *https://commons.wikimedia.org/wiki/File:Challenger_explosion.jpg*
Also used on pages 76-77

pp. 18-19
'Bell X-1'. Hoover, Robert A. NASA, *https://sk.wikipedia.org/wiki/Bell_X-1#/media/S%C3%BAbor:Bell_X-1_color.jpg*
'Geostationary Airship Satellite'. Kcida10. *https://en.wikipedia.org/wiki/High-altitude_platform_station#/media/File:Geostationary_airship_satellite.jpeg*
'Buk-M1-2 SAM system'. ,:Ajvol:. *https://en.wikipedia.org/wiki/Buk_missile_system#/media/File:Buk-M1-2_9A310M1-2.jpg*
'HySICS Flight Balloon'. NASA. *https://www.nasa.gov/image-feature/goddard/hysics-flight-balloon/* Also used on pages 20-21 (Journey into outer space, From Mesosphere to the Thermosphere)
Untitled. Free-photos. *https://pixabay.com/photos/stars-nightsky-milky-way-darkness-1246590/* Also used on pages 108-109, 110-111, 112-113
'High-altitude balloon'. Shelton, George. NASA, *https://www.nasa.gov/images/content/726453main_Weather%20Balloon-Full.jpg*
'Mount Pinatubo'. Harlow, Dave. USGS, *http://vulcan.wr.usgs.gov/Volcanoes/Philippines/Pinatubo/images.html*
'Helios Prototype'. NASA. *https://www.nasa.gov/centers/armstrong/news/FactSheets/FS-068-DFRC.html*
Untitled. Tabble. *https://pixabay.com/illustrations/paper-planes-aircraft-send-flat-1023097/*
'Red Bull Stratos exhibit'. Smithsonian. *https://commons.wikimedia.org/wiki/File:RedBullStratosNASM.jpg*
'Ye-155R3'. Grunherz. *https://en.wikipedia.org/wiki/Mikoyan-Gurevich_MiG-25#/media/File:Ye-155.png*
Untitled. Clker. *https://pixabay.com/vectors/petri-dish-bacteria-laboratory-311960/*
Untitled. Myriams-Fotos. *https://pixabay.com/photos/hot-air-balloon-balloon-aircraft-3542903/*

pp. 20-21 (Journey into outer space, From Mesosphere to the Termosphere)
'Big-60'. NASA. *https://www.nasa.gov/feature/goddard/2018/touchdown-nasa-s-football-stadium-sized-scientific-balloon-takes-flight*
Untitled. Zajcsik. *https://pixabay.com/photos/lightning-storm-night-firebird-342341/*
Untitled. OpenClipart-Vectors *https://pixabay.com/vectors/space-shuttle-atlantis-nasa-156012/* Also used on pages 24-25 (Journey into outer space, Domain of the satellites)
'Astronaut Bruce McCandless on First-ever Untethered Spacewalk'. NASA. *https://www.nasa.gov/image-feature/astronaut-bruce-mccandless-on-first-ever-untethered-spacewalk/* Also used on pages 74-75, 76-77, 78-79
'Project Mercury – Little Joe Test'. NASA. *https://archive.org/details/C-1959-52201*
Untitled. Nicman. *https://pixabay.com/photos/monkey-primates-capuchin-2139295/*

pp. 22-23 (Journey into outer space, Atop the ocean of air)
Untitled. 12019. *https://pixabay.com/photos/aurora-borealis-northern-lights-2647474/*
'X-15'. NASA. *https://www.dfrc.nasa.gov/Gallery/Photo/X-15/HTML/EC88-0180-1.html*
'Art Concept – Apollo VIII – Command Module (CM) – Re-Entry Orientation'. NASA. *https://images.nasa.gov/details-S68-55292.html*
Untitled. SankerG. *https://pixabay.com/photos/monkey-primate-wildlife-macaque-3243666/*
'HARP 16-inch (410 mm) gun'. US Department of Defense. *https://en.wikipedia.org/wiki/Project_HARP#/media/File:Project_Harp.jpg*
'Alan Shepard suited up'. NASA. *https://www.nasa.gov/topics/history/milestones/spacesuit.html* Also used on pages 72-73
'Mercury spacecraft'. NASA. *https://www.nasa.gov/mission_pages/mercury/missions/spacecraft.html* Also used on pages 72-73
Untitled. Noel_Bauza. https://pixabay.com/photos/aurora-polar-lights-northen-lights-1185464/n Also used on pages 72-73, 74-75, 76-77, 78-79, 80-81
Untitled. Free_Photos. https://pixabay.com/photos/milky-way-galaxy-night-sky-stars-984050/ Also used on pages 72-73, 74-75, 76-77, 78-79, 80-81, 86-87, 108-109, 108-109, 110-111, 112-113
Untitled. O12. https://pixabay.com/photos/milky-way-starry-sky-night-sky-star-2695569/ Also used on pages 72-73, 74-75, 76-77, 78-79, 80-81, 86-87, 108-109, 110-111, 112-113, 113-115, 134-135
Untitled. Free_Photos. https://pixabay.com/photos/sky-stars-constellations-astronomy-828648/ Also used on pages 72-73, 74-75, 76-77, 78-79, 80-81, 86-87, 110-111, 112-113, 134-135
'John Glenn in his spacesuit'. NASA. *https://commons.wikimedia.org/wiki/File:Mercury_6,_John_H_Glenn_Jr.jpg*
'Friendship 7'. Balon Greyjoy. *https://commons.wikimedia.org/wiki/File:Friendship_7_National_Air_and_Space_Museum.jpg* https://pixabay.com/photos/stars-nightsky-milky-way-darkness-1246590/ Also used on pages 108-109
'Sputnik 1'. NASA *https://en.wikipedia.org/wiki/Sputnik_1#/media/File:Sputnik_asm.jpg* Also used on pages 28-29, 72-73, 128-129
'International Space Station'. NASA. *http://spaceflight.nasa.gov/gallery/images/station /crew-27/hires/*

iss027e036636.jpg Also used on pages 28-29, 76-77, 80-81
'Skylab'. NASA. https://www.nasa.gov/content/40-years-ago-skylab-paved-way-for-international-space-station Also used on pages 26-27 (from artificial spacecrafts to asteroids), 74-75
'Yuri Gagarin'. NASA. https://www.nasa.gov/topics/history/features/gagarin/gagarin2.html Also used on pages 72-73
'Vostok Spacecraft'. Pascal. https://commons.wikimedia.org/wiki/File:Vostok_spacecraft_replica.jpg Also used on pages 72-73
1963 Soviet Union 10 kopeks stamp. Valentina Tereshkova. USSR, https://commons.wikimedia.org/wiki/File:Soviet_Union-1963-Stamp-0.10._Valentina_Tereshkova.jpg
DOS-4. NASA. http://heasarc.gsfc.nasa.gov/Images/misc_missions/salyut-4.gif Also used on pages 28-29, 74-75
'Mir Space Station viewed from Endeavour during STS-89'. NASA. https://en.wikipedia.org/wiki/Mir#/media/File:Mir_Space_Station_viewed_from_Endeavour_during_STS-89.jpg Also used on pages 28-29, 76-77

pp. 24-25 (Journey into outer space, Domain of the satellites)
'WISE Spacecraft'. NASA. https://en.wikipedia.org/wiki/Wide-field_Infrared_Survey_Explorer#/media/File:WISE_artist_concept_(PIA17254,_crop).jpg
'Hubble Space Telescope'. NASA. https://en.wikipedia.org/wiki/File:HST-SM4.jpeg Also used on pages 28-29, 76-77, 128-129
'Iridium Satellite'. Kozuch. https://en.wikipedia.org/wiki/Iridium_satellite_constellation#/media/File:Iridium_Satellite.jpg (under CC BY 2.0)
'COBE Satellite'. NASA/Cobe Science Team. https://map.gsfc.nasa.gov/media/990295/index.html Also used on pages 28-29, 76-77
'BADR-1'. Suparco. http://www.suparco.gov.pk/assets/images/badr-1.gif
'Alouette-1'. NASA. http://history.nasa.gov/SP-4402/ch2.htm
'A Black Brant XII being launched from Wallops Flight Facility'. NASA/Wallops. https://en.wikipedia.org/wiki/Sounding_rocket#/media/File:Black_Brant.jpg
'Echo-II'. NASA. https://en.wikipedia.org/wiki/Project_Echo#/media/File:Echo_II.jpg
'Romanian stamp from 1959 with Laika'. Romania. https://en.wikipedia.org/wiki/Laika#/media/File:Posta_Romana_-_1959_-_Laika_120_B.jpg
Untitled. Colin00B. https://pixabay.com/illustrations/helium-atom-animation-540560/
'Blue Screen of Death'. Praseodymium. https://en.wikipedia.org/wiki/Blue_Screen_of_Death#/media/File:Windows_XP_BSOD.png
'Ilustration about the collision of CERISE'. NASA. https://en.wikipedia.org/wiki/Cerise_(satellite)#/media/File:Cerise_sat_broke.jpg
'Gemini spacecraft'. NASA. https://www.nasa.gov/image-feature/from-mercury-mark-ii-to-project-gemini
'Bumper 1948 Rocket'. miyeon. http://absfreepic.com/free-photos/download/bumper-1948-rocket-4752x3168_90633.html Also used on pages 86-87

pp. 26-27
'Carina Nebula'. ESO. https://commons.wikimedia.org/wiki/File:ESO_-_The_Carina_Nebula_(by).jpg (under CC BY 4.0) Also used on pages 28-29, 51-52, 56-57, 58-59, 62-63, 64-65, 66-67 68-69, 92-93, 94-95, 96-97, 98-99, 100-101
'Pale Blue Dot'. NASA. http://visibleearth.nasa.gov/view.php?id=52392
'Jupiter's South Pole'. NASA. https://www.nasa.gov/image-feature/jupiters-south-pole
'Ultima Thule'. NASA/Johns Hopkins University Applied Physics Laboratory. https://en.wikipedia.org/wiki/(486958)_2014_MU69#/media/File:UltimaThule_CA06_color_vertical.png Also used on pages 128-129
'Mosaic image od asteroid Bennu'. NASA. https://en.wikipedia.org/wiki/101955_Bennu#/media/File:BennuAsteroid.jpg Also used on pages 28-29, 30-31
'Heliospheric Current Sheet'. NASA. http://lepmfi.gsfc.nasa.gov/mfi/hcs/hcs_shape.html
'V838 Monocerotis Light Echo'. NASA. https://en.wikipedia.org/wiki/V838_Monocerotis#/media/File:V838_Monocerotis_expansion.jpg
'A rosette for the VLT'. European Southern Observatory. https://www.eso.org/public/images/potw1847a/ (under CC-BY 4.0)
'The Milky Way Galaxy'. Gilli, Bruno. European Southern Observatory https://www.eso.org/public/images/potw1847a/ (under CC-BY 4.0).
'Messier 15'. NASA, ESA. http://www.spacetelescope.org/static/archives/images/large/heic1321a.jpg (under CC-BY 3.0) Also used on pages 110-111
'Streaks and stripes'. NASA, ESA. https://www.spacetelescope.org/images/potw1748a/ (under CC-BY 3.0)
'Large Scale Structure in the Local Universe: The 2MASS Galaxy Catalog' Jarret, Thomas. IPAC/Caltech, http://www.acgc.uct.ac.za/~jarrett/lss/index.html
'NGC-1275'. NASA. https://en.wikipedia.org/wiki/NGC_1275#/media/File:NGC_1275_Hubble.jpg
'Simulated distribution of matter in the universe within a region over 50 million light years across'. Pontzen, Andrew; Governato, Fabio. NASA, https://commons.wikimedia.org/wiki/File:Large-scale_structure_of_light_distribution_in_the_universe.jpg (CC BY 2.0)
'Sharpest ever view of the Andromeda galaxy'. NASA, ESA. https://www.spacetelescope.org/images/heic1502a/ (CC BY 4.0)

pp. 28-29
'Artist's concept of NuSTAR on orbit'. NASA. https://en.wikipedia.org/wiki/NuSTAR#/media/File:NuSTAR_illustration_(transparent_background).png
'Rosetta spacecraft'. IanShazell. https://commons.wikimedia.org/wiki/File:Rosetta.jpg
'Tesla Roadster'. US Department of Energy. http://www.publicdomainfiles.com/show_file.php?id=14020027212538
'OSIRIS-REx'. NASA. https://commons.wikimedia.org/wiki/File:OSIRIS-REx_spacecraft.png Also used on pages 78-79
'Orbiting Carbon Observatory 2'. NASA. https://en.wikipedia.org/wiki/Orbiting_Carbon_Observatory_2#/media/File:Orbiting_Carbon_Observatory-2_artist_rendering_(PIA18374).jpg
'Mars Reconnaissance Orbiter'. Waste, Corby. NASA, https://photojournal.jpl.nasa.gov/catalog/PIA04916 Also used on pages 76-77
'Artist's conception of OAO-1 in orbit'. NASA. https://en.wikipedia.org/wiki/Orbiting_Astronomical_Observatory#/media/File:Orbiting_Astronomical_Observatory.jpg Also used on pages 72-73
'ExoMars Trace Gas Orbiter'. NASA. https://commons.wikimedia.org/wiki/File:ExoMars_Trace_Gas_Orbiter.jpg
'Illustration of the Terra satellite'. NASA. https://directory.eoportal.org/web/eoportal/satellite-missions/t/terra
'UARS1'. NASA. https://commons.wikimedia.org/wiki/File:UARS_1.jpg
'Extreme Ultraviolet Explorer'. NASA. https://en.wikipedia.org/wiki/Extreme_Ultraviolet_Explorer#/media/File:EUVE_Photo.gif
'Illustration of the Terra satellite'. NASA. https://directory.eoportal.org/web/eoportal/satellite-missions/t/terra
'Artist illustration of the Chandra X-ray Observatory'. NASA/CXC/NGST. https://en.wikipedia.org/wiki/Chandra_X-ray_Observatory#/media/File:Chandra_artist_illustration.jpg Also used on pages 76-77
'NASA's Compton Gamma Ray Observatory'. Cameron, Ken. NASA, https://en.wikipedia.org/wiki/Compton_Gamma_Ray_Observatory#/media/File:CGRO_s37-96-010.jpg
'Messenger'. NASA. Also used on pages 78-79 https://www.nasa.gov/directorates/heo/scan/services/missions/solarsystem/MESSENGER.html
'NASA's Swift Gamma Ray Burst Detecting Satellite'. NASA E/PO. https://en.wikipedia.org/wiki/Neil_Gehrels_Swift_Observatory#/media/File:NASA_Swift_Gamma-Ray_Burst_Mission_(transparent).png Also used on pages 84-85
'A transparent image of the Juno spacecraft'. NASA. https://en.wikipedia.org/wiki/Juno_(spacecraft)#/media/File:Juno_spacecraft_model_1.png
'Sputnik 2'. NASA. https://commons.wikimedia.org/wiki/File:Sputnik2_vsm.jpg
'Gemini spacecraft being prepared in the shop'. NASA. http://www.astronautix.com/graphics/g/gemprefl.jpg
'Kepler space observatory'. NASA/JPL-Caltech/Wendy Stenzel https://en.wikipedia.org/wiki/Kepler_space_telescope#/media/File:Kepler_spacecraft_artist_render_(crop).jpg
'Space Shuttle Atlantis'. NASA. http://spaceflight.nasa.gov/gallery/images/station/crew-13/html/iss013e81234.html Also used on pages 74-75
'Space Shuttle Atlantis'. NASA. https://www.nasa.gov/sites/default/files/thumbnails/image/569718main_fd3predock_full.jpg Also used on pages 74-75
'Mariner 4'. NASA. https://nssdc.gsfc.nasa.gov/image/spacecraft/mariner04.gif .
'Crew Dragon – Pad Abort Vehicle'. SpaceX. https://www.flickr.com/photos/spacex/16661791299/
'A full-scale mockup of Bigelow Aerospace's Space Station Alpha'. Ingalls, Bill. NASA, https://en.wikipedia.org/wiki/Bigelow_Aerospace#/media/File:Bigelow_Aerospace_facilities.jpg
'MAVEN'. NASA. https://en.wikipedia.org/wiki/MAVEN#/media/File:MAVEN_spacecraft_model.png
'James Webb Space Telescope rendering'. NASA. https://en.wikipedia.org/wiki/James_Webb_Space_Telescope#/media/File:James_Webb_Space_Telescope_2009_top.jpg Also used on pages 78-79
'A transparent image of the New Horizons spacecraft'. NASA. https://en.wikipedia.org/wiki/New_Horizons#/media/File:New_Horizons_Transparent.png Also used on pages 78-79, 128-129
'Artist's conception of the Spitzer Space Telescope'. NASA/JPL-Caltech http://www.spitzer.caltech.edu/images/3072-SIRTF-Spitzer-Rendered-against-an-Infrared-100-Micron-Sky Also used on pages 76-77
'Artist concept of NASA's Lunar Reconnaissance Orbiter'. NASA. https://en.wikipedia.org/wiki/Lunar_Reconnaissance_Orbiter#/media/File:Lunar_Reconnaissance_Orbiter_001.jpg
'Artist's impression of a Navstar-2F satellite in orbit'. USAF. https://commons.wikimedia.org/wiki/File:Navstar-2F.jpg Also used on pages 76-77
'Cassini-Huygens'. NASA/JPL. https://en.wikipedia.org/wiki/Cassini%E2%80%93Huygens#/media/File:Cassini_Saturn_Orbit_Insertion.jpg
'Earth Radiation Budget Satellite'. NASA. https://en.wikipedia.org/wiki/Earth_Radiation_Budget_Satellite#/media/File:ERBS.jpg
'Pascomsat Gridsphere'. USAF. https://en.wikipedia.org/wiki/PasComSat#/media/File:OV1-8_PASCOMSAT_Gridsphere.jpg
'Voyager spacecraft model'. NASA. https://en.wikipedia.org/wiki/Voyager_1#/media/File:Voyager_spacecraft_model.png Also used on pages 74-75, 76-77, 78-79, 80-81, 130-131
'Galileo probe, Io and Jupiter'. NASA. https://photojournal.jpl.nasa.gov/catalog/PIA18176 Also used on pages 76-77
'The Apollo 15 Service Module as viewed from the Apollo Lunar Module'. NASA. https://en.wikipedia.org/wiki/Apollo_command_and_service_module#/media/File:Apollo_CSM_lunar_orbit.jpg Also used on pages 72-73, 74-75, 82-83
'Apollo Lunar Module diagram'. NASA. https://commons.wikimedia.org/wiki/File:Lunar_Module_diagram.jpg Also used on pages 74-75
'Mars Exploration Rovers: Spirit & Opportunity - Artist's Concept'. NASA/JPL-Caltech Also used on pages 76-77 https://mars.nasa.gov/resources/22356/mars-exploration-rovers-spirit-opportunity-artists-concept/
'Space Exploration Vehicle'. NASA. https://www.nasa.gov/exploration/multimedia/galleries/mmsev_sev_compare.html
'Curiosity 3D model'. NASA. https://mars.nasa.gov/msl/multimedia/interactives/learncuriosity/index-2.html
'Orion Sevice Module'. NASA. http://www.nasa.gov/multimedia/imagegallery/image_feature_2426c.html
'Infrared Astronomical Satellite'. NASA. https://www.jpl.nasa.gov/missions/infrared-astronomical-satellite-iras/ Also used on pages 74-75, 128-129
'A transparent image of the Dawn spacecraft'. NASA. https://sk.wikipedia.org/wiki/S%C3%BAbor:Dawn_spacecraft_model.png Also used on pages 78-79
'Artist depiction of Magellan at Venus'. NASA. https://en.wikipedia.org/wiki/Magellan_(spacecraft)#/media/File:Magellan_-_artist_depiction.png Also used on pages 76-77
'Artist's rendition of the Pioneer 10 spacecraft at Jupiter'. NASA. https://www.nasa.gov/centers/ames/missions/archive/pioneer.html Also used on pages 74-75
'Asteroid Ida and its moon'. NASA. https://www.jpl.nasa.gov/spaceimages/details.php?id=PIA00136 Also used on pages 30-31
'The nucleus of Comet Tempel 1'. NASA. https://en.wikipedia.org/wiki/Comet_nucleus#/media/File:Tempel_1_(PIA02127).jpg Also used on pages 30-31

pp. 30-31
Untitled. skeeze. https://pixabay.com/sk/photos/new-york-city-manhattan-ostrov-550174/
'Phobos'. NASA/JPL-Caltech/University of Arizona. http://photojournal.jpl.nasa.gov/catalog/PIA10368 Also used on pages 80-81
'Deimos-MRO'. NASA/JPL-Caltech/University of Arizona https://commons.wikimedia.org/wiki/File:Deimos-MRO.jpg
'951 Gaspra'. NASA. https://commons.wikimedia.org/wiki/File:951_Gaspra.jpg
'Flow-like features on Helene's leading hemisphere'. NASA. Also used on pages 76-77 https://en.wikipedia.org/wiki/Helene_(moon)#/media/File:PIA12758_Helene_crop.jpg
Untitled. PellissierJP. https://pixabay.com/photos/background-texture-metal-scratches-1172581/ Also used on pages 46-47, 48-49, 50-51, 82-83, 86-87, 134-135
'Metal texture'. PellissierJP.https://all-free-download.com/free-photos/download/texture-metal-metallic_236076.html Also used on pages 46-47, 48-49, 50-51, 82-83, 86-87, 134-135
Untitled. kjpargeter. https://www.freepik.com/free-photo/grunge-wall-texture_988115.htm
'New England'. NASA. https://visibleearth.nasa.gov/view.php?id=94002
'Isle of Man'. NASA. https://visibleearth.nasa.gov/view.php?id=76972
'Core of comet 81P/Wild'. NASA. https://en.wikipedia.org/wiki/81P/Wild#/media/File:Wild2_3.jpg
'Six different views of Eros in approximate natural color from NEAR-Shoemaker'. NASA/JPL/JHUAPL Also used on pages 74-75 https://en.wikipedia.org/wiki/433_Eros#/media/File:PIA02475_Eros'_Bland_Butterscotch_Colors.jpg
'Flyby of comet Hartley 2'. NASA. https://en.wikipedia.org/wiki/103P/Hartley#/media/File:PIA13602-16epoxi.gif
'Methone'. NASA/JPL-Caltech/Space Science Institute. https://en.wikipedia.org/wiki/Methone_(moon)#/media/File:Methone_PIA14633.jpg
'Calypso Close Up'. NASA. https://solarsystem.nasa.gov/resources/14920/calypso-close-up/
'Telesto Cassini Closeup'. NASA. https://en.wikipedia.org/wiki/Telesto_(moon)#/media/File:Telesto_cassini_closeup.jpg
'Processed image of Nix'. NASA/JHU-APL/SwRI. https://en.wikipedia.org/wiki/Nix_(moon)#/media/File:Nix_best_view-true_color.jpg
'Prometheus'. NASA/JPL/Space Science Institute. https://en.wikipedia.org/wiki/Prometheus_(moon)#/media/File:Prometheus_12-26-09b.jpg
'Approximately true-color mosaic of Saturn's moon Hyperion'. NASA/JPL/Space Science Institute. http://photojournal.jpl.nasa.gov/catalog/PIA07761 Also used on pages 32-33
'Earth from the International Space Station' NASA. https://earthobservatory.nasa.gov/
'Mathilde'. NASA. http://nssdc.gsfc.nasa.gov/imgcat/html/object_page/nea_19970627_mos.html
'Epimetheus'. NASA/JPL/Space Science Institute. https://en.wikipedia.org/wiki/Epimetheus_(moon)#/media/File:PIA09813_Epimetheus_S._polar_region.jpg
'Vesta'. NASA/JPL-Caltech. https://en.wikipedia.org/wiki/List_of_Solar_System_objects_by_size#/media/File:Vesta_full_mosaic.jpg
'NASA Visible Earth: Hawaii' NASA https://eoimages.gsfc.nasa.gov/images/imagerecords/82000/82975/hawaii_tmo_2014026_lrg.jpg
'Successive images of a rotating Ida'. NASA/JPL/USGS. https://en.wikipedia.org/wiki/243_Ida#/media/File:243_Ida_rotation.jpg
'Janus'. NASA/JPL/Space Science Institute. https://en.wikipedia.org/wiki/Janus_(moon)#/media/File:PIA12714_Janus_crop.jpg
'Pandora'. NASA/JPL/Space Science Institute. https://solarsystem.nasa.gov/moons/saturn-moons/pandora/in-depth/
'Satellite image of Malta'. NASA. https://commons.wikimedia.org/wiki/File:Satelite_image_of_Malta).jpg

pp. 32-33
'Africa and Europe from a Million Miles Away'. NASA. https://www.nasa.gov/image-feature/africa-and-europe-from-a-million-miles-away Also used on pages 86-87, 92-93, 94-95, 98-99, 100-101, 116-117
'Great Britain and Ireland'. NASA. https://commons.wikimedia.org/wiki/File:MODIS_-_Great_Britain_and_Ireland_-_2012-06-04_during_heat_wave.jpg
'Makemake'. Solarsystemscope. https://www.solarsystemscope.com/textures/ (under CC BY 4.0) Also used on pages 34-35, 36-37, 38-39, 40-41, 42-43, 110-111, 112-113, 114-115
'Haumea'. Solarsystemscope. https://www.solarsystemscope.com/textures/ (under CC BY 4.0) Also used on pages 34-35, 36-37, 38-39, 40-41, 42-43, 110-111, 112-113, 114-115
'Ceres'. Solarsystemscope. https://www.solarsystemscope.com/textures/ (under CC BY 4.0) Also used on pages 34-35, 36-37, 38-39, 40-41, 42-43, 110-111, 112-113, 114-115
'Eris'. Solarsystemscope. https://www.solarsystemscope.com/textures/ (under CC BY 4.0) Also used on pages 34-35, 36-37, 38-39, 40-41, 42-43, 110-111, 112-113, 114-115
'Sun'. Solarsystemscope. https://www.solarsystemscope.com/textures/ (under CC BY 4.0) Also used on pages 34-35, 36-37, 38-39, 40-41, 42-43, 72-73, 74-75, 76-77, 78-79, 80-81, 92-93, 110-111, 112-113, 114-115
'Moon'. Solarsystemscope. https://www.solarsystemscope.com/textures/ (under CC BY 4.0) Also used on pages 34-35, 36-37, 38-39, 40-41, 42-43, 72-73, 74-75, 76-77, 78-79, 80-81, 96-97,98-99 114-115, 114-115, 116-117, 122-123
Untitled. dunc. https://pixabay.com/photos/full-moon-moon-bright-sky-space-496873/ Also used on pages 34-35, 36-37, 38-39, 40-41, 42-43, 72-73, 74-75, 76-77, 78-79, 80-81, 96-97, 110-111, 112-113, 114-115, 122-123
'Venus'. Solarsystemscope. https://www.solarsystemscope.com/textures/ (under CC BY 4.0) Also used on pages 34-35, 36-37, 38-39, 40-41, 42-43, 72-73, 74-75, 76-77, 78-79, 80-81, 94-95, 110-111, 112-113, 114-115, 116-117, 132-133
'Mercury'. Solarsystemscope. https://www.solarsystemscope.com/textures/ (under CC BY 4.0) Also used on pages 34-35, 36-37, 38-39, 40-41, 42-43, 72-73, 74-75, 76-77, 78-79, 80-81, 92-93, 110-111, 112-113, 114-115, 116-117, 118-119
'Mars'. Solarsystemscope. https://www.solarsystemscope.com/textures/ (under CC BY 4.0) Also used on pages 34-35, 36-37, 38-39, 40-41, 42-43, 72-73, 74-75, 76-77, 78-79, 80-81, 110-111, 112-113, 114-115, 116-117, 130-131, 132-133
'Largest known Kuiper Belt objects'. NASA. https://www.nasa.gov/images/content/606923main_comets2_lg.jpg
'Largest known trans-Neptunian objects'. NASA. https://en.wikipedia.org/wiki/File:EightTNOs.png
'Kleopatra'. NASA. https://en.wikipedia.org/wiki/216_Kleopatra#/media/File:Kleopatra.jpg
'Pluto by LORRI and Ralph'. NASA/JHUAPL/SWRI. https://commons.wikimedia.org/wiki/File:Pluto_by_LORRI_and_Ralph,_13_July_2015.jpg Also used on pages 40-41, 128-129
'Global color mosaic of Triton'. NASA/JPL/USGS. https://photojournal.jpl.nasa.gov/catalog/PIA00317 Also used on pages 40-41
'Charon'. NASA/JHUAPL.SWRI. https://commons.wikimedia.org/wiki/File:Charon_in_Color_(HQ).jpg Also used on pages 128-129
Untitled. Skitterphoto. https://pixabay.com/sk/photos/plame%C5%88-ohe%C5%88-inferno-oran%C5%BEov%C3%A1-726268/
'Dione'. NASA/JPL/Space Science Institute. https://en.wikipedia.org/wiki/Dione_(moon)#/media/File:Dionean_Linea_PIA08256.jpg
'Ariel'. NASA. https://en.wikipedia.org/wiki/Ariel_(moon)#/media/File:Ariel_HiRes.jpg
'Oberon'. NASA. https://en.wikipedia.org/wiki/Oberon_(moon)#/media/File:Voyager_2_picture_of_Oberon.jpg Also used on pages 128-129
'Tethys'. NASA/JPL/Space Science Institute. https://en.wikipedia.org/wiki/Tethys_(moon)#/media/File:PIA18317-SaturnMoon-Tethys-Cassini-20150411.jpg
'Umbriel'. NASA. https://en.wikipedia.org/wiki/Umbriel_(moon)#/media/File:PIA00040_Umbrielx2.47.jpg
'Mimas'. NASA/JPL-Caltech/Space Science Institute. https://en.wikipedia.org/wiki/Mimas_(moon)#/media/File:Mimas_Cassini.jpg
'Titania'. NASA/JPL. https://en.wikipedia.org/wiki/Titania_(moon)#/media/File:Titania_(moon)_color,_edited.jpg Also used on pages 128-129
'Rhea'. NASA/JPL/Space Science Institute. http://photojournal.jpl.nasa.gov/catalog/PIA07763
'Enceladus'. NASA/JPL. https://en.wikipedia.org/wiki/Enceladus#/media/File:PIA17202_-_Approaching_Enceladus.jpg Also used on pages 80-81
'Iapetus'. NASA/JPL/Space Science Institute https://en.wikipedia.org/wiki/Iapetus_(moon)#/media/File:Iapetus_as_seen_by_the_Cassini_probe_-_20071008.jpg
'Miranda'. NASA/JPL-Caltech. https://en.wikipedia.org/wiki/Miranda_(moon)#/media/File:PIA18185_Miranda's_Icy_Face.jpg

'Front and back of Larisso'. NASA. https://en.wikipedia.org/wiki/Larissa_(moon)#/media/File:Larissa.jpg
'Proteus'. NASA. https://en.wikipedia.org/wiki/Proteus_(moon)#/media/File:Proteus_(Voyager_2).jpg
'Titan Infrared Mosaic'. NASA. https://photojournal.jpl.nasa.gov/catalog/PIA19658
'An image of Jupiter taken by the Hubble Space Telescope'. NASA. https://en.wikipedia.org/wiki/Jupiter#/media/File:Jupiter_and_its_shrunken_Great_Red_Spot.jpg Also used on pages 34-35, 36-37, 38-39, 40-41, 42-43, 72-73, 110-111, 112-113, 118-119
'Size comparison of several objects with potential for dwarf planet status'. Kornmesser, Martin. IAU/NASA, https://en.wikipedia.org/wiki/2_Pallas#/media/File:Iau_dozen.jpg
'Eros, Vesta and Ceres size comparison'. NASA. https://en.wikipedia.org/wiki/File:Eros,_Vesta_and_Ceres_size_comparison.jpg Also used on pages 72-73, 128-129
'Dawn'. NASA/JPL-Caltech/UCLA. https://en.wikipedia.org/wiki/Ceres_(dwarf_planet)#/media/File:PIA19562-Ceres-DwarfPlanet-Dawn-RC3-image19-20150506.jpg Also used on pages 72-73, 128-129
'Amalthea'. Hamilton, Calvin J. NASA, http://solarviews.com/cap/jup/amalthea.htm
'Saturn's moon Pan'. NASA/JPL-Caltech. https://commons.wikimedia.org/wiki/File:PIA21436.jpg
'Atlas'. NASA/JPL-Caltech. https://commons.wikimedia.org/wiki/File:Atlas_2017-04-12_raw_preview.jpg

pp. 34-35

'Uranus'. NASA/JPL-Caltech. https://en.wikipedia.org/wiki/Uranus#/media/File:Uranus2.jpg Also used on pages 36-37, 38-39, 40-41, 42-43, 74-75, 110-111, 112-113, 114-115, 128-129, 130-131, 132-133
'Neptune'. Solarsystemscope. https://www.solarsystemscope.com/textures/ (under CC BY 4.0) Also used on pages 36-37, 38-39, 40-41, 42-43, 76-77, 110-111, 112-113, 114-115, 128-129, 132-133
'Neptune Full'. NASA/JPL. https://en.wikipedia.org/wiki/Neptune#/media/File:Neptune_Full.jpg Also used on pages 36-37, 38-39, 40-41, 42-43, 76-77, 110-111, 112-113, 114-115, 128-129, 132-133
'Gas Giant textures by Dagohbert'. Dagohbert. https://www.deviantart.com/dagohbert/art/Gas-Giant-Textures-427702908 (Under CC BY 3.0) Also used on pages 36-37, 38-39, 40-41, 42-43, 110-111, 112-113, 114-115
'Jupiter'. Solarsystemscope. https://www.solarsystemscope.com/textures/ (under CC BY 4.0) Also used on pages 36-37, 38-39, 40-41, 42-43, 72-73, 74-75, 78-79, 80-81, 110-111, 112-113, 114-115
'Saturn'. Solarsystemscope. https://www.solarsystemscope.com/textures/ (under CC BY 4.0) Also used on pages 36-37, 38-39, 40-41, 42-43, 74-75, 78-79, 80-81, 110-111, 112-113, 114-115, 130-131
'TRAPPIST 1'. NASA, https://exoplanets.nasa.gov/blog/1469/all-these-worlds-are-yours/ Also used on pages 110-111, 112-113, 114-115
'Exoplanet WaterWorlds'. NASA, https://commons.wikimedia.org/wiki/File:NASA-Exoplanet-WaterWorlds-20180817.jpg Also used on pages 110-111, 112-113, 114-115
'An artist's impression of 10 hot Jupiter exoplanets'. NASA, https://exoplanets.nasa.gov/system/news_items/main_images/238_Dec14.jpg Also used on pages 110-111, 112-113, 114-115

pp. 36-37

'Lagoon Nebula, visible light view'. NASA, https://www.nasa.gov/feature/goddard/2018/lagoon-nebula-visible-light-view
Untitled. OpenClipart-vectors. https://pixabay.com/vectors/alphabet-word-images-decoration-1295152/
'Titan in true color'. NASA. http://photojournal.jpl.nasa.gov/catalog/PIA14602 Also used on pages 40-41, 76-77

pp. 40-41

'Center of Messier 1'. NASA. https://www.nasa.gov/sites/default/files/thumbnails/image/crab-nebula-xlarge_web.jpg

pp. 44-45

'Sun Releases M5.2-class Flare'. NASA/SDO. https://www.nasa.gov/content/goddard/sun-release-m52-class-flare-feb-3-2014 Also used on pages 46-47, 48-49, 50-51, 56-57, 60-61, 92-93, 108-109, 110-111, 112-113, 118-119, 120-121, 122-123, 124-125, 126-127, 128-129, 130-131, 132-133, 134-135
'Sun Erupts With Significant Flare'. NASA/SDO. https://www.nasa.gov/sites/default/files/thumbnails/image/sept10x8blend1311714k.jpg Also used on pages 46-47, 48-49, 50-51, 92-93, 108-109, 110-111, 112-113, 118-119, 124-125, 126-127, 128-129, 130-131, 132-133, 134-135
'NASA's SDO Sees Sun Emit Mid-Level Flare'. NASA/SDO. https://solarsystem.nasa.gov/resources/768/nasas-sdo-sees-sun-emit-mid-level-flare-oct-1/ Also used on pages 46-47, 48-49, 50-51, 56-57, 60-61, 92-93, 108-109, 110-111, 112-113, 118-119, 120-121, 122-123, 124-125, 126-127, 128-129, 130-131, 132-133, 134-135
'The Sun flares with activity'. NASA/SDO. https://eoimages.gsfc.nasa.gov/images/imagerecords/76000/76998/20120123_032721_2048_0304.jpg Also used on pages 46-47, 48-49, 50-51, 56-57, 60-61, 92-93, 108-109, 110-111, 112-113, 118-119, 120-121, 122-123, 124-125, 126-127, 128-129, 130-131, 132-133, 134-135
'Map of the full Sun by STEREO and SDO spacecraft'. NASA/STEREO/SDO/GSFC. .https://en.wikipedia.org/wiki/Sun#/media/File:Map_of_the_full_sun.jpg Also used on pages 46-47, 48-49, 50-51, 92-93, 108-109, 110-111, 112-113, 118-119, 122-123, 124-125, 126-127, 128-129, 130-131, 132-133, 134-135
'Sun's Quiet Corona'. NASA/SDO. http://www.nasa.gov/multimedia/imagegallery/image_feature_2485.html Also used on pages 46-47, 48-49, 50-51, 108-109, 110-111, 112-113, 118-119, 120-121, 122-123, 124-125, 126-127, 128-129, 130-131, 132-133, 134-135
'Sun - SDO'. NASA/Solar and Heliospheric Observatory. http://www.starnetlibraries.org/wp-content/uploads/2016/08/sun-nasa-sdo.jpg Also used on pages 46-47, 48-49, 50-51, 56-57, 108-109, 110-111, 112-113, 118-119, 122-123, 124-125, 126-127, 130-131, 132-133, 134-135

pp. 52-53

'Cygnus X-1'. NASA/CXC/M.Weiss. https://www.nasa.gov/sites/default/files/cygx1_ill_0.jpg Also used on pages 110-111, 112-113, 114-115, 124-125, 128-129
'Protoplanetary Disk'. NASA. https://commons.wikimedia.org/wiki/File:Protoplanetary_disk.jpg Also used on pages 110-111, 112-113, 114-115, 124-125, 128-129
'Artist's conception of circumstellar disk around AA Tauri'. NASA. http://www.spitzer.caltech.edu/Media/releases/ssc2008-06/ssc2008-06c.shtml Also used on pages 114-115, 124-125
'Protoplanetary Disk'. NASA. http://photojournal.jpl.nasa.gov/catalog/PIA03243 Also used on pages 114-115, 124-125
'Artist's conception of the dust and gas surrounding a newly formed planetary system'. NASA. https://commons.wikimedia.org/wiki/File:NASA-ExocometsAroundBetaPictoris-ArtistView-2.jpg Also used on pages 114-115
'Protoplanetary Disk'. NASA. https://commons.wikimedia.org/wiki/File:Ssc2009-11b.jpg Also used on pages 114-115

pp. 54-55

'Stingray Nebula'. Bobrowsky, Matt. NASA, https://commons.wikimedia.org/wiki/File:Stingraynebula.jpg Also used on pages 128-129
'NGC 7027'. Latter, William B. NASA, https://commons.wikimedia.org/wiki/File:NGC_7027HSTFull.jpg
'Eskimo Nebula'.NASA, ESA, Andrew Fruchter (STScI), and the ERO team (STScI + ST-ECF) https://commons.wikimedia.org/wiki/File:Ngc2392.jpg Also used on pages 126-127
'NGC 2022'. NASA. https://commons.wikimedia.org/wiki/File:NGC2022.jpg
'NGC 6210'. ESA/Hubble and NASA. https://commons.wikimedia.org/wiki/File:NGC_6210_HST.tif
'NGC 2440'. NASA, ESA, and K. Noll (STScI). https://commons.wikimedia.org/wiki/File:NGC_2440_by_HST.jpg
'Cat's Eye Nebula'. NASA. https://commons.wikimedia.org/wiki/File:Catseye-big.jpg Also used on pages 126-127
'Ghost of Jupiter Planetary Nebula'. B. Balick (U. Washington) et al., WFPC2, HST, NASA. https://apod.nasa.gov/apod/ap970331.html Also used on pages 126-127
'NGC 6886'. ESA/Hubble & NASA. https://commons.wikimedia.org/wiki/File:NGC_6886.jpg (Under CC-BY 4.0)
'M1-92'. ESA/Hubble & NASA. https://en.wikipedia.org/wiki/M1-92#/media/File:Minkowski_92.jpg (Under CC-BY 3.0)
'NGC 6818'. NASA. https://en.wikipedia.org/wiki/NGC_6818#/media/File:Ngc6818.jpg
'NGC 6826'. NASA. https://commons.wikimedia.org/wiki/File:NGC_6826HSTFull.jpg
'NGC 6741'. ESA/Hubble and NASA. https://sk.wikipedia.org/wiki/NGC_6741#/media/S%C3%BAbor:NGC_6741.tif
'NGC 6918'. ESA/Hubble and NASA. https://sk.wikipedia.org/wiki/NGC_3918#/media/S%C3%BAbor:Planetary_nebula_NGC_3918.jpg
'NGC 3132'. NASA. http://www.spacetelescope.org/images/opo9839a/
'NGC 5307'. NASA. http://hubblesite.org/newscenter/archive/releases/2007/33/image/e/
'HEN2-47'. NASA. https://commons.wikimedia.org/wiki/File:Hen2-47.jpg
'MZ-3'. NASA, ESA & the Hubble Heritage Team. https://commons.wikimedia.org/wiki/File:Ant_Nebula.jpg
'M2-9'. ESA/Hubble. https://en.wikipedia.org/wiki/M2-9#/media/File:The_Twin_Jet_Nebula.jpg (under CC BY 4.0)
'Red Rectangle Nebula'. NASA. http://antwrp.gsfc.nasa.gov/apod/ap040513.html
'Saturn Nebula'. NASA. https://www.spacetelescope.org/images/opo9738g/
'Homunculus Nebula'. Smith, Nathan. NASA. https://commons.wikimedia.org/wiki/File:EtaCarinae.jpg Also used on pages 128-129
'Retina Nebula'. NASA and the Hubble Heritage Team. https://commons.wikimedia.org/wiki/File:Retinanebel.jpg
'Spirograph Nebula'. NASA and the Hubble Heritage Team. https://commons.wikimedia.org/wiki/File:Spirograph_Nebula_-_Hubble_1999.jpg
'The Hourglass Nebula'. NASA. https://www.spacetelescope.org/images/opo9607a/ Also used on pages 126-127
'Fine Ring Nebula'. ESO. https://commons.wikimedia.org/wiki/File:Fine_Ring_Nebula.jpg (CC BY 3.0)
'Helix Nebula'. NASA. https://svs.gsfc.nasa.gov/30792 Also used on pages 126-127

pp. 56-57

'NGC 7293'. NASA, ESA, and C.R. O'Dell. https://commons.wikimedia.org/wiki/File:NGC7293_(2004).jpg
'Pillars of Creation'. NASA, Jeff Hester, and Paul Scowen (Arizona State University). https://commons.wikimedia.org/wiki/File:Eagle_nebula_pillars.jpg
'Horsehead Nebula up Close'. O'Donnell, Dylan. https://commons.wikimedia.org/wiki/File:Horsehead_Nebula_up_Close.jpg
'The Pencil Nebula'. ESO. https://www.wikidata.org/wiki/Q840805#/media/File:The_Pencil_Nebula,_a_strangely_shaped_leftover_from_a_vast_explosion.jpg (Under CC BY 3.0)
'NGC 6357'. NASA, ESA and Jes?s Maz Apellÿniz. https://en.wikipedia.org/wiki/NGC_6357#/media/File:EmissionNebula_NGC6357.jpg
'Butterfly Nebula'. NASA, ESA and the Hubble SM4 ERO Team. https://en.wikipedia.org/wiki/NGC_6302#/media/File:NGC_6302_Hubble_2009.full.jpg Also used on pages 126-127
'Fox Fur Nebula'. Donatiello, Giuseppe. https://commons.wikimedia.org/wiki/Category:NGC_2264#/media/File:Fox_Fur_Nebula_(30611917930).jpg
'Cassiopeia A'. NASA/JPL-Caltech. https://commons.wikimedia.org/wiki/File:Cassiopeia_A_Spitzer_Crop.jpg Also used on pages 110-111, 112-113, 122-123, 124-125, 128-129, 130-131
'Tycho's Supernova' NASA/CXC/Rutgers/J.Warren & J.Hughes http://chandra.harvard.edu/photo/2005/tycho/ Also used on pages 112-113, 122-123, 126-127
'Kepler's Supernova'. NASA/ESA/JHU/R.Sankrit & W.Blair. https://en.wikipedia.org/wiki/Kepler%27s_Supernova#/media/File:Keplers_supernova.jpg
'NGC 2818'. NASA, ESA and the Hubble Heritage Team. https://commons.wikimedia.org/wiki/File:NGC_2818_by_the_Hubble_Space_Telescope.jpg
'Southern Owl Nebula'. ESO. https://commons.wikimedia.org/wiki/File:Eso1532a.jpg Also used on pages 120-131
'Ring Nebula'. NASA and the Hubble Heritage Team. https://commons.wikimedia.org/wiki/File:M57_The_Ring_Nebula.JPG Also used on pages 68-69
'Eye of Sauron Nebula'. ESA/Hubble and NASA https://commons.wikimedia.org/wiki/File:Hubble_Observes_Glowing,_Fiery_Shells_of_Gas.jpg (Under CC BY 4.0)
'Cotton Candy Nebula'. NASA. https://en.wikipedia.org/wiki/Cotton_Candy_Nebula#/media/File:Cottoncandynebula.jpg
'NGC 2261'. NASA, https://apod.nasa.gov/apod/ap991020.html
'Flame Nebula'. X-ray: NASA/CXC/PSU/K.Getman, E.Feigelson, M.Kuhn & the MYStIX team; Infrared: NASA/JPL-Caltech. https://en.wikipedia.org/wiki/Flame_Nebula#/media/File:NASA-FlameNebula-NGC2024-20140507.jpg
'Frosty Leo Nebula'. NASA, ESA and the Hubble Heritage Team. https://en.wikipedia.org/wiki/Frosty_Leo_Nebula#/media/File:Frosty_Leo_Nebula.jpg (Under CC BY 4.0)
'Boomerang Nebula'. NASA, ESA and the Hubble Heritage Team. https://en.wikipedia.org/wiki/Boomerang_Nebula#/media/File:Boomerang_nebula.jpg
'Bubble Nebula'. NASA, ESA and the Hubble Heritage Team. https://en.wikipedia.org/wiki/NGC_7635#/media/File:The_Bubble_Nebula_-_NGC_7635_-_Heic1608a.jpg (Under CC BY 3.0)
'Cone Nebula'. ESA/Hubble. https://www.spacetelescope.org/images/heic0206f/ (under CC by 4.0)
'Orion Nebula'. NASA. https://commons.wikimedia.org/wiki/File:Orion_Nebula_(M42)_part_HST_4800px.jpg Also used on pages 12-13, 124-125
'Dumbbell Nebula'. ESO/I. Appenzeller, W. Seifert, O. Stahl https://www.eso.org/public/images/eso9846a/ (under CC BY 3.0)
'NGC 7129'. NASA. http://www.spitzer.caltech.edu/Media/releases/ssc2004-02/release.shtml
'Crab Nebula'. NASA, ESA, J. Hester and A. Loll https://commons.wikimedia.org/wiki/File:Crab_Nebula.jpg Also used on pages 126-127
'G1.9+0.3'. NASA/CXC/NCSU/K.Borkowski. http://chandra.harvard.edu/photo/2013/g19/g19_xray.tif

pp. 58-59

'G2920+1.8'. NASA/CXC/SAO. http://www.nasa.gov/chandra/multimedia/chandra-15th-anniversary-g292.html
'RCW 103'. NASA/CXC/Penn State/G.Garmire. http://chandra.harvard.edu/photo/2007/rcw103/
'Heart and Soul Nebulae'. NASA/JPL-Caltech/UCLA. https://commons.wikimedia.org/wiki/Category:IC_1848#/media/File:Heart_and_Soul_nebulae.jpg
'Puppis A'. NASA/CXC/IAFE. http://chandra.harvard.edu/photo/2014/puppisa/ Also used on pages 126-127
'North America Nebula'. NASA/JPL https://commons.wikimedia.org/wiki/File:Changing_Face_of_the_North_America_Nebula.jpg
'N49'. NASA. http://chandra.harvard.edu/photo/2010/n49/
'Rosette Nebula'. O'Donnell, Dylan. https://commons.wikimedia.org/wiki/File:NGC-2237-deography-valentinesday-2018-HaRGB-Rosette-Mosaic.jpg
'Lagoon Nebula'. ESO/S.Guisard https://commons.wikimedia.org/wiki/File:Lagoon_Nebula_(ESO).jpg
'A Swarm of Ancient Stars'. NASA, The Hubble Heritage Team. https://en.wikipedia.org/wiki/Globular_cluster#/media/File:A_Swarm_of_Ancient_Stars_-_GPN-2000-000930.jpg Also used on pages 110-111, 112-113
'IC 2118'. NASA. https://commons.wikimedia.org/wiki/File:Reflection.nebula.arp.750pix.jpg
'Eagle Nebula from ESO'. ESO. https://commons.wikimedia.org/wiki/File:Eagle_Nebula_from_ESO.jpg
'Omega Centauri'. ESO/INAF-VST/OmegaCAM. https://www.eso.org/public/images/eso1119b/ (Under CC BY 4.0) Also used on pages 110-111
'NGC 346 in Small magellanic cloud' NASA, ESA and A. Nota. https://commons.wikimedia.org/wiki/File:NGC_346_in_Small_magellanic_cloud.jpg Also used on pages 118-119
'Simeis 147'. Donatiello, Giuseppe; Stone, Tim. WISE https://commons.wikimedia.org/wiki/File:Simeis_147_(28254062070).jpg
'RCW 103'. NASA. http://chandra.harvard.edu/photo/2016/rcw103/
'SN-185'. NASA/CXC/NCSU/K.Borkowski. https://en.wikipedia.org/wiki/SN_185#/media/File:RCW_86.jpg
'Pacman Nebula'. NASA. https://en.wikipedia.org/wiki/NGC_281#/media/File:NGC_281_from_Chandra.jpg
'Messier 54'. ESA/Hubble & NASA. Also used on pages 60-61 https://en.wikipedia.org/wiki/Messier_54#/media/File:Messier_54_HST.jpg (under CC BY 3.0)
'Ghost Head Nebula'. ESA, NASA, & Mohammad Heydari-Malayer. https://commons.wikimedia.org/wiki/File:NGC2080.jpg
'Monkey Head Nebula'. NASA. https://commons.wikimedia.org/wiki/File:Hubble_Celebrates_24th_Anniversary_with_Infrared_Image_of_Nearby_Star_Factory_(13225104285).jpg
'47 Tucanae'. NASA, ESA, Digitized Sky Survey. https://en.wikipedia.org/wiki/47_Tucanae#/media/File:Hubble_finds_evidence_of_multiple_stellar_populations_in_globular_cluster_47_Tucanae.jpgAlso used on pages 60-61, 110-111, 130-131, 132-133
'Cygnus Loop'. Ultraviolet image of the Cygnus Loop Nebula. https://commons.wikimedia.org/wiki/File:Ultraviolet_image_of_the_Cygnus_Loop_Nebula_crop.jpg Also used on pages 126-127
'Tarantula Nebula'. Gendler, Robert; Colombari,Robert; NASA, European Southern Observatory. https://apod.nasa.gov/apod/ap160226.html (under CC BY 3.0)
'Messier 9'. NASA/ESA. https://commons.wikimedia.org/wiki/File:Globular_cluster_Messier_9_(captured_by_the_Hubble_Space_Telescope).tif Also used on pages 60-61

pp. 60-61

'I Zwicky 18'. Gendler, Robert; Colombari,Robert; NASA, European Southern Observatory. https://apod.nasa.gov/apod/ap160226.html
'ESO 540'. ESA/Hubble & NASA. https://en.wikipedia.org/wiki/ESO_540-2z``d030#/media/File:ESO_540-030.jpg
'Henize 2-10'. NASA, CXC. http://chandra.harvard.edu/photo/2011/he210/
'Tucana Dwarf galaxy by Hubble space telescope'. NASA. https://en.wikipedia.org/wiki/Tucana_Dwarf#/media/File:Tucana_Dwarf_Hubble_WikiSky.jpg
'NGC 2419'. ESA/Hubble. https://commons.wikimedia.org/wiki/File:NGC2419_-_HST_-_Potw1908a.jpg (under CC by 4.0)
'Carina Nebula'. ESO. http://www.eso.org/public/images/eso1250a/
'IC-10'. NASA, NOAO. https://commons.wikimedia.org/wiki/File:IC10_BVHa.jpg
'NGC 604'. Yang, Hui. NASA, https://commons.wikimedia.org/wiki/File:Nursery_of_New_Stars_-_GPN-2000-000972.jpg Also used on pages 68-69
'Leo A'. Dionatiello, Giuseppe. https://en.wikipedia.org/wiki/Leo_II_(dwarf_galaxy)#/media/File:LG_Leo_II_(26422781005).jpg
'NGC 1569'. NASA/A. Aloisi https://en.wikipedia.org/wiki/NGC_1569#/media/File:Starburst_in_a_Dwarf_Irregular_Galaxy.jpg Also used on pages 108-109
'WLM Galaxy'. ESO/VST/Omegacam Local Group Survey. https://www.eso.org/public/images/eso1610a/
'Sombrero galaxy'. NASA/ESA and The Hubble Heritage Team. https://commons.wikimedia.org/wiki/File:M104_ngc4594_sombrero_galaxy_hi-res.jpg Also used on pages 64-65, 66-67, 108-109
'UGC 5497'. ESA/Hubble & NASA. http://annesastronomynews.com/photo-gallery-ii/galaxies-clusters/ugc-5497/
'Fornax Dwarf'. ESO/Digitized Sky Survey 2. https://en.wikipedia.org/wiki/Fornax_Dwarf#/media/File:Fornax_dwarf_galaxy.jpg Also used on pages 62-63, 108-109
'PGC 51017'. ESA/Hubble https://commons.wikimedia.org/wiki/File:An_intriguing_young-looking_dwarf_galaxy.jpg (under CC by 4.0)
'Antlia Dwarf'. NASA. https://en.wikipedia.org/wiki/Antlia_Dwarf#/media/File:Antlia_Dwarf_PGC_29194_Hubble_WikiSky.jpg Also used on pages 62-63, 108-109
'UGC 4879'. ESA/Hubble https://en.wikipedia.org/wiki/UGC_4879#/media/File:A_mysterious_hermit.jpg (under CC BY 4.0)
'Small Magellanic Cloud'. ESA/Hubble https://upload.wikimedia.org/wikipedia/commons/7/7a/Small_Magellanic_Cloud_%28Digitized_Sky_Survey_2%29.jpg (under CC BY 4.0) Also used on pages 108-109

pp. 62-63

'NGC 3310'. NASA and The Hubble Heritage Team. https://commons.wikimedia.org/wiki/File:NGC_3310.jpg Also used on pages 108-109
'NGC 7742'. Hubble Heritage Team (AURA/STScI/NASA/ESA). https://commons.wikimedia.org/wiki/File:Seyfert_Galaxy_NGC_7742.jpg Also used on pages 108-109
'NGC 7217'. NASA. https://en.wikipedia.org/wiki/NGC_7217#/media/File:NGC_7217_Hubble.jpg Also used on pages 108-109
'Spindle Galaxy'. NASA, ESA, and The Hubble Heritage Team http://hubblesite.org/newscenter/archive/releases/2006/24/image/a Also used on pages 108-109
'NGC 7793'. NASA/JPL-Caltech/R. Kennicutt https://commons.wikimedia.org/wiki/File:NGC_7793SpitzerFull.jpg Also used on pages 66-67

'Milky Way'. Risinger, Nick. https://en.wikipedia.org/wiki/File:Milky_Way_Galaxy.jpg used on pages 64-65, 66-67, 108-109, 110-111, 112-113, 116-117, 118-119, 130-131, 132-133
'Andromeda Galaxy'. NASA/JPL-Caltech https://commons.wikimedia.org/wiki/File:Andromeda_galaxy_2.jpg Also used on pages 64-65, 66-67, 108-109, 110-111, 116-117, 118-119, 132-133
'Centaurus A'. ESO. http://www.eso.org/public/images/eso0903a/ (under CC BY 4.0)
'NGC 3982'. NASA, ESA, and the Hubble Heritage Team Also used on pages 108-109 https://commons.wikimedia.org/wiki/File:NGC_3982_galaxy_hubble.jpg
'Antennae Galaxies'. NASA, ESA, and the Hubble Heritage Team https://commons.wikimedia.org/wiki/File:Antennae_galaxies_xl.jpg Also used on pages 66-67, 108-109, 110-111, 120-121, 130-131
'Silverado Galaxy'. NASA, ESA, and the Hubble Heritage Team https://commons.wikimedia.org/wiki/File:NGC_3370_Hi.jpg Also used on pages 108-109
'NGC 55'. ESO. https://commons.wikimedia.org/wiki/File:Irregular_Galaxy_NGC_55_(ESO_0914a).jpg Also used on pages 108-109
'NGC 4945'. ESO. http://www.eso.org/public/images/eso0931a/ Also used on pages 108-109
'NGC 2207+IC2163'. NASA, ESA, and the Hubble Heritage Team https://commons.wikimedia.org/wiki/File:NGC2207%2BIC2163.jpg Also used on pages 108-109
'Triangulum Galaxy taken by VLT Survey Telescope'. ESO. http://www.eso.org/public/images/eso1424a/ Also used on pages 108-109
'Messier 96'. ESO/Oleg Maliy. http://www.eso.org/public/images/potw1143a/ Also used on pages 108-109
'NGC 2976'. NASA, ESA, and J. Dalcanton and B. Williams. https://commons.wikimedia.org/wiki/File:The_Outer_Regions_of_NGC_2976.jpg
'Maffei 2'. NASA. http://www.spitzer.caltech.edu/images/3463-sig10-025-Maffei-2-The-Hidden-Galaxy Also used on pages 108-109
'Flocculent spiral NGC 2841'. NASA, ESA and The Hubble Heritage Team. https://www.spacetelescope.org/images/heic1104a/ Also used on pages 108-109
'Large Magellanic Cloud'. ESO/S. Brunier. https://commons.wikimedia.org/wiki/File:Magellanic_Clouds_%E2%80%95_Irregular_Dwarf_Galaxies.jpg
'NGC 1672'. NASA, ESA and The Hubble Heritage Team. https://commons.wikimedia.org/wiki/File:NGC_1672_HST.jpg Also used on pages 108-109
'NGC 1512'. NASA. https://commons.wikimedia.org/wiki/File:NGC_1512.jpg Also used on pages 108-109
'Cigar Galaxy'. NASA, ESA and The Hubble Heritage Team. https://commons.wikimedia.org/wiki/File:M82_HST_ACS_2006-14-a-large_web.jpg Also used on pages 66-67, 108-109, 110-111, 112-113
'M82 Chandra HST Spitzer'. NASA/JPL-Caltech/STScI/CXC/UofA/ESA/AURA/JHU. https://en.wikipedia.org/wiki/File:M82_Chandra_HST_Spitzer.jpg Also used on pages 66-67, 108-109, 110-111, 112-113
'NGC 1300'. NASA, ESA and The Hubble Heritage Team. https://commons.wikimedia.org/wiki/File:Hubble2005-01-barred-spiral-galaxy-NGC1300.jpg Also used on pages 66-67, 108-109
'Black Eye Galaxy'. NASA and The Hubble Heritage Team. https://commons.wikimedia.org/wiki/File:Blackeyegalaxy.jpg Also used on pages 66-67, 108-109
'NGC 6745'. NASA Goddard Space Flight Center NASA-GSFC. https://commons.wikimedia.org/wiki/File:NGC_6745.jpg Also used on pages 108-109

pp. 64-65
'NGC 4565 and 4562'. Jschulman555. https://commons.wikimedia.org/wiki/File:NGC_4565_and_4562.jpg (under CC BY 3.0) Also used on pages 66-67, 108-109
'Condor Galaxy'. ESO/VLT. https://en.wikipedia.org/wiki/NGC_6872#/media/File:NGC_6872.png (under CC BY 4.0).
'NGC 5257 & 5258'. NASA and The Hubble Heritage Team. https://commons.wikimedia.org/wiki/File:Hubble_Interacting_Galaxy_NGC_5257_(2008-04-24).jpg Also used on pages 108-109
'Cartwheel Galaxy'. NASA, ESA, and K. Borne. https://commons.wikimedia.org/wiki/File:Cartwheel-galaxy.jpg Also used on pages 108-109
'NGC 4921'. NASA, ESA and K. Cook https://commons.wikimedia.org/wiki/File:NGC_4921_by_HST.jpg Also used on pages 108-109
'NGC 3314'. NASA, ESA and The Hubble Heritage Team. https://commons.wikimedia.org/wiki/File:Hubble_view_of_NGC_3314_-_Heic1208a.tif Also used on pages 108-109
'Pinwheel Galaxy'. European Space Agency & NASA https://commons.wikimedia.org/wiki/File:M101_hires_STScI-PRC2006-10a.jpg (under CC BY 3.0) Also used on pages 66-67, 88-89, 108-109, 112-113, 128-129
'Hoag's Object'. NASA and The Hubble Heritage Team. https://commons.wikimedia.org/wiki/File:Hoag%27s_object.jpg Also used on pages 108-109, 116-117
'NGC 6744'. ESO. http://www.eso.org/public/images/eso1118a/. (under CC BY 3.0) Also used on pages 108-109, 110-11
'ARP 256'. ESA/Hubble. https://en.wikipedia.org/wiki/Arp_256#/media/File:Crash_in_progress_Arp_256.jpg. (Under CC BY 3.0) Also used on pages 108-109, 110-111
'Sunflower Galaxy'. NASA. https://commons.wikimedia.org/wiki/File:Messier_63_GALEX_WikiSky.jpg Also used on pages 108-109
'NGC 1232'. ESO/IDA/Danish 1.5 m/R.Gendler and A. Hornstrup. https://commons.wikimedia.org/wiki/File:Spiral_Galaxy_NGC_1232_(wallpaper).jpg (under CC BY 4.0) Also used on pages 66-67, 108-109
'NGC 4696'. NASA. https://commons.wikimedia.org/wiki/File:NGC_4696_hubble.jpg Also used on pages 108-109
'Seyfert's Sextet'. NASA. https://commons.wikimedia.org/wiki/File:Seyfert_Sextet_full.jpg Also used on pages 108-109
'Virgo A'. NASA. https://commons.wikimedia.org/wiki/File:Messier_87_Hubble_WikiSky.jpg Also used on pages 66-67, 108-109
'NGC 1566'. ESA/Hubble. https://upload.wikimedia.org/wikipedia/commons/6/62/Potw1422a.jpg (under CC BY 3.0) Also used on pages 66-67, 108-109
'Messier 86'. NASA. https://commons.wikimedia.org/wiki/File:Messier_86_Hubble_WikiSky.jpg Also used on pages 108-109
'Messier 49'. NASA, ESA, STScI and J. Blakeslee. https://commons.wikimedia.org/wiki/File:M49.png Also used on pages 108-109
'NGC 1448'. ESO. https://commons.wikimedia.org/wiki/File:Supernovae_in_NGC_1448_(captured_by_the_Very_Large_Telescope).jpg (under CC BY 4.0) Also used on pages 66-67, 108-109
'AM 0644-741'. NASA, ESA and The Hubble Heritage Team. https://commons.wikimedia.org/wiki/File:AM_0644-741.jpg Also used on pages 108-109
'HCG 90'. NASA, ESA and The Hubble Heritage Team. https://commons.wikimedia.org/wiki/File:HCG_90HST.jpg Also used on pages 108-109
'Abell 2199'. NASA/JPL-Caltech/SDSS/NOAO. https://commons.wikimedia.org/wiki/File:Monster_Galaxies_Lose_Their_Appetite_With_Age_03.jpg Also used on pages 108-109, 110-111
'NGC 1409 & 1410'. NASA, William C. Keel. https://commons.wikimedia.org/wiki/File:NGC_1409HSTFull.jpg Also used on pages 108-109
'NGC 6240'. NASA, ESA, and The Hubble Heritage Team Also used on pages 108-109, 110-111. https://commons.wikimedia.org/wiki/File:Hubble_revisits_tangled_NGC_6240.jpg (under CC BY 4.0) Also used on pages 108-109
'NGC 634'. ESA/Hubble & NASA. https://commons.wikimedia.org/wiki/File:A_Perfect_Spiral_with_an_Explosive_Secret.jpg. (Under CC BY 3.0) Also used on pages 108-109
'Abell 383'. NASA. https://commons.wikimedia.org/wiki/File:Abell_383.jpg (under CC BY 4.0) Also used on pages 108-109, 110-111, 112-113
'NGC 4526'. NASA/ESA. https://upload.wikimedia.org/wikipedia/commons/a/a2/SN1994D.jpg (under CC BY 3.0) Also used on pages 108-109
'NGC 2768'. ESA/Hubble. https://commons.wikimedia.org/wiki/File:Dusty_detail_in_elliptical_galaxy_NGC_2768.jpg Also used on pages 108-109
'NGC 1316'. NASA, ESA and The Hubble Heritage Team. https://commons.wikimedia.org/wiki/File:Ngc1316_hst.jpg Also used on pages 66-67, 108-109
'UGC 8335'. NASA, ESA, and the Hubble Heritage. https://commons.wikimedia.org/wiki/File:Hubble_Interacting_Galaxy_UGC_8335_(2008-04-24).jpg Also used on pages 108-109
'ESO 77-14'. NASA, ESA, and the Hubble Heritage. https://commons.wikimedia.org/wiki/File:Hubble_Interacting_Galaxy_ESO_77-14_(2008-04-24).jpg Also used on pages 108-109
'Messier 106'. NASA, ESA, and the Hubble Heritage. https://commons.wikimedia.org/wiki/File:Messier_106_visible_and_infrared_composite.jpg Also used on pages 108-109
'UGC 9618'. NASA, ESA, and the Hubble Heritage. https://en.wikipedia.org/wiki/Arp_302#/media/File:Hubble_Interacting_Galaxy_UGC_9618_(2008-04-24).jpg Also used on pages 108-109
'Arp 148'. NASA, ESA, and the Hubble Heritage. https://commons.wikimedia.org/wiki/File:Hubble_Interacting_Galaxy_Arp_148_(2008-04-24).jpg Also used on pages 108-109
'NGC 1275'. NASA, ESA, and the Hubble Heritage. https://commons.wikimedia.org/wiki/File:NGC_1275_Hubble.jpg Also used on pages 108-109
'NGC 3190'. NASA. https://commons.wikimedia.org/wiki/File:NGC_3190_Hubble_mosaic.jpg Also used on pages 108-109
'NGC 4145'. NASA. http://www.nasa.gov/mission_pages/spitzer/multimedia/spitzer-20090805.html Also used on pages 108-109
'NGC 3981'. ESO. https://commons.wikimedia.org/wiki/File:A_Galactic_Gem_NGC_3981.tif Also used on pages 108-109
'Hercules A'. NASA, ESA, and the Hubble Heritage team. https://en.wikipedia.org/wiki/Hercules_A#/media/File:A_Multi-Wavelength_View_of_Radio_Galaxy_Hercules_A.jpg (under CC BY 3.0) Also used on pages 108-109
'NGC 1532'. ESO. https://commons.wikimedia.org/wiki/File:ESO_-_Ngc1532_gendler_(by).jpg (under CC BY 4.0) Also used on pages 108-109
'NGC 6744'. ESO. http://www.eso.org/public/images/eso1118a/ Also used on pages 108-109
'NGC 2441'. ESA/Hubble https://en.wikipedia.org/wiki/NGC_2441#/media/File:A_curious_supernova_in_NGC_2441.jpg (CC BY 4.0) Also used on pages 108-109
'NGC 4622'. NASA and the Hubble Heritage team. https://commons.wikimedia.org/wiki/File:NGC_4622HSTFull.jpg
'NGC 1097'. NASA, ESA, and the Hubble Heritage. https://commons.wikimedia.org/wiki/File:NGC_1097_center_Hubble.jpg Also used on pages 66-67, 108-109
'Tadpole Galaxy'. NASA. https://commons.wikimedia.org/wiki/File:UGC_10214HST.jpg Also used on pages 108-109
'Bode's Galaxy'. NASA, ESA, and the Hubble Heritage team https://commons.wikimedia.org/wiki/File:Messier_81_HST.jpg Also used on pages 108-109
'Cygnus A, an active galactic nucleus'. NASA/NRAO. https://astronomynow.com/2016/10/25/hotspots-in-cygnus-a-an-active-galactic-nucleus/
'Stephan's Quintet'. NASA, ESA, and the Hubble SM4 ERO Team. https://en.wikipedia.org/wiki/Stephan%27s_Quintet#/media/File:Stephan's_Quintet_Hubble_2009.full_denoise.jpg
'Mice Galaxies'. NASA. https://commons.wikimedia.org/wiki/File:Merging_galaxies_NGC_4676_(captured_by_the_Hubble_Space_Telescope).jpg Also used on pages 108-109
'IC 1101'. NASA/ESA/Hubble Space Telescope. https://en.wikipedia.org/wiki/File:IC_1101_in_Abell_2029_(hst_06228_03_wfpc2_f702w_pc).jpg Also used on pages 108-109, 110-111, 112-113, 130-131, 132-133, 134-135

pp. 66-67
'NGC 7090'. NASA. https://commons.wikimedia.org/wiki/File:NGC_7090_spiral_galaxy_by_Hubble_Space_Telescope.jpg
'Messier 96'. ESO/Oleg Maliy. https://commons.wikimedia.org/wiki/Messier_96#/media/File:NGC_3368_ESO.jpg (under CC by 3.0)
'Whirlpool Galaxy'. NASA and ESA. https://commons.wikimedia.org/wiki/File:Messier51_sRGB.jpg Also used on pages 112-113, 128-129
'Cosmic Horseshoe'. ESA/Hubble & NASA. https://commons.wikimedia.org/wiki/File:A_Horseshoe_Einstein_Ring_from_Hubble.JPG

pp. 68-69
'Pleiades'. NASA, ESA, AURA/Caltech, Palomar Observatory. https://en.wikipedia.org/wiki/Pleiades#/media/File:Pleiades_large.jpg Also used on pages 114-115, 122-123
'Messier 58'. SIRTF/NASA. http://www.spitzer.caltech.edu/images/2265-sig06-003-NGC-4579
'A swarm of ancient stars'. NASA, The Hubble Heritage Team. https://commons.wikimedia.org/wiki/File:A_Swarm_of_Ancient_Stars_-_GPN-2000-000930.jpg

pp. 70-71
'Great Observatories Unique Views of the Milky Way'. NASA. https://images.nasa.gov/details-PIA12348.html Also used on pages 64-65, 66-67, 68-69, 70-71, 92-93,94-95, 96-97, 98-99, 100-101, 102-103
'Expedition 46 Soyuz Launch to the International Space Station'. NASA. https://www.nasa.gov/image-feature/expedition-46-soyuz-launch-to-the-international-space-station
'Space Shuttle Atlantis landing at KSC'. NASA/Chuck Luzier. https://commons.wikimedia.org/wiki/File:Space_Shuttle_Atlantis_landing_at_KSC_following_STS-122.jpg Also used on pages 78-79
'Falcon Heavy Demo Mission'. SpaceX. https://commons.wikimedia.org/wiki/File:Falcon_Heavy_Demo_Mission_(40126461851).jpg
'Volkov during a Russian EVA'. NASA. http://spaceflight.nasa.gov/gallery/images/station/crew-28/html/iss028e020718.html
'Launch of Bumper 2'. NASA/U.S. Army. https://en.wikipedia.org/wiki/RTV-G-4_Bumper#/media/File:Bumper8_launch-GPN-2000-000613.jpg
'Robert H. Goddard'. Goddard, Esther C. https://commons.wikimedia.org/wiki/File:Goddard_and_Rocket.jpg
'James Webb Space Telescope Revealed'. NASA. https://en.wikipedia.org/wiki/File:James_Webb_Space_Telescope_Revealed_(26832090085).jpg
'Aldrin walks on the surface of the Moon during Apollo 11'. NASA. https://en.wikipedia.org/wiki/Buzz_Aldrin#/media/File:Aldrin_Apollo_11_original.jpg
'Arecibo Observatory'. Broad, David. https://commons.wikimedia.org/wiki/File:Arecibo_radio_telescope_observatory_Puerto_Rico_-_panoramio_(9).jpg (under CC BY 3.0)
'Very Large Telescope'. ESO/José Francisco Salgado https://es.wikipedia.org/wiki/Archivo:ESO_Very_Large_Telescope.jpg (under CC BY 3.0)
'Parkes Radio Telescope'. CSIRO. https://commons.wikimedia.org/wiki/File:CSIRO_ScienceImage_4350_CSIROs_Parkes_Radio_Telescope_with_moon_in_the_background.jpg (under CC BY 3.0)
'Hooker Telescope'. Doc Searls, https://www.flickr.com/photos/docsearls/28792930640/ (under CC BY 2.0)
'Mars Pathfinder Lander preparations'. NASA. https://commons.wikimedia.org/wiki/File:Mars_Pathfinder_Lander_preparations.jpg
'Soyuz spacecraft'. NASA. https://commons.wikimedia.org/wiki/File:Soyuz_TMA-7_spacecraft2.jpg

pp. 72-73
'Hubble's Panoramic View of a Turbulent Star-making Region'. NASA, ESA, D. Lennon and E. Sabbi. http://hubblesite.org/image/2942/news_release/2012-01 Also used on pages 74-75, 76-77, 78-79, 80-81, 82-83
Untitled. Fehmi2029. https://cdn.pixabay.com/photo/2017/04/07/13/56/night-sky-2211032_960_720.jpg Also used on pages 86-87
'Earth from the ISS'. NASA. https://www.nasa.gov/sites/default/files/thumbnails/image/iss043e194350_0.jpg
'Fruit flies'. Skeeze. https://pixabay.com/photos/fruit-flies-mexican-female-insects-520905/
'Space Dogs Veterok and Ugoljok'. Teknisha Museet. https://en.wikipedia.org/wiki/File:Rymdhundarna_Veterok_och_Ugoljok_(16493301588).jpg (Under CC by 2.0)
'Ham Retrieval'. NASA. https://commons.wikimedia.org/wiki/File:Ham_Retreival_GPN-2000-001004.jpg
'Explorer 6 paddles up'. NASA. https://en.wikipedia.org/wiki/Explorer_6#/media/File:Explorer_6_paddles_up.jpg
'Explorer 1'. NASA. https://commons.wikimedia.org/wiki/File:Explorer1.jpg
'Luna 1'. Posta Shqiptare. http://rammb.cira.colostate.edu/dev/hillger/Albania.C68.jpg
'Laika'. Posta Romana. https://commons.wikimedia.org/wiki/File:Posta_Romana_-_1959_-_Laika_120_B.jpg
'Ham Retrieval' NASA. https://commons.wikimedia.org/wiki/File:Ham_Retreival_GPN-2000-001004.jpg
'OSCAR 1 Satellite'. Smithsonian Institution. https://commons.wikimedia.org/wiki/File:OSCAR_1_satellite-01.jpg
'Mariner 2'. NASA-JPL. https://en.wikipedia.org/wiki/Mariner_2#/media/File:Mariner_2.jpg
'Alexey Leonov'. Soviet Union. https://commons.wikimedia.org/wiki/File:Soviet_Union-1965-Stamp-0.10._Voskhod-2._First_Spacewalk.jpg
'Valentina Tereshkova'. Soviet Union. https://commons.wikimedia.org/wiki/File:Soviet_Union-1963-Stamp-0.10._Valentina_Tereshkova.jpg
'Mariner 5'. NASA. https://commons.wikimedia.org/wiki/File:Mariner_5.jpg
'Luna 10'. NASA. https://commons.wikimedia.org/wiki/File:Luna-10.jpg
'Jumping Salute'. NASA. https://moon.nasa.gov/resources/103/jumping-salute/
'Mariner 7'. NASA. https://commons.wikimedia.org/wiki/File:Mariner_6-7.png
'Luna 9'. Posta Romana. https://media4.allnumis.com/19416/160-lei-statia-automata-luna-9_19416_86432536dd2e7aL.jpg
'Surveyor 1'. NASA. https://nssdc.gsfc.nasa.gov/nmc/spacecraft/display.action?id=1966-045A
'Venera 4'. Armael. https://commons.wikimedia.org/wiki/File:Venera_4_(MMA_2011)_(1).JPG

pp. 74-75
'Model of a lander Venera'. Montgomery, Don S. US NAVY, https://commons.wikimedia.org/wiki/File:Cut-away_model_of_a_Soviet_communications_satellite.JPEG
'Mariner 10'. NASA. https://en.wikipedia.org/wiki/Mariner_10#/media/File:Mariner_10_transparent.png
'SAS 2'. NASA. https://en.wikipedia.org/wiki/Small_Astronomy_Satellite_2#/media/File:SAS_2.gif
'Mariner 9'. NASA. https://commons.wikimedia.org/wiki/File:Mariner09.jpg
'Lunar Roving Vehicle'. NASA. https://commons.wikimedia.org/wiki/File:Apollo15LunarRover.jpg
'Luna 24'. NASA. https://commons.wikimedia.org/wiki/File:Luna24-mission-nasa.jpg
'Viking Lander Model'. NASA. https://www.nasa.gov/multimedia/imagegallery/image_feature_2055.html
'Lunokhod mission diagram'. NASA. https://en.wikipedia.org/wiki/Lunokhod_programme#/media/File:Lunokhod-mission.jpg
'Helios spacecraft'. NASA/Max Planck. https://commons.wikimedia.org/wiki/File:Helios_-_testing.png
'Pioneer Venus orbiter'. NASA. https://commons.wikimedia.org/wiki/File:Pioneer_Venus_orbiter.jpg
'Artist's impression of a GPS-IIR satellite in orbit'. US Government. https://commons.wikimedia.org/wiki/File:GPS-IIR.jpg

pp. 76-77
'Genesis Spacecraft'. NASA. https://www.jpl.nasa.gov/news/news.php?feature=6199
'CubeSat'. NASA. https://commons.wikimedia.org/wiki/File:CubeSat_picture-02.png
'Artist's concept of Cassini diving between Saturn and its innermost ring'. NASA. https://solarsystem.nasa.gov/missions/cassini/mission/grand-finale/overview/ Also used on pages 78-79
'NEAR Shoemaker'. NASA. https://sk.wikipedia.org/wiki/NEAR_Shoemaker
'2001 Mars Odyssey'. NASA. https://commons.wikimedia.org/wiki/File:2001_mars_odyssey_wizja.jpg
'Mars Express'. NASA/JPL/Corby Waste. http://marsprogram.jpl.nasa.gov/express/gallery/artwork/marsis-radarpulses.html

pp. 78-79
'Mars 2020 Rover'. NASA/JPL-Caltech. https://mars.nasa.gov/resources/mars-2020-rover-artists-concept/
'SpaceX Dragon'. NASA. https://commons.wikimedia.org/wiki/File:COTS2Dragon.6.jpg
'Kepler Space Telescope'. NASA. https://commons.wikimedia.org/wiki/File:Kepler_Space_Telescope.jpg
'Herschel Space Telescope'. NASA. https://commons.wikimedia.org/wiki/File:Herschel_Space_Observatory.jpg
'Venus Express in Orbit'. Mirecki, Andrzej. https://commons.wikimedia.org/wiki/File:Venus_Express_in_orbit.jpg
'Stardust'. NASA. https://en.wikipedia.org/wiki/Stardust_(spacecraft)#/media/File:Stardust_-_Concepcao_artistica.jpg
'Hayabusa'. JGarry. https://en.wikipedia.org/wiki/Hayabusa#/media/File:Hayabusa_hover.jpg
'Phoenix landing'. NASA/JPL/Corby Waste. https://en.wikipedia.org/wiki/File:Phoenix_landing.jpg
'Dawn'. NASA. https://www.jpl.nasa.gov/news/news.php?feature=4425
'SLS'. NASA. https://www.jpl.nasa.gov/edu/images/activities/sls.jpg
Untitled. OpenClipart-Vectors. https://pixabay.com/vectors/drop-water-rain-tear-teardrop-147190/
'InSight'. NASA. https://nssdc.gsfc.nasa.gov/nmc/spacecraft/display.action?id=INSIGHT

pp. 80-81
'Dart'. NASA/JHUAPL. https://cdn.mos.cms.futurecdn.net/x7UyNZyK9muEwufSE7o8Fi-970-80.jpg'

'Europa Clipper'. NASA. https://commons.wikimedia.org/wiki/File:Europa_Clipper_transparent.png
'Sample return concept'. NASA/JPL. https://en.wikipedia.org/wiki/Mars_sample-return_mission#/media/File:Mars_sample_returnjpl.jpg
'Deep Space Gateway'. NASA/JPL. https://commons.wikimedia.org/wiki/File:Orion_visiting_Deep_Space_Gateway.jpg
'Deep Space Transport'. NASA/JPL. https://nvite.jsc.nasa.gov/presentations/b2/D1_Mars_Connolly.pdf
'Orion spacecraft'. NASA/JPL. https://commons.wikimedia.org/wiki/File:Orion_with_ATV_SM.jpg
'Wide Field Infrared Survey Telescope'. NASA/JPL. https://commons.wikimedia.org/wiki/File:WFIRSTRender10_light.jpg
'Parker Solar Probe'. NASA/JHUAPL. https://commons.wikimedia.org/wiki/File:Parker_Solar_Probe_insignia.png
'Different designs for the ATLAST telescope'. Postman, Marc. Space Telescope Science Institute, https://en.wikipedia.org/wiki/Advanced_Technology_Large-Aperture_Space_Telescope#/media/File:Atlast_concepts_all3.jpg

pp. 82-83
'A complete F-1 engine assembly as pictured in 1968'. https://archive.org/details/MSFC-6862832 Also used on pages 86-87
'Shell buckling test'. NASA. https://www.nasa.gov/sites/default/files/_0fd8664_0.jpg Also used on pages 86-87
'Building a big rocket fuel tank'. NASA. https://www.nasa.gov/sites/default/files/img_0505a_0.jpg Also used on pages 86-87
'SLS liquid hydrogen tank'. NASA. https://www.nasa.gov/centers/marshall/michoud/welders-complete-sls-liquid-hydrogen-tank Also used on pages 86-87
Untitled. OpenClipart-Ve561ctors. https://pixabay.com/vectors/steel-tubes-metallic-metal-pipes-576434/ Also used on pages 86-87

pp. 84-85
'Southern California and Baja California'. NASA. https://commons.wikimedia.org/wiki/File:ISS-55_Southern_California_and_Baja_California,_Mexico.jpg

pp. 86-87
'Hall Thruster'. NASA. https://www1.grc.nasa.gov/space/sep/
'Fuel rods'. Japo. https://commons.wikimedia.org/wiki/File:JETE-control_rod.jpg
'Electron gun'. Liftarn. https://commons.wikimedia.org/wiki/File:Egunr.jpg
Untitled. ChristopherMeinersmann. https://pixabay.com/photos/satellite-dish-to-listen-radio-2801490/
'Canberra Deep Space Communication Complex'. NASA. https://commons.wikimedia.org/wiki/File:Canberra_Deep_Dish_Communications_Complex_-_GPN-2000-000502.jpg https://www.nasa.gov/exploration/systems/mpcv/gallery/abort_crew/ed08-0230-362.html===
'Orion Crew vehicle mockup'. https://www.nasa.gov/exploration/systems/mpcv/gallery/parachute_testing/jsc2012e031604.html
'Boilerplate Orion Crew module'. NASA. https://www.nasa.gov/exploration/systems/mpcv/gallery/abort_crew/ed08-0230-362.html

pp. 88-89
Untitled. Ildigo. https://pixabay.com/photos/chicago-cloud-gate-2520867/
Untitled. Free-Photos. https://pixabay.com/photos/ocean-waves-water-sea-blue-nature-918897/
Untitled. MustangJoe. https://pixabay.com/photos/beach-dominican-republic-caribbean-1236581/
Untitled. quangle. https://pixabay.com/photos/sunrise-boat-rowing-boat-nobody-1014713/
Untitled. jplenio. https://pixabay.com/photos/road-nevada-clouds-bird-landscape-3856796/
Untitled. HarmonyCenter. https://pixabay.com/photos/cumulus-clouds-dramatic-white-blue-499176/
Untitled. diego_torres. https://pixabay.com/photos/sunset-field-poppy-sun-nature-815270/
Untitled. kordi_vahle. https://pixabay.com/photos/beach-north-sea-sea-sunset-water-2179624/
'A star-forming region in the Large Magellanic Cloud'. ESA/Hubble. https://en.wikipedia.org/wiki/Star#/media/File:Starsinthesky.jpg
'The pillars of creation'. NASA. https://www.nasa.gov/image-feature/the-pillars-of-creation Also used on pages 92-93

pp. 90-91
'Impact craters on the surface of Venus'. NASA/JPL. https://en.wikipedia.org/wiki/Venus#/media/File:PIA00103_Venus_-_3-D_Perspective_View_of_Lavinia_Planitia.jpg
'Venera 13 surface views'. NASA. https://nssdc.gsfc.nasa.gov/photo_gallery/photogallery-venus.html
'False-colour image of Maat Mons'. NASA/JPL. https://en.wikipedia.org/wiki/Venus#/media/File:Maat_Mons_on_Venus.jpg
'Han Solo on Mercury'. NASA/Johns Hopkins University Applied Physics Laboratory/Carnegie Institution of Washington. https://www.dailyedge.ie/nasa-messenger-captures-photo-hans-solo-1093251-Sep2013/
'Lava-flooded craters and large expanses of smooth volcanic plains on Mercury'. NASA/JHU/APL. https://en.wikipedia.org/wiki/MESSENGER#/media/File:MESSENGER_-_BV_Microsymposium49.jpg
Untitled. Afrikit. https://pixabay.com/photos/ethiopia-ethiopian-desert-desert-634222/
'McMurdo Dry Valleys'. Rejcek, Peter. NSF, https://www.nsf.gov/news/news_images.jsp?cntn_id=242559&org=NSF

'Aerial view of the Amazon rainforest'. Jorge.kike.medina. https://en.wikipedia.org/wiki/Amazon_rainforest#/media/File:Campo12Foto_2.JPG
'Ice confirmed at the Moon's poles'. NASA. https://www.jpl.nasa.gov/news/news.php?feature=7218
'Apollo 11 seismic experiment'. NASA. https://moon.nasa.gov/resources/13/apollo-11-seismic-experiment
'Curiosity succesfully drills "Duluth"'. NASA. https://mars.nasa.gov/resources/21876/first-drilled-sample-on-mars-since-2016/?site=insight
'Curiosity Self-Portrait at "Okoruso" Drill Hole'. NASA. https://www.jpl.nasa.gov/spaceimages/details.php?id=PIA20602
'NASA's Curiosity rover spots a bed of Earth-like pebbles'. NASA/JPL-Caltech/MSSS. https://mars.nasa.gov/msl/multimedia/raw/?rawid=2356MH0007210010804588C00_DXXX&s=2356
'A South Polar Pit or an Impact Crater?'. NASA. https://www.jpl.nasa.gov/spaceimages/details.php?id=PIA21636
'Curiosity Mars Rover Checks Odd-looking Iron Meteorite'. NASA/JPL-Caltech/MSSS. https://mars.nasa.gov/news/curiosity-mars-rover-checks-odd-looking-iron-meteorite/
'Mystery Lines on Mars Carved By Water'. NASA/JPL-Caltech/Univ. of Arizona. https://www.space.com/12543-mars-mystery-slopes-salt-water.html
'Mare Tranquillitatis pit crater'. NASA/GSFC/Arizona State University https://en.wikipedia.org/wiki/Lunar_lava_tube#/media/File:Mare_Tranquillitatis_pit_crater.jpg
'Earth – Apollo 11'. NASA. https://nssdc.gsfc.nasa.gov/imgcat/html/object_page/a11_h_44_6552.html

pp. 92-93
'Mercury MESSENGER Global DEM 665m v2'. NASA/USGS Astrogeology Science Center https://astrogeology.usgs.gov/search/map/Mercury/Topography/MESSENGER/Mercury_Messenger_USGS_DEM_Global_665m_v2
'Mercury MESSENGER MDIS Basemap Enhanced Color Global Mosaic 665m'. NASA/USGS Astrogeology Science Center/Johns Hopkins University Applied Physics Laboratory https://astrogeology.usgs.gov/search/map/Mercury/Messenger/Global/Mercury_MESSENGER_MDIS_Basemap_EnhancedColor_Mosaic_Global_665m
'Mercury MESSENGER MDIS Color Global Mosaic 665m v3'. NASA/Arizona State University. https://astrogeology.usgs.gov/search/map/Mercury/Messenger/Global/Mercury_MESSENGER_MDIS_ClrMosaic_global_665m_v3
'Carina Nebula'. NASA, ESA, and the Hubble SM4 ERO Team. https://en.wikipedia.org/wiki/Nebula#/media/File:Hs-2009-25-e-full_jpg.jpg
'Sunspots at solar maximum and minimum'. NASA, ESA, and the Hubble SM4 ERO Team. https://earthobservatory.nasa.gov/images/37575/sunspots-at-solar-maximum-and-minimum
'First Global Topographic Map of Mercury Released'. NASA. https://www.usgs.gov/news/first-global-topographic-map-mercury-released

pp. 94-95
'Venus Real Color'. NASA or Ricardo Nunes. https://en.wikipedia.org/wiki/Venus#/media/File:Venus-real_color.jpg Also used on pages 130-131, 132-133
'Hemispheric View of Venus'. NASA. https://solarsystem.nasa.gov/resources/486/hemispheric-view-of-venus/
'Combined Magellan radar map and altimetry map of Venus surface'. NASA. http://lasp.colorado.edu/~eparvier/astr3720/notes/3_04_03_images/venuscyl5.jpg
Untitled. qimono. https://pixabay.com/photos/clouds-sky-blue-nature-weather-2085112/

pp. 96-97
'Crafting the Blue Marble'. NASA. https://earthobservatory.nasa.gov/blogs/elegantfigures/2011/10/06/crafting-the-blue-marble/ Also used on pages 116-117, 118-119, 122-123, 124-125, 126-127, 132-133
'Bathymetry'. NASA. https://visibleearth.nasa.gov/view.php?id=73963
'Age of Oceanic Lithosphere'. NOAA. https://upload.wikimedia.org/wikipedia/commons/e/e7/2008_age_of_oceans_plates.jpg

pp. 100-101
'Mars Height Map'. NASA. https://marsoweb.nas.nasa.gov/globalData/images/fullscale/MOLA_cylin.jpg Also used on pages 114-115

pp. 102-103
'Big Bang'. Cedric Sorel. https://commons.wikimedia.org/wiki/File:Big_bang.jpg
'Protogalaxies'. NASA. https://commons.wikimedia.org/wiki/File:Stellar_Fireworks_Finale.jpg
'Glow from probable Population III Stars'. NASA. www.spitzer.caltech.edu/images/1505-ssc2005-22a1-Fiery-First-Stars
'Artist's rendering of a quasar'. ESO/M. Kornmesser. http://www.eso.org/public/images/eso1122a/ (under CC BY 4.0)
'A2744 YD4'. ESO/M. Kornmesser. http://www.eso.org/public/images/eso1708a/ (under CC BY 4.0)
'Protoplanetary disk'. ESO/L. Calçada. https://commons.wikimedia.org/wiki/File:Artist%E2%80%99s_Impression_of_a_Baby_Star_Still_Surrounded_by_a_Protoplanetary_Disc.jpg (under CC BY 4.0)
'Artist's depiction of a collision between two planetary bodies'. NASA/JPL-Caltech. http://www.nasa.gov/multimedia/imagegallery/image_feature_1454.html

'Most distant gamma-ray burst'. ESO/A. Roquette. http://www.eso.org/public/news/eso0917/ (under CC BY 3.0)
'Accretion disk binary system'. NASA. https://commons.wikimedia.org/wiki/File:Accretion_Disk_Binary_System.jpg
'Formation of Large Scale Structure'. Kravtsov, Andrey. Anatoly, Klypin. http://astro.wku.edu/astr106/structure/filaments.html (Under CC BY 3.0)
'NASA's Hubble Shows Milky Way is Destined for Head-On Collision'. NASA; ESA; Z. Levay and R. van der Marel, STScI; T. Hallas, and A. Mellinger.
'Geological Time Spiral'. USGS. https://commons.wikimedia.org/wiki/File:Geological-time-spiral_ja.png

pp. 104-105
'Dark Clouds of the Carina Nebula'. NASA. https://www.nasa.gov/multimedia/imagegallery/image_feature_1146.html Also used on pages 53-54, 104-105, 106-107, 108-109, 110-111, 112-113, 114-115, 116-117, 118-119, 120-121, 122-123, 124-125, 126-127, 128-129, 130-131, 132-133, 134-135
Untitled. noonexy. https://pixabay.com/illustrations/abstract-digital-art-fractal-2352668/ Also used on pages 134-135.
Untitled. DeltaWorks. https://pixabay.com/illustrations/fractal-sphere-maru-digital-art-662891/
Untitled. DeltaWorks. https://pixabay.com/illustrations/fractal-digital-art-669722/
Untitled. DeltaWorks. https://pixabay.com/illustrations/fractal-sphere-maru-digital-art-662894/ Also used on pages 106-107
Untitled. DeltaWorks. https://pixabay.com/illustrations/fractal-graphics-design-digital-art-169358/ Also used on pages 106-107
Untitled. darksouls1. https://pixabay.com/illustrations/fractal-light-fractal-light-chaos-1764082/
Untitled. dawnydawny. https://pixabay.com/illustrations/fractal-texture-background-design-2573303/ Also used on pages 132-133, 134-135
Untitled. darksouls1. https://pixabay.com/illustrations/fractal-light-light-fractal-neon-1820202/ Also used on pages 106-107, 108-109, 110-111, 132-133, 134-135
Untitled. TheDigitalArtist. https://pixabay.com/illustrations/fractal-big-bang-universe-bang-big-1073403/ Also used on pages 106-107, 108-109, 110-111, 132-133, 134-135
'Calabi-yau manifold'. Jbourjai. https://commons.wikimedia.org/wiki/File:Calabi_yau.jpg

pp. 106-107
'Cosmic Microwave Background'. NASA/WMAP. https://map.gsfc.nasa.gov/media/121238/ilc_9yr_moll4096.png Also used on pages 128-129

pp. 108-109
Untitled. Maxpixel. https://www.maxpixel.net/Connections-Texture-Network-Structure-Pattern-3354142
'Abell 2744'. NASA. https://commons.wikimedia.org/wiki/File:Heic1401a-Abell2744-20140107.jpg Also used on pages 110-111, 112-113
'Abell 1689'. ESA/Hubble https://en.wikipedia.org/wiki/Abell_1689#/media/File:New_Hubble_view_of_galaxy_cluster_Abell_1689.jpg (under CC BY 4.0) Also used on pages 110-111, 112-113
'Galaxies Gone Wild!'. NASA, ESA and the Hubble Heritage Project. https://commons.wikimedia.org/wiki/File:Galaxies_Gone_Wild!.jpg

pp. 110-111
'Hubble Finds an Einstein Ring'. ESA/Hubble & NASA; Acknowledgment: Judy Schmidt. https://www.nasa.gov/image-feature/goddard/2018/hubble-finds-an-einstein-ring (under CC BY 4.0)
Untitled. G4889166, https://pixabay.com/illustrations/star-field-stars-abstract-2294797/ Also used on pages 112-113, 114-115, 116-117, 118-119, 120-121, 122-123, 124-125, 126-127, 128-129, 130-131, 132-133, 134-135
'A smiling lens'. NASA. https://commons.wikimedia.org/wiki/File:HST-Smiling-GalaxyClusterSDSS-J1038%2B4849-20150210.jpg
Untitled. geralt, https://pixabay.com/photos/nerves-cells-star-dendrites-sepia-2926087/
Untitled. geralt, https://pixabay.com/photos/nerves-network-nervous-system-line-2728138/
'SNR 0509-67.5'. NASA. https://www.nasa.gov/mission_pages/chandra/multimedia/photo10-173.html
'Hubble eXtreme Deep Field'. NASA; ESA; G. Illingworth, D. Magee, and P. Oesch https://en.wikipedia.org/wiki/Hubble_Ultra-Deep_Field#/media/File:Hubble_Extreme_Deep_Field_(full_resolution).png Also used on pages 134-135
'Gn-z11'. NASA/ESA. https://upload.wikimedia.org/wikipedia/commons/4/42/Distant_galaxy_GN-z11_in_GOODS-N_image_by_HST.jpg

pp. 112-113
Untitled. 189748. https://pixabay.com/illustrations/fractal-fractal-art-969517/
Untitled. ColiN00B. https://pixabay.com/illustrations/galaxy-spiral-galaxy-fog-space-2961606/
Untitled. PeterDargatz. https://pixabay.com/photos/moon-super-moon-space-science-sky-416973/ Also used on pages 116-117, 118-119, 122-123, 132-133
'Fermi Bubbles'. NASA's Goddard Space Flight Center. http://astronomy.com/sitefiles/resources/image.aspx?item={64386BCA-C111-48AC-9BFE-628720848241}
'Asteroids'. rOEN911. https://www.deviantart.com/roen911/art/Asteroids-png-434091174 Also used on pages 114-115

pp. 114-115
'Bok Globules'. NASA and The Hubble Heritage Team. https://en.wikipedia.org/wiki/Bok_globule#/media/File:Bok_globules_in_IC2944.jpg
'Carina Nebula'. NASA, ESA, and the Hubble SM4 ERO Team. https://en.wikipedia.org/wiki/Nebula#/media/File:Hs-2009-25-e-full_jpg.jpg
Untitled. Arcturian. https://pixabay.com/illustrations/lava-cracked-background-fire-656827/
Untitled. Skeeze. https://pixabay.com/photos/lava-volcanic-crust-molten-window-1524277/ Also used on pages 116-117
Untitled. 8385. https://pixabay.com/illustrations/volcano-lava-landscape-glow-lake-1728164/
'Lunar Grail Mission'. NASA/GSFC/JPL/Colorado School of Mines/MIT. https://commons.wikimedia.org/wiki/File:14-236-LunarGrailMission-OceanusProcellarum-Rifts-Overall-20141001.jpg
'Halobacteria'. NASA. https://commons.wikimedia.org/wiki/File:Halobacteria.jpg Also used on pages 116-117, 132-133
'Tectonic plate boundaries'. USGS. https://commons.wikimedia.org/wiki/Category:Subduction_diagrams#/media/File:Tectonic_plate_boundaries2.png Also used on pages 132-133
'Banded Iron Formation at the Fortescue Falls'. Churchard, Graeme. https://commons.wikimedia.org/wiki/File:Banded_iron_formation_Dales_Gorge.jpg (under CC BY 2.0)
Untitled. Barni1. https://pixabay.com/photos/red-caps-australia-stromatolit-1105915/
Untitled. PamperedwithNature. https://pixabay.com/photos/tide-pool-ocean-hawaii-tide-beach-2546969/ Also used on pages 116-117
Untitled. Koenraadboshoff. https://pixabay.com/photos/sea-ocean-rocky-beach-waves-2430638/ Also used on pages 116-117

pp. 116-117
'Escherichia Coli'. Rocky Mountain Laboratories, NIAID, NIH https://en.wikipedia.org/wiki/File:EscherichiaColi_NIAID.jpg
'Alba Mons'. NASA/Arizona State University. https://en.wikipedia.org/wiki/Alba_Mons#/media/File:Alba_MOLA.jpg
Untitled. 4311868. https://pixabay.com/sk/photos/pl%C3%A1%C5%BE-pobre%C5%BEia-pacifiku-oce%C3%A1n-2089936/
Untitled. Pezibear. https://pixabay.com/photos/ice-snow-crystals-nature-frozen-1997288/
'Gravitational Waves'. R.Hurt/JPL-CALTECH. https://www.jpl.nasa.gov/news/news.php?feature=6975 Also used on pages 118-119, 128-129
'Black holes don't make a big splash'. Swinburne Astronomy Productions
https://www.jpl.nasa.gov/news/news.php?feature=3942 (under CC BY 3.0) Also used on pages 118-119, 128-129
Untitled. PxHere. https://pxhere.com/sk/photo/612264
Untitled. FreePhotos. https://pixabay.com/photos/rocky-mountain-cliff-landscape-1149298/
Untitled. derwiki. https://pixabay.com/photos/glacier-argentina-south-america-583419/
Untitled. AlexAntropov86. https://pixabay.com/illustrations/meteor-asteroid-space-disaster-3129573/ Also used on pages 122-123, 130-131, 132-133
Untitled. Pexels. https://pixabay.com/photos/crater-volcano-hawaii-1846775/
Untitled. doctor-a. https://pixabay.com/photos/volcano-hawaii-lava-cloud-ash-2262295/ Also used on pages 122-123
'Australia satellite plane'. Stöckli, Reto. NASA, https://en.wikipedia.org/wiki/File:Australia_satellite_plane.jpg Also used on pages 118-119, 120-121, 122-123, 124-125, 130-131, 132-133
'Sahara'. NASA/GSFC. https://visibleearth.nasa.gov/view.php?id=57230 Also used on pages 118-119, 120-121, 122-123, 124-125, 130-131, 132-133
Untitled. Clker-free-vector-images. https://pixabay.com/vectors/bacteriophage-virus-phage-medicine-309497/
Untitled. argzombies. https://pixabay.com/illustrations/mitochondria-cell-biology-science-3016868/
Untitled. franknowledge. https://pixabay.com/illustrations/animal-cell-biology-eukaryote-1608621/
Untitled. Msaeedsalem. https://pixabay.com/photos/rocks-granite-limestone-stones-2485461/
'Valles Marineris'. NASA/JPL-Caltech. https://en.wikipedia.org/wiki/Valles_Marineris#/media/File:VallesMarinerisHuge.jpg
Untitled. KatinkavomWolfenmond. https://pixabay.com/photos/gem-crystal-amethyst-stone-quartz-3328161/
'Fryxelsee'. National Science Foundation. https://en.m.wikipedia.org/wiki/File:Fryxellsee.jpg
Untitled. maxos_dim. https://pixabay.com/sk/photos/sopka-lava-pary-island-913954/ Also used on pages 118-119
Granite style linoleum floor texture. Photos public domain. http://www.photos-public-domain.com/2013/08/29/granite-style-linoleum-floor-texture/
Untitled. Funky_noodle. https://pixabay.com/sk/photos/fire-mountain-lanzarote-hora-lava-3641907/
Untitled. dric. https://pixabay.com/photos/volcano-lava-rash-science-fiction-2876292/

pp. 118-119
Untitled. Pxhere. https://pxhere.com/sk/photo/612256
Untitled. sasint. https://pixabay.com/sk/photos/vulk%C3%A1n-geografia-zobrazen%C3%AD-1807514/
Untitled. tommygbeatty. https://pixabay.com/sk/photos/kilauea-sopka-hawaii-n%C3%A1rodn%C3%BD-park-3088675/